DISCRETE MATHEMATICS
WITH
COMPUTER SCIENCE APPLICATIONS

DISCRETE MATHEMATICS
WITH
COMPUTER SCIENCE APPLICATIONS

Romualdas Skvarcius
Boston University

William B. Robinson
The Foxboro Company

The Benjamin/Cummings Publishing Company, Inc.
Menlo Park, California • Reading, Massachusetts •
Don Mills, Ontario • Wokingham, U.K. • Amsterdam •
Sydney • Singapore • Tokyo • Mexico City • Bogota •
Santiago • San Juan

Sponsoring Editor: Alan Apt
Production Editor: Larry Olsen
Cover Designer: Michael Rogondino
Book Designer: Lisa Mirsky
Artists: Carol Verbeeck, Janet Hayes
Cover Art: Morgan O'Hara
Title: August 1980

A color coded accounting of the artist's time during the month of August 1980. Twenty four concentric circles represent the hours of the day. Each wedge represents a single day. Colors indicate: creative time (red); social time (green); educational time (blue); time spent on survival needs (yellow).

Reprinted with corrections, May 1986.

Library of Congress Cataloging in Publication Data

Skvarcius, Romualdas.
 Discrete mathematics with computer science applications.

 Includes index.
 1. Electronic data processing—Mathematics.
I. Robinson, William B. II. Title.
QA76.9.M35S58 1986   001.64   85-13237
ISBN 0-8053-7044-7

BCDEFGHIJ-DO-8 9876

The Benjamin/Cummings Publishing Company, Inc.
2727 Sand Hill Road
Menlo Park, California 94025

Dedicated to
Dianne and Linda
and to all our children

Preface

The purpose of this text is to introduce discrete mathematics and its computer science applications to computer science students so that they will have an adequate set of mathematical tools for problem solving in their advanced courses. The intended audience is freshmen and sophomore students who are taking a concentration in computer science courses. We assume only that the students using the text have had a course in high school algebra.

We have selected a set of topics that we regard as both necessary and comprehensible for beginning students. Our central theme is that discrete mathematics provides a useful set of tools for modeling problems in computer science, just as calculus and continuous mathematics provide a useful set of tools for modeling in the physical sciences and engineering.

This text meets the recommendations for a course on discrete structures made by the curriculum committees of the ACM and the IEEE Computer Society. It also satisfies the requirements for a course in applied algebra set forth by the Committee on the Undergraduate Program in Mathematics of the MAA.

Chapter Pedagogy

We have attempted to make the style of the text as friendly and as conversational as possible. The book has several features designed to reinforce the concepts and techniques in the text to support the student in using the book in a self-instructional manner.

Examples and Self Tests

We have included numerous Examples for each major concept presented in the text. Each section contains a set of Self-Test questions. Students should try to answer the Self Test questions as they read the text. Answers to Self Tests are given at the end of each chapter, providing immediate feedback to the student.

Chapter Exercises

A comprehensive set of Exercises follows each section. Each chapter also has a robust set of Review Problems, which can be used to supplement the section Exercises. Solutions to the odd-numbered section Exercises and Review Problems are at the end of the text. A complete set of solutions to all Exercises and Review Problems is available in the Instructor's Guide. Some Review Problems are actually programming problems, which can be assigned in conjunction with a programming course or as optional projects. More difficult Review Problems are preceded by an asterisk.

Computer Science Applications

In chapters 2 through 10 we have included a discussion of a significant area of computer science to which the concepts presented in the chapter can be applied. These Application sections are optional and can be assigned as additional readings. Our hope is that these Applications will convince the student that the chapter material is not just part of an idle mathematics requirement but is of practical use in everyday computer science problem solving. The Applications and the associated mathematical topics are:

Application	Mathematical Topics	Chapter
Expert Systems	Logic and Sets	2
Relational Databases	Relations and Functions	3
Worst Case Analysis of Sorting Algorithms	Combinatorics	4
Languages and Parsing	Undirected Graphs	5
Routing Algorithms for Communications Networks	Directed Graphs	6
Circuit Design	Boolean Algebra	7
Coding Theory	Group Theory	8
Problem Solving	Finite State Automata	9
Average Case Analysis of Sorting Algorithms	Probability	10

Glossary

A Glossary of major terms used in the text appears at the end for easy reference.

How to Use This Book

Chapter Structure

We have kept chapters as independent as possible so that several different courses could be given from the text, depending on the inclination of the instructor. The relationships among chapters are shown in the following tree.

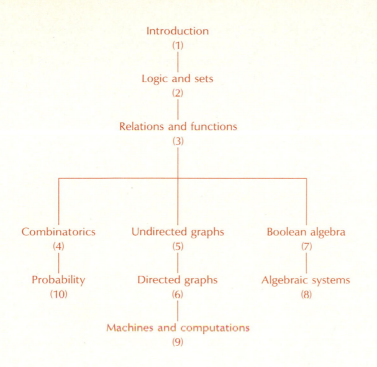

If an instructor chooses to downplay the representation of finite state automata as directed graphs, the material on machines and computations can be discussed independently of the chapter on directed graphs without too much difficulty.

We have successfully used the material from the first seven chapters plus chapter ten in a one-semester course. To fulfill the requirements of an applied algebra course, the material from Chapter 8 can be used in place of Chapter 10. If a course is to be oriented toward language theory or the theory of computation, then Chapter 9 should be included instead.

Topic Coverage

Foundations.   Chapter 1 introduces discrete mathematics and the idea of modeling, the concept of an algorithm, and a simple language for expressing algorithms. Chapter 2 covers the basics of logic and sets that will be used throughout the remainder of the text. In addition, the important Principle of Mathematical Induction is presented. The chapter contains an optional section on correctness of algorithms. The Application section discusses the use of logic and sets in expert systems. Chapter 3 introduces the concepts of relations and functions and their many properties. The Application section examines operations on n-ary relations as they are used in database management systems. The first three chapters can be covered quickly when the students have a reasonable background in the foundations of mathematics.

Combinatorics.   Chapter 4, Combinatorics, covers the basic counting rules for finite sets, including permutations and combinations. An optional section shows how these can be used in analysis of algorithms. The Application section covers issues involved in analyzing sorting algorithms.

Graphs.   Chapter 5 introduces Undirected Graphs, covering paths and circuits, connectivity, and trees. The Application section shows the application of graphs to the syntax of formal languages. Chapter 6 completes the study of graphs by focusing on Directed Graphs. In particular, topological sorting and various path problems are shown to have algorithmic solutions. The Application section shows how routing problems in communications networks can be modeled and solved by using directed graphs.

Boolean Algebra.   Chapter 7, Boolean Algebra, begins with the standard theoretical results on Boolean expressions and their representations, covers minimization through Karnaugh maps, gives examples from switching theory, and concludes with an Application section showing how Boolean logic can be used in the design of a 2-bit adder.

Abstract Algebra.   Chapter 8 introduces the student to abstract algebraic systems, and, through examples, shows how algebras can be used to unify much of the material from earlier parts of the text. Structures considered are semigroups, monoids, and groups, as well as methods for building new structures from existing ones by means of product and quotient operations. The Application section ends the chapter with a discussion of group codes for error detection and correction.

Models of Computation.   Chapter 9 is concerned with abstract models of computation. Two ends of the spectrum of computational power are defined: finite state automata and Turing machines. The role of each type of machine as language recognizers and as function computers is discussed. The Application section illustrates the use of finite automata as problem-solving models.

Probability.   The final chapter is an introduction to discrete probability theory. Only very basic topics are treated, including uniform probability spaces, conditional probabilities and Bayes' Theorem, the binomial distribution, and expected values. The Application section deals with average case analysis of two sorting algorithms.

Acknowledgments

We wish to thank the students at Boston University who have helped us "debug" the material in the text by using early drafts when we found no other available text suitable. We also express our appreciation to the reviewers, who have contributed many helpful suggestions on the structure and content of the text:

Michael Townsend, Columbia University

Margaret Cozzens, Northeastern University

Douglas Campbell, Brigham Young University

Udi Manber, University of Wisconsin

Anthony Evans, Wright State University

Peter Linz, University of California at Davis

David Cantor, University of California at Los Angeles

Richard Pollock, New York University

Andreas Blass, University of Michigan

Anthony Ralston, State University of New York at Buffalo

Chris Brown, University of Rochester

Jim Calhoun, University of Iowa

Fred S. Roberts, Rutgers University

Kirby Baker, University of California at Los Angeles

Donald Friesen, Texas A&M University

Romualdas Skvarcius
William B. Robinson

Brief Contents

Contents

4   COMBINATORICS   128

8   ALGEBRAIC SYSTEMS   306

9 MACHINES AND COMPUTATIONS   358

DISCRETE MATHEMATICS
WITH
COMPUTER SCIENCE APPLICATIONS

1 Introduction to Discrete Mathematics

Many concepts and techniques from mathematics are useful for solving problems in computer science. Our purpose in this book is to discuss the nature of discrete mathematics and explore the relationship between mathematics and computer science. In this chapter, we present the basic concepts underlying mathematics as a set of tools and techniques and the general modeling methodology for applying mathematics to real problems. We conclude the chapter with the definition of a simple algorithmic language, which we will use later for algorithm development.

1.1   What Is Discrete Mathematics?

It is not easy to give a precise definition of the word *discrete* as it applies to mathematics. In general usage, the word *discrete* means "made up of distinct parts." Discrete mathematics is often distinguished from other mathematics by contrasting it with continuous mathematics. The term *continuous*, as used here, means "without interruption and without abrupt changes." Some of the more familiar areas of continuous mathematics are calculus and differential equations.

Many of the topics and techniques of discrete and continuous mathematics may be the same. For instance, in both discrete and continuous mathematics, we are often concerned with collections of objects and the structures that are defined on them. These collections of objects are called **sets**. However, continuous mathematics concerns itself mostly with sets that are analogous to the set of real numbers. As a consequence, these sets usually have continuous representations, in the geometric sense. They are often intervals of a real line or regions of a plane. Discrete mathematics studies sets whose elements are separate and geometrically disconnected from each other. Moreover, most sets of interest in discrete mathematics are finite or countable, whereas those in continuous mathematics are uncountable.

Similar distinctions are possible within areas other than set theory. For example, as we will see later, relations and functions are studied in discrete mathematics and in continuous mathematics. However, the objects on which these relations and functions are defined are different.

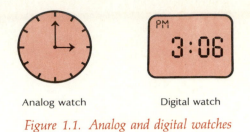

Analog watch   Digital watch

Figure 1.1. Analog and digital watches

The terms *discrete* and *continuous* have their counterparts in computer science. Computers may be classified as *digital* or *analog* computers. The classification is based on the type of data the computers are designed to process. Data may be obtained either as a result of *counting* or through the use of some *measuring* instrument. Data that are obtained through counting are discrete data, and computers designed to process such data are called digital computers. Data that are obtained through measurement are called continuous data and can be processed directly by analog computers. As an example of digital versus analog devices, consider the two watches in Figure 1.1. The watch with the dial and sweep hands is an analog device. It displays time along a continuous scale (the markings on the dial) to a degree of accuracy determined by the accuracy of the mechanism and by our ability to read the dial. The digital watch counts pulses generated at a fixed frequency and converts these counts to numbers that represent hours, minutes, and seconds. The digital watch works in discrete quantities.

The study of discrete mathematical systems has grown in importance as the use and influence of digital computers have increased. The modern digital computer is basically a finite discrete system, and many of its properties can be understood and illustrated by modeling them within the framework of discrete mathematical systems. Thus, a major reason for the study of discrete mathematics is to acquire the tools and techniques we need to design and understand computer systems. Discrete mathematics provides a framework to understand the problems of designing digital microprocessor systems and programs.

1.2   Tools, Techniques, and Methodologies

In this section, we discuss the general relationship of tools, techniques, and methodologies. Let us begin with an example from the sport of tennis.

What are the tools a tennis player needs to play the game? Even if you don't play, you can probably come up with a list that includes racket, balls, court, net, rules, and tennis whites. But, given all these tools, one can't yet play the game without mastering some of the basic techniques—the ways that one uses the tools effectively. The techniques of tennis include the forehand stroke, the backhand stroke, the serve, the volley, the overhead smash, and the lob. Having the right tools and techniques, you might think yourself ready to play the game. Perhaps you could play a respectable game at this point, but any tennis player's game is enhanced when the player learns

TABLE 1.1
Tools, Techniques, and
Methodology of Tennis

Tools	Tennis shoes Racket Balls Court Tennis whites (optional)
Techniques	Forehand stroke Backhand stroke Serve Volley Overhead smash Lob
Methodology	Singles strategy Doubles strategy Playing a left-hander

some of the methodology of the game. What we mean by *methodology* is the set of concepts that brings cohesion to the tools and techniques, showing when, how, and why to use them in various situations. The methodology of tennis includes strategies for singles play, for doubles play, for playing a left-handed player, and so on. Table 1.1 contains a summary of the tools, techniques, and methodologies of tennis.

Similarly, mathematics may be viewed as a tool, or rather a collection of tools, for problem solving in many disciplines. Once we have the mathematical tools, we often can use standard techniques for applying the tools in particular situations. The general methodology for using mathematics for problem solving is known as mathematical modeling. Figure 1.2 shows how the modeling process is used.

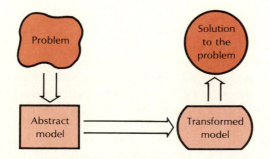

Figure 1.2. The modeling process

This diagram is a model of the modeling process. You have probably encountered numerous models for real objects, such as model airplanes and trains. Such models reflect the actual structure and function of the full-sized object. We call such models **concrete** models. When we see a mathematical model, we may not immediately recognize the real object that it represents because we have selected some of the properties and ignored others. We call these models **abstract** models. Consider the following simple example, which is typical of a class of problems in algebra:

> Jack has twelve apples, which cost him ten cents apiece. How many apples must he sell at twenty cents apiece before he begins to make a profit?

How do we go about solving this problem? We might begin by writing some equations to provide us with a model that describes the problem.

$$\text{income} = (\text{sale price}) \cdot (\text{number sold})$$
$$\text{expenses} = (\text{cost}) \cdot (\text{number bought})$$

The problem then becomes to find the number sold for which income equals or exceeds expenses. We might let x be the number of apples Jack needs to sell to break even. Then we need to solve the equation

$$(\text{sale price}) \cdot (x) = (\text{cost}) \cdot (\text{number bought})$$

Substituting the values from our problem, we have

$$20x = 10 \cdot 12$$

The equation is our model, although there is not an apple in sight. In fact, this particular model might well fit numerous problems that have nothing to do with apples. It is how we *interpret* the model that makes it fit our problem. Our next step is to apply some formal rules to the equation to find the value of x. We divide both sides of the equation by 20; this yields the transformed model

$$x = 6$$

The step back to our original problem is now trivial. Jack must sell six apples before he begins to make a profit.

Although the example may seem to be a laborious way of achieving a simple result, it does illustrate the elements of the modeling process shown in Figure 1.2.

We consider one more example of the modeling process, this one a little more substantial. The details of the example may not be as clear to you at this point as the details of the apples example because we have not yet presented any graph theory.

Table 1.2 contains mileage distances between six towns called A, B, C, D, E, and F. For towns that are not directly connected, the table entry is a dash. The problem is to find a road network of minimal total length that connects all the towns.

TABLE 1.2
Distances Between Towns (miles)

	A	B	C	D	E
B	5				
C	—	5			
D	10	10	20		
E	—	20	—	20	
F	—	—	30	—	10

The table itself is already a model of the real-world situation. We can further transform the table to represent the information geometrically. The geometric model is shown in Figure 1.3(a) as a graph. A **graph** consists of a set of points, called **vertices**, and lines connecting the points, called **edges**. We represent the towns as vertices, and the existing roads as edges.

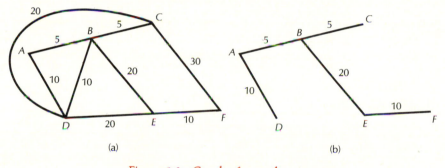

Figure 1.3. *Graph of a road system*

Now that we have a suitable mathematical model for our problem, we can attempt to find a solution in terms of that model. The task is to reduce the graph to another graph of minimal total length that connects all the vertices. By inspection, we can begin by including those roads between towns that are separated by the least distance. There must be a path between any pair of towns, but there must not be any roads that would cause a loop to form, since that leads to extra paths. More formally, we can approach the problem by attempting to find a solution in the form of an algorithm. An **algorithm** is a precise and unambiguous sequence of instructions that leads to the solution of a problem in a finite amount of time. To be useful, an algorithm must also give a solution to a specific class of problems. To find our graph of minimal total length, we perform the following steps:

1. Arbitrarily select any vertex, and connect it to the nearest adjacent vertex.
2. Scan the unconnected vertices. Locate the unconnected vertex nearest to one of the connected vertices, and connect these two vertices. (If more than one vertex

is the shortest distance from the connected vertices, select any one of them arbitrarily.) Repeat this process until all the vertices are connected.

A result of applying this algorithm to the graph of Figure 1.3(a) is shown in Figure 1.3(b). It is quite straightforward to obtain the road network of minimal total length connecting all the towns.

This approach of first obtaining a mathematical model for a problem and then formulating an algorithm in terms of that model is commonly used as a problem-solving technique in computer science. Not all problems we will consider will be solved by devising algorithms for their solution, but many will.

So far we have talked about discrete mathematics as a set of tools and techniques for problem solving. This is precisely the theme we will develop in the remainder of the text. Because the fields of discrete mathematics are so broad and the possibilities for applications so vast, we will only be able to consider some of the methodologies for applying what we learn here. The mathematical topics we will study are not new. They certainly predate computer science. Taken together, they form a very useful set of tools for solving many computer science problems.

Exercise 1.1

1. Assuming that the graph in Figure 1.4 represents a road network connecting a set of towns, find a road network of minimal total length that connects all the towns. Is this minimal network unique?

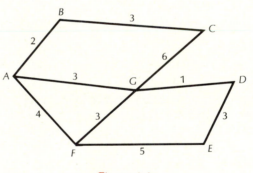

Figure 1.4.

1.3   An Algorithmic Language

Computer scientists often state solutions to problems in terms of algorithms. In this section, we present the details of the language used to describe the algorithms presented in this text. Students who have taken or are taking a programming course should skim this section quickly to familiarize themselves with the notation used in this book. Students who have not yet taken a programming course should proceed through this section at a slower pace.

What we need is a language that will allow for clear expression of algorithms. There are four alternatives: (1) natural languages (such as English), (2) computer programming languages, (3) flowcharts, and (4) pseudocode. Let us examine each of these alternatives.

A natural language such as English is appealing for use in problem solving because it would not require us to learn a new language. With some simple algorithms, such as the one of the previous section, it is adequate, but English has some major deficiencies in the description of algorithms. First, the English language can be ambiguous, and algorithms must not be ambiguous. Second, using English will probably result in a description of the algorithm that is too long and too difficult to understand.

The second way to describe algorithms is to write a complete computer program that implements the algorithm. However, there are some strong reasons for avoiding this method as well. First, computer languages include details that are necessary for a computer to execute the program, and such details often tend to hide the essence of the algorithm from a human reader. Furthermore, while designing the algorithm, we do not want to be burdened with specific details, such as punctuation, of a particular computer language. Finally, there is no universal language that all computer scientists use. For instance, if we described the algorithms in Pascal, but you knew only FORTRAN, you would have to learn Pascal before you could understand the algorithm's descriptions.

Flowcharts, the third alternative, are not regarded as good tools for defining structured algorithms. The basic trouble with flowcharts is that they allow too much freedom in describing control structures, hence they often result in poorly structured algorithms. Some aspects of algorithms, such as loops and conditional tests, may be difficult to identify in the flowchart, especially if the algorithm is large or complex.

Pseudocode

Pseudocode consists of a small set of structured language elements similar to a programming language together with English-like descriptions of the actions of the algorithm. Such an approach allows us to describe algorithms in an unambiguous manner and yet remain free of the details of an actual programming language. Since the main elements of our language are similar to those of actual programming languages, it should be relatively straightforward to translate these algorithms into programs in any language you desire.

Our pseudocode is based on the programming language Pascal, but is more informal than Pascal. In this language, all algorithms will take the following general form:

begin
    {Statements to perform operations
    as well as statements designed
    to control the order in which
    statements are to be executed}
end.

The building block of the algorithmic language is the **statement**. There are two basic types of statements in our algorithmic language, assignment statements and control statements.

Assignment Statements

The most commonly used statement in the description of algorithms is the **assignment statement**, which causes a value to be assigned to a variable. The assignment statement has the form

$$variable\ name \leftarrow expression$$

When executed, the expression on the right is evaluated, and the resulting value is assigned to the variable on the left, replacing any value the variable may have had previously. For example, given a variable A that takes integer values, the assignment statement

$$A \leftarrow A + 1$$

increments the value of A by 1. (*Note*: The symbol $\leftarrow$ is often called the assignment operator.)

Example 1.1 illustrates a simple algorithm that computes the sum of two numbers, *First* and *Second*, and assigns the sum to *Sum*.

EXAMPLE 1.1

{Algorithm to compute the sum of two numbers.}
{The numbers are *First* and *Second*. The   }
{result is assigned to *Sum*.   }

begin
    Input *First* and *Second*
    Sum $\leftarrow$ *First* + *Second*
end.

The text between braces is called a comment. Here it simply describes the function of the algorithm.

Sometimes we will use informal expressions on the right of the assignment, or we will describe some action in such terms as "select an arbitrary vertex." These statements should always be sufficiently clear to make the description of the algorithm unambiguous.

Control Statements

Control statements are used to structure the flow of control through the algorithm. That is, they are used to determine the order in which the statements of the algorithm are to be executed. There are three types of control statements: (1) *sequence* or *compound* statements, (2) *conditional* statements, and (3) *iterative* statements.

We give definitions and examples of these statement types below, along with flowchart representations of each. This is one place where a flowchart is useful to illustrate what each statement does. To make the control statements easier to read, we will use certain *keywords* and will highlight them in the text of the algorithm in **boldface** type.

Compound Statements.   A **compound statement** or a **sequence** of statements is a list of statements that is to be executed as a single unit in the order given (see Figure 1.5). The list of statements between the **begin** and **end** in Example 1.1 makes up a sequence. In other words, we can think of the entire algorithm as constituting a single compound statement. We will see shortly that compound statements are useful when we want to treat a group of statements as if they were a single statement.

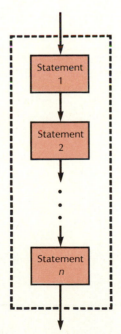

Figure 1.5. A sequence structure

Conditional Statements.   A **conditional statement** is a statement that allows the selection of one alternative between two. We will use two forms of the conditional statement:

if condition **then**
  Statement

if condition **then**
  Statement 1
else
  Statement 2

Figure 1.6 shows flowcharts for the conditional statements. The condition in the **if–then** statement must have a value of *true* or *false*. When the value is *true*, then the statement following the **then** is executed. Otherwise, the statement following the **if–then** statement is executed. Example 1.2 is an algorithm that computes the absolute value of a number.

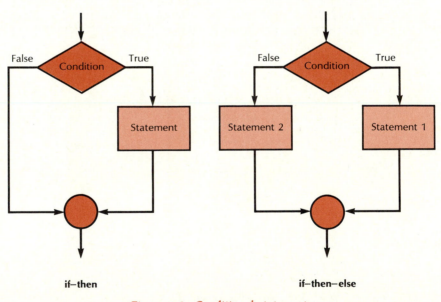

Figure 1.6. Conditional statements

EXAMPLE 1.2

{Algorithm to compute the absolute value of *n*.}
{The result is in *abs*.   }

```
begin
     Input n
     if n < 0 then
         n ← −n
     abs ← n
     Output abs
end.
```

The statement $n \leftarrow -n$ will be executed only if the condition $n < 0$ is *true*. If the condition is *false*, the statement following **then** is skipped. Either way, the next statement to be executed is $abs \leftarrow n$.

Likewise, the condition in the **if−then−else** statement must have a value of *true* or *false*. When the condition value is *true*, Statement 1 is executed. Otherwise, Statement 2 is executed. Either way, the next statement to be executed will be the one following the **if−then−else** statement. Example 1.3 illustrates the use of the **if−then−else** statement.

EXAMPLE 1.3

{Algorithm to find the larger value of}
{the values of a and b. The variable   }
{c holds the larger of the two values.}

```
begin
     Input a
     if  a > b then
         c ← a
     else
         c ← b
     Output c
end.
```

Iterative Statements.   There are four forms of **iterative statements**, which are often referred to as **loops**.

(1) **for** variable ← initial value **to** final value **do**
         Statement

(1a) **for** all elements of a set **do**
         Statement

(2) **while** *expression* **do**
    Statement

(3) **repeat**
      Statement 1
      Statement 2
          $\vdots$
      Statement *n*
    until condition

Figure 1.7 shows flowchart representations of the iterative statements. Note the arrow from the bottom of each flowchart to the junction at the top, giving each statement its characteristic *loop* structure. The statements inside the loop, which are repeated for each iteration of the loop, are collectively called the *body* of the loop.

   The **for** loop (top) executes the statement in the body a fixed number of times. When the **for** statement is encountered, the initial and final values are determined, and the loop counter variable shown in the diagram is set to the *initial value*. After the statement has been executed, the loop counter is increased by one. When the loop counter is greater than the *final value*, the loop terminates, and the statement following the **for** statement is executed. Example 1.4 shows a **for** loop that finds the sum of the first ten integers. Note that the statement constituting the body of the **for** loop may be a compound statement.

EXAMPLE 1.4

{Algorithm to sum the first ten positive}
{integers and store the result in the   }
{variable *sum*   }

begin
      sum ← 0
      for *i* ← 1 **to** 10 **do**
          sum ← *sum* + *i*
      Output *sum*
end.

We will use the **for** statement in (1a) when working with sets. This form of the **for** statement is similar to the previous one. The number of iterations equals the number of elements in the set. The loop is executed once for each element.

   The **while** and **repeat . . . until** loops, in contrast to the **for** loop, cause the statement or statements in their bodies to execute an indeterminate number of times. When a **while** statement is encountered, the condition is evaluated to see whether it is *true* or *false*. If it is *true*, then the statement following the **do** is executed. The con-

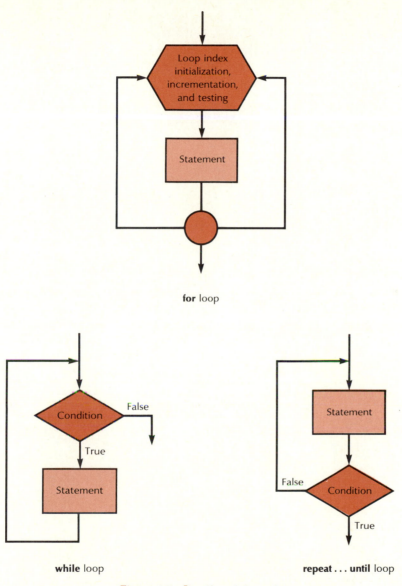

Figure 1.7. Iterative statements

dition is tested again, and as long as its value is *true*, the body of the loop will be executed. When the condition becomes *false*, the statement following the **end** will be executed. Thus, it is possible that the body of the **while** loop will not be executed even once.

The **repeat . . . until** loop will always execute its internal statements at least once. After the first execution, the condition after **until** will be evaluated. If it is *true*, the statement following the loop will be executed. Otherwise, the statements in the body of the loop will be executed again. This cycle repeats until the condition becomes *true*.

Observe that we have used the phrases *until the condition becomes true* and *while the condition remains true*. It is imperative in the **while** and **repeat . . . until** loops that some internal action *change* the value of the condition controlling the loop. In Examples 1.5 and 1.6, we elaborate on Example 1.4 by computing the sum of the first n positive odd integers.

EXAMPLE 1.5

{Algorithm to compute the sum of the first   }
{n positive odd integers and output the result}

begin
    $sum \leftarrow 0$
    $i \leftarrow 1$
    Input n
    while $i \leq 2n$ **do**
        begin
            $sum \leftarrow sum + i$
            $i \leftarrow i + 2$
        end
    Output sum
end.

EXAMPLE 1.6

{Algorithm to compute the sum of the first   }
{n positive odd integers and output the result}

begin
    $sum \leftarrow 0$
    $i \leftarrow 1$
    Input n
    repeat
        $sum \leftarrow sum + i$
        $i \leftarrow i + 2$
    until $i > 2n$
    if $n = 0$ **then**
        $sum \leftarrow 0$
    Output sum
end.

Note the following things about the above examples: First, the condition in the **while** loop in Example 1.5 is the opposite of the condition in the **repeat . . . until** loop in Example 1.6, although both algorithms do the same thing. This is typical of the translation between the two equivalent forms of iteration. Second, since we wanted to repeat more than one statement, we used a compound statement with the **while** loop. This is not required with the **repeat** loop because the keywords **repeat** and **until** delimit the body of the loop. Finally, we have not bothered to check that the variable n has met our criterion of being positive. You should try to figure out the consequences of reading a negative integer in both versions of the algorithm.

As a final example of an algorithm description, Example 1.7 presents the graph algorithm using pseudocode.

EXAMPLE 1.7

{Algorithm to reduce a graph to a graph}
{of minimal total length.   }

begin
    $v \leftarrow$ any vertex
    $u \leftarrow$ nearest adjacent vertex to v
    Connect v and u
    while unconnected vertices remain **do**
        begin
            $u \leftarrow$ one of the unconnected vertices that is
                nearest to one of the connected vertices
            Connect u to the nearest connected vertex
        end
end.

The control statements are the ones we described in our pseudocode, except that we have stated conditions, such as the one for the **while** loop, informally rather than as formal conditional expressions. We will do this whenever such informal statements make the algorithm more readable, even if the translation of such an algorithm into a computer program would require more work.

Some Comments About Algorithms

Although we will present various algorithms in this text, we will omit most of the details that would make it possible to convert these algorithms immediately to computer programs. For example, we will usually not be concerned with *data structures*. We also will not be concerned with algorithm design issues. Although such issues are clearly important in computer science, they are the proper subject of programming and data structures texts.

However, we will be concerned with algorithm *correctness* and *efficiency*. We will encounter many of our algorithms in the context of a formal mathematical treatment of a topic, such as graph theory. In that context, the algorithm will appear, in most cases, as a theorem and will be proven to be correct. Correctness of an algorithm is a nontrivial issue, as can be seen by carefully considering the algorithm in Example 1.7. Although the correctness of this algorithm seems quite plausible at first, it is less plausible after some reflection. How do we know, for instance, that the overall optimal solution will be obtained by making an optimal selection at each intermediate stage? This, as we shall see later, is rarely the case with algorithms for seeking optimal solutions.

In some cases, we will encounter more than one algorithm as a possible solution to a problem. In such situations, we will need some means of comparing them. A common means of comparing algorithms is to compare their running times. This is a difficult topic, but we will discuss some of the tools needed to make such comparisons at various points in the text.

Exercises 1.2

1. Modify the algorithm in Example 1.5, which computes the sum of the first n odd integers, so that it also finds the sum of the squares of the first n odd integers.

2. a. What is the final value of the variable *sum* in Examples 1.5 and 1.6 if the value of n is negative?
   b. Modify the algorithms in Examples 1.5 and 1.6 to bypass the loops when the value of n is negative.

3. What do the following algorithms do?

   a. **begin**
      count ← 10
      sum ← 0
     while *count* > 0 **do**
       begin
        sum ← 0
        count ← *count* − 1
       end
    end.

   b. **begin**
      first ← 1
      Output *first*
      second ← 1
      Output *second*
      next ← *first* + *second*
     while *next* < 15 **do**
       begin
        Output *next*
        first ← *second*
        second ← *next*
        next ← *first* + *second*
       end
    end.

SUMMARY

1. **Discrete mathematics** is characterized by the study of objects that are disconnected or "discretely" represented.

2. The study of discrete mathematical systems has grown in importance as the use and influence of digital computers have increased.

3. A modern digital computer is basically a finite discrete system, and many of its properties can be understood and illustrated by modeling them within the framework of discrete mathematical systems.

4. Mathematics may be viewed as a collection of tools for problem solving in many disciplines. That is, mathematics can provide us with a set of concepts that can serve as a vehicle for finding solutions to problems in other disciplines.

5. An **algorithm** is a precise and unambiguous sequence of instructions that lead to a problem's solution in a finite amount of time. To be useful, an algorithm must also give a solution to a specific class of problems.

6. **Pseudocode** consists of a small set of structured language elements similar to a programming language, together with English-like descriptions of the actions of the algorithm.

7. The **assignment statement** has the form

$$variable\ name \leftarrow expression$$

When executed, the expression on the right is evaluated, and the resulting value is assigned to the variable named on the left, replacing any value the variable may have had previously.

8. **Control statements** are used to structure the flow of control through the algorithm.

9. There are three types of control statements:
   a. the **sequence**, or **compound**, statement
   b. the **conditional** statement
   c. the **iterative** statement

10. A **compound** statement, or a sequence of statements, is a list of statements that are to be executed as a single unit in the order given.

11. A **conditional** statement is a statement that allows the selection of one alternative between two.

12. An **iterative** statement allows for the repetition of one or more statements.

13. There are three forms of iterative statements, which are often referred to as loops. The three iterative statements are:
   a. the **for** loop
   b. the **while** loop
   c. the **repeat . . . until** loop

14. The **for** loop executes a statement or a group of statements a fixed number of times.

15. The **while** and **repeat . . . until** loops cause a statement or group of statements to execute an indeterminate number of times while some condition remains true or until some condition becomes true.

REVIEW PROBLEMS

1. What will be the output of the following algorithm if the input value for n is 3?

> **begin**
> > $f \leftarrow 1$
> > Input n
> > **for** $i \leftarrow 1$ **to** n **do**
> > > $f \leftarrow f \cdot i$
> >
> > Output f
>
> **end.**

2. Consider the following algorithm.

> **begin**
> > Input x and y
> > **while** $x \neq y$ **do**
> > > **begin**
> > > > **if** $x < y$ **then**
> > > > > $u \leftarrow x$
> > > >
> > > > **else**
> > > > > $u \leftarrow y$
> > > >
> > > > **if** $x > y$ **then**
> > > > > $v \leftarrow x$
> > > >
> > > > **else**
> > > > > $v \leftarrow y$
> > > >
> > > > $x \leftarrow u$
> > > > $y \leftarrow v - u$
> > >
> > > **end**
> >
> > Output x
>
> **end.**

Find the output of the algorithm, given the following input values. (The first value is assigned to x, the second to y.

a. 3, 6   c. 3, 7
b. 9, 12   d. 12, 30

3. Find the flaw in the following loop.

```
begin
    count ← 1
    while count ≠ 9 do
        begin
            sum ← sum + x
            count ← count + 2
        end
end.
```

4. Find the output of the following algorithm.

```
begin
    for i ← 1 to 5 do
        for j ← 1 to 3 do
            Output i + j
end.
```

5. The following algorithm is supposed to compute the average of n values. What is wrong with it?

```
begin
    count ← 0
    sum ← 0
    Input n
    while count < n do
        begin
            Input (x)
            sum ← sum + x
            count ← count + 1
        end
    average ← sum / n
end.
```

6. Modify the algorithm in problem 5 so that it correctly computes the average of n numbers.

7. Develop an algorithm for measuring exactly 4 liters of water if only a 5-liter container and a 3-liter container are available.

8. Develop an algorithm to compute the roots of a quadratic equation.

9. Develop an algorithm that inputs n integers and counts the number of negative values, positive values, and zeros.

*10. Consider the following algorithm as an alternative to the algorithm in Example 1.7. Then trace it (show its execution) for the graph in Figure 1.3.

{Algorithm to reduce a graph to a graph of minimal total length.}

```
begin
    Order the edges according to decreasing lengths.
    m ← number of vertices
    remaining ← number of edges
    current ← 1
    while remaining > m − 1 do
        begin
            {Examine ordered edges in turn}
            if current edge does not disconnect
              graph then
                begin
                    Delete current edge
                    remaining ← remaining − 1
                end
            current ← current + 1
        end
end.
```

2   Logic and Sets

The use of logic is essential in any discipline that is a formal science. Within computer science, logic is directly applicable to the design and verification of algorithms, to the design of circuits (which are often called *logic circuits*), and in certain problem-solving approaches called logic programming. The second part of this chapter contains an introduction to sets. Sets and the operations on sets will be found in various forms throughout the remainder of the text.

2.1   Logic and Propositions

Logic consists of a set of rules for drawing inferences. Notice that we did not conclude the previous sentence with a phrase such as *about topic X*. Formal logic, or symbolic logic, is a method that we can apply to any set of concepts. For example, plane geometry presents a set of concepts and axioms from which numerous theorems are derived using logical rules. Similarly, in algebra, given definitions and properties of real numbers can be used to derive other, less evident properties of real numbers. The two subject areas have different content, but the same rules of logic can be applied to both. Hence, the form of logic can be separated from the content of the subject to which it is applied. In our discussion of logic, we will devote attention first to propositional logic and then to predicate logic. Then we will look at methods of establishing proofs. The last topic in the section concerns mathematical induction, which is an extremely important method of proving statements in computer science applications.

Propositions

In an axiomatic system, we assume that certain key statements, usually about a class of objects, are *true*. These statements are called **axioms**. It is possible to make many statements that are meaningful within the axiomatic system, but whose truth value we do not know. For instance, in plane geometry we can assert "Every right triangle is an equilateral triangle." Some statements will be *true* and some (such as the example) will be *false*. For any such statement, we can apply the rules of logic in an attempt to derive the truth value of the statement from the axioms. If we find that the statement is *true*, we call the statement a *theorem*. The sequence of logical steps leading from the axioms to the statement is called a *proof* of the theorem.

We define a **proposition** to be a statement that has a **truth value**; that is, the statement is either *true* or *false*. Here are some simple examples of propositions that are *true*:

100 > 99

Harry Truman was a U.S. President.

The moon is not made of green cheese.

Here are some examples of propositions that are *false*:

7 > 42

There are 1200 members of the U.S. Congress.

Some sentences we will not allow as propositions because they do not have a truth value. These sentences are known as *paradoxes* because it is impossible to assign a truth value to them without also concluding that the sentence has the opposite truth value. Consider the following sentence:

This sentence is false.

If we say that the sentence is *true*, then we have to conclude, because of what it says, that it is also *false*. However, if we say that the sentence is *false*, it must be *true*. There are numerous other examples of sentences involving this vicious circle, including such "truisms" as "Never say never" and "There is an exception for every rule."

There are other kinds of sentences to which we might assign a certain truth value but to which others might assign the opposite value. For instance, consider the following sentence:

Robert Heinlein is a better author than Arthur C. Clarke.

There may be grounds for believing that the statement is *true*, but one is likely to get an argument from avid Clarke fans. Assigning a truth value to the statement requires a subjective judgment, and such judgments are not of real concern in logic. However, nothing prevents us from including such a subjective sentence as part of our assumptions. Then we have to remember that changing our assumptions can have a drastic effect on the logical consequences.

Logical Operations

Propositions can be built from other propositions through the use of logical operations. Some logical operations are called **connectives** because they are used to connect two propositions. The connectives are **and**, **or**, and **if–then**. We also use the logical operator **not**, which is not a connective because it affects only one proposition. In logic, these operators are defined precisely so that the result is always a compound proposition with a value of *true* or *false*. Furthermore, the truth value of the result depends only on the truth value of the components. Once we know the truth value of our initial propositions, we can always determine the truth value of propositions constructed from them, no matter how complex such propositions become.

We start with **not**. In logic, just as in English, the use of **not** acts to reverse the truth value of the statement with which it is used. The operation of applying **not** to a proposition is called **negation**. We can state the definition of negation as follows: The negation of a true proposition results in a false one; the negation of a false proposition results in a true one. This is summarized in the table below. A table such as this one, which precisely and concisely defines the meaning of **not**, is called a **truth table**. In the table, *true* is represented as T, and *false* is represented as F.

P	not P
T	F
F	T

Truth table for **not**

In the table, the letter P represents a proposition, and **not** P represents the negation of that proposition. In the case that P is the proposition "$7 > 5$," **not** P stands for "7 is not greater than 5," or "$7 \leq 5$." Note also that P is *true* while **not** P is *false*.

EXAMPLE 2.1

Show that the negation of a negation has the same truth value as the original proposition; that is, **not** (**not** P) has the same truth value as P.

Solution

We construct the following table:

P	not P	not (not P)
T	F	T
F	T	F

Examining the first and last columns of the truth table reveals that P and **not** (**not** P) have the same truth value.

We say that two propositions are **logically equivalent** if they have the same truth value. Example 2.1 shows that the proposition P is logically equivalent to the proposition **not** (**not** P).

Next, consider the logical connective **and**. Suppose that P stands for the proposition "7 > 5," and Q stands for the proposition "8 is even." What do we say about the truth value of P **and** Q? Since P and Q are both *true*, we expect that the compound proposition P **and** Q is also *true*. The proposition P **and** Q is called the **conjunction** of P and Q. We define **and** by assigning truth values for all the remaining combinations of the truth values of P and Q. The definition is shown in the truth table for **and** below.

P	Q	P **and** Q
T	T	T
T	F	F
F	T	F
F	F	F

*Truth table for **and***

EXAMPLE 2.2

Construct the truth table for the compound proposition P **and** (**not** Q) and determine the truth value when P is *true* and Q is *false*.

Solution

We construct the following truth table:

P	Q	**not** Q	P **and** (**not** Q)
T	T	F	F
T	F	T	T
F	T	F	F
F	F	T	F

Examining the second row of the truth table reveals that P **and** (**not** Q) is *true* when P is *true* and Q is *false*.

The connective **or** has two meanings in English. One is called the *exclusive or* and is sometimes expressed as "either one or the other, but not both." The second is called the *inclusive or* and is sometimes expressed in legal documents as "and/or." In mathematics and in logic, **or** *always* means the *inclusive or*. The truth table for **or** is shown below. The compound proposition P **or** Q is often called the **disjunction** of P and Q.

P	Q	P **or** Q
T	T	T
T	F	T
F	T	T
F	F	F

*Truth table for **or***

The only time P **or** Q is *false* is when both P and Q are *false*.

EXAMPLE 2.3

Let P be the proposition "Today is Friday," and let Q be the proposition "Yesterday it rained." Describe the following propositions in words.

1. P **and** Q
2. P **or** Q
3. **not** P

Solution

1. P **and** Q means "Today is Friday and yesterday it rained."
2. P **or** Q means "Today is Friday or it rained yesterday."
3. **not** P means "Today is not Friday."

EXAMPLE 2.4

Verify, by means of truth tables, that the propositions R = **not** (P **and** Q) and S = (**not** P) **or** (**not** Q) are logically equivalent.

Solution

We have to construct two truth tables. The truth table for R is as follows:

P	Q	P **and** Q	R
T	T	T	F
T	F	F	T
F	T	F	T
F	F	F	T

Look at the fourth row, where P is *false* and Q is *false*. As the third column shows, (P **and** Q) is *false*. By definition of **not**, R is *true*. Now look at the following truth table for S:

P	Q	**not** P	**not** Q	S
T	T	F	F	F
T	F	F	T	T
F	T	T	F	T
F	F	T	T	T

Look at the fourth row here as well. Since P and Q are both *false*, **not** P and **not** Q are both *true*. Hence, by the definition of **or**, S must be *true*.

The logical equivalence in Example 2.4 is half of a theorem known as **DeMorgan's Laws**. The other half is obtained by interchanging the **and** and the **or**. You will be asked to confirm this in the exercises.

The remaining basic connective is called **implication** and uses the phrase *if . . . then*. Assume that P and Q are propositions. The truth table below defines the truth value of the implication "If P, then Q." Another phrase for "If P, then Q" is "P implies Q." Implication is written symbolically as P $\Rightarrow$ Q.

P	Q	P $\Rightarrow$ Q
T	T	T
T	F	F
F	T	T
F	F	T

Truth table for implication

The third column of the truth table for implication contains the truth values of "P implies Q" as determined from the truth values of P and Q. The first and second rows are probably what you expected, but the third and fourth rows often surprise people when they first look at them. Let's consider an example where P is *false*, which is often called the *vacuous case*.

Let P be the proposition "The moon is made of green cheese." We have agreed that P is *false*. Let Q be the proposition "Ludwig van Beethoven was the King of France." According to the truth table, the compound proposition "P implies Q" is *true*! Thus the following proposition is a true statement:

> If the moon is made of green cheese, then Ludwig
> van Beethoven was the King of France.

Do not despair! The following proposition is also a true statement:

> If the moon is made of green cheese, then Ludwig
> van Beethoven was not the King of France.

In other words, as soon as we discover that the moon is not made of green cheese, we no longer care whether Beethoven was King of France. We are willing to accept the two implications as *vacuously true*. We note that the only way the proposition $P \Rightarrow Q$ can be *false* is if P is *true* and Q is *false*.

EXAMPLE 2.5

Determine if the following sentences are logically equivalent:

1. If it is sunny Saturday and my homework is done, I am going cycling.
2. If it is sunny Saturday, then, if my homework is done, I am going cycling.

Solution

Define P, Q, and R as follows:

P: It is sunny Saturday.
Q: My homework is done.
R: I am going cycling.

Then the first sentence is (P **and** Q) $\Rightarrow$ R, and the second sentence is $P \Rightarrow (Q \Rightarrow R)$. We have eight possible truth value combinations for P, Q, and R. We construct the following truth tables:

P	Q	R	P **and** Q	(P **and** Q) $\Rightarrow$ R
T	T	T	T	T
T	T	F	T	F
T	F	T	F	T
T	F	F	F	T
F	T	T	F	T
F	T	F	F	T
F	F	T	F	T
F	F	F	F	T

P	Q	R	Q $\Rightarrow$ R	P $\Rightarrow$ (Q $\Rightarrow$ R)
T	T	T	T	T
T	T	F	F	F
T	F	T	T	T
T	F	F	T	T
F	T	T	T	T
F	T	F	F	T
F	F	T	T	T
F	F	F	T	T

The last columns of the truth tables are identical, so we conclude that the two sentences are logically equivalent.

SELF TEST 1

1. Assume that P and Q are both true propositions and R and S are both false propositions. Determine whether each of the following is *true* or *false*.
   a. P **and** R   c. P ⇒ Q
   b. P **or** S   d. R ⇒ (**not** S)
2. Construct the truth table for the compound proposition P **or** (**not** Q). Determine the truth value of P **or** (**not** Q) when P is *true* and Q is *false*.
3. Let P be the proposition "The sunset is red," and let Q be the proposition "Tomorrow will be clear." Describe each of the following propositions in words.
   a. P **and** Q
   b. P **or** Q
   c. P ⇒ Q
4. Determine whether the following sentences are logically equivalent.
   a. It is false that the cat is sleeping and the dog is barking.
   b. The cat is not sleeping or the dog is not barking.

Exercises 2.1

1. Let P, Q, and R be propositions defined as follows:

   P: I am thirsty.
   Q: My glass is empty.
   R: It is three o'clock.

   Write each of the following propositions as expressions involving P, Q, and R.
   a. I am thirsty and my glass is not empty.
   b. It is three o'clock and I am thirsty.
   c. If it is three o'clock, then I am thirsty.
   d. If I am thirsty, then my glass is empty.
   e. If I am not thirsty, then my glass is not empty.

2. Construct the truth table for (**not** P) **or** Q. Compare this with the truth table for P $\Rightarrow$ Q.

3. Verify the other half of DeMorgan's Laws; namely, **not** (P **or** Q) is logically equivalent to (**not** P) **and** (**not** Q).

4. Construct a truth table for each of the following compound propositions.

   a. P **and** (Q **or** R)
   b. P **or** (Q **and** R)
   c. P **or** (Q **or** R)
   d. (P **or** Q) **or** R
   e. (P **and** Q) $\Rightarrow$ R
   f. R $\Rightarrow$ (**not** P $\Rightarrow$ Q)
   g. **not** P $\Rightarrow$ **not** (R **and** Q)

5. Which of the propositions in exercise 4 are logically equivalent?

6. Show that the proposition **not** Q $\Rightarrow$ **not** P is logically equivalent to the proposition P $\Rightarrow$ Q.

2.2   Predicate Logic

In our discussion of propositions and the principles for building new propositions from existing ones, we have so far dealt only with simple declarative sentences. How can we form more complex (and more useful) propositions? In general, we are interested in making statements that include properties of objects or relationships among them. For instance, we might want to say, "There exists an integer n such that $n^2 =$ 25." In this sentence we are *quantifying* the existence of an object (some integer) satisfying a particular property (its square is 25). The property is an example of a **predicate**, or a statement about an object. We use the notation P(x) to indicate that the predicate concerns properties of the object x. There are two **predicate quantifiers**: **for all** and **there exists**. We will use $\forall$ to denote "for all," and $\exists$ to denote "there exists."

There are often some subtleties involved in using quantifiers. For instance, if we have one plate that all students must share, we say

$$\exists \text{ one plate } \forall \text{ students.}$$

But, if each student has a separate plate, we would say

$$\forall \text{ students } \exists \text{ one plate.}$$

In English, the sentences "For all students there exists one plate" and "There exists one plate for all students" appear to mean the same thing and would probably be accepted as such. However, in the formalism we have defined, their meaning is quite distinct, as you can see from Figure 2.1.

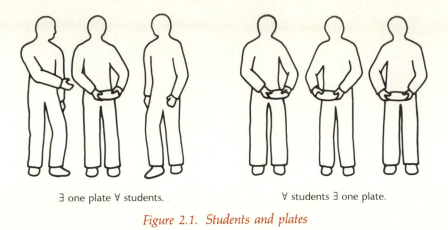

∃ one plate ∀ students.   ∀ students ∃ one plate.

Figure 2.1. Students and plates

EXAMPLE 2.6

Assume that x is an integer. Let $P(x)$ be the predicate "$x = x^2$." What is the truth value of each of the following propositions?

1. $\forall x\, P(x)$
2. $\exists x$ such that $P(x)$

   Solution

1. Here the question is whether *every* integer equals its own square. A little reflection shows that if $x = 2$, then $x^2 = 4$. Thus the proposition is *false*.
2. If $x = 1$, then $x^2 = 1$, so this proposition is *true*.

EXAMPLE 2.7

Let x and y be positive integers. Express the negation of each of the following propositions.

1. $\forall x\, P(x)$
2. $\exists y$ such that $Q(y)$

   Solution

1. If a predicate $P(x)$ is not *true* for all x, there exists some x such that $P(x)$ is *false*. Then **not** $P(x)$ is *true*. We are saying

$$\textbf{not}\ (\forall x\, P(x))$$

   is the same as

$$\exists x \text{ such that } (\textbf{not}\ P(x))$$

2. Similarly, we can see that for a proposition $Q(y)$,

$$\textbf{not } (\exists y \text{ such that } Q(y))$$

is the same as

$$\forall y \text{ } (\textbf{not } Q(y))$$

---

1. Assume that x is an integer. Let $P(x)$ be the predicate "$2x = x^2$." What is the truth value of each of the following propositions?
   a. $\forall x \text{ } P(x)$
   b. $\exists x$ such that $P(x)$
2. Let x and y be positive integers. Express the negation of the following proposition.

$$\forall x \text{ } \exists y \text{ such that } y^2 > x$$

Exercises 2.2

1. Use appropriate capital letters to write each of the following predicates in symbolic form.
   a. There exists an integer x such that $x = x/2$.
   b. All cats drink milk.
   c. For each integer x, $x < x + 1$.
2. Let x and y be integers. Rewrite the expression

$$\textbf{not } (\exists x \text{ such that } \forall y \text{ } y > x)$$

by eliminating the operator **not**.
3. Write English sentences that are the negations of each of the following sentences.
   a. All dogs bark.
   b. Some birds fly.
   c. No cat likes to swim.
   d. No computer science major does not know mathematics.
4. Use convenient symbols to express each of the propositions in exercise 3 symbolically. Then express each negation symbolically.

2.3   Proofs

Having established some of the basic tools for propositions and predicates, we are ready to use these tools for constructing logical arguments. Logical arguments are the same thing as *proofs*. Many students find formal proofs tedious, but they are an integral part of mathematics, and the logical arguments that we will present apply to the design and verification of algorithms as well.

We typically encounter the need for formal proofs in situations where we are given some simple or compound propositions that we assume to be true. We want to use a logical argument to get from these assumptions to a conclusion.

One of the most commonly encountered proofs in mathematics involves proving an implication. For example, consider a well-known theorem from plane geometry.

> The Pythagorean Theorem says that in a right triangle with hypotenuse C and sides A and B, $C^2 = A^2 + B^2$. Let $P(x)$ be the proposition "x is a right triangle with hypotenuse C and sides A and B." Let $Q(x)$ be the proposition "For triangle x, $A^2 + B^2 = C^2$." Then the Pythagorean Theorem is the implication $\forall x\, P(x) \Rightarrow Q(x)$. To prove the theorem is to prove this implication.

Remember that an implication $P \Rightarrow Q$ is *true* when both P and Q are *true* and when P is *false*. There are several standard methods of proving that an implication $P \Rightarrow Q$ is *true*.

1. Assume P is *true*, and show that Q must be *true*. This is often called a **direct argument**.
2. Assume Q is *false*, and show that P must be *false*. This approach is called proving the **contrapositive**.
3. Show that P is always *false*, so that the implication must be vacuously *true*.
4. Show that Q is always *true*.
5. Show that if P is *true* and Q is *false*, then a contradiction results.

We will not consider proofs involving situations where P is always *false* or where Q is always *true* because such situations are generally of little interest.

Direct Proofs

Direct arguments are probably familiar to you from algebra and plane geometry. Let us consider an example.

EXAMPLE 2.8

Recall that the absolute value of a number x, denoted $|x|$, is x when x is positive or zero and is $-x$ when x is negative. Give a direct proof for the proposition "If $|x| > |y|$, then $x^2 > y^2$."

Solution

We first observe that $|x|^2 > |y|^2$, since $|x| > |y|$. Also $|z|^2 = z^2$ for any number z. The conclusion follows.

Contrapositive Proofs

The second kind of argument, contraposition, is based on the fact that **not** $Q \Rightarrow$ **not** P is logically equivalent to $P \Rightarrow Q$ (see exercise 6 on page 73). In some instances, a contrapositive proof is easier to begin.

EXAMPLE 2.9

Let n be a positive integer. Give a contrapositive proof that, if n is a prime number different from 2, then n is odd.

Solution

The contrapositive form of the statement "If n is a prime number different from 2, then n is odd" is "If n is even, then either $n = 2$ or n is not a prime number." We suppose that n is even. Then $n = 2 \cdot p$ for a positive integer, $p < n$. Now, either $p = 1$ or $p > 1$. If $p = 1$, then $n = 2$. If $p > 1$, then n is not prime because n is divisible by p and $p > 1$. In either case, we have proven the contrapositive. Therefore, the initial implication is *true*.

You might wonder how to decide which of the two methods to apply in a given situation. Practice and experience will give you intuition as to which will be the better method, based on the characteristics of the problem.

Proofs by Contradiction

A **proof by contradiction** is a proof of an implication that shows that joining the assumption "Q is *false*" together with the premise "P is *true*" leads to a contradiction. To see how a proof by contradiction works, we will give another proof of the implication from Example 2.9.

EXAMPLE 2.10

Let n be a positive integer. Give a proof by contradiction that if n is a prime number different from 2, then n is odd.

Solution

Let $P(n)$ be the statement "n is a prime number different from 2," and let $Q(n)$ be the statement "n is odd." To carry out the proof by contradiction, we need to assume that n is a prime number different from 2 and that n is even. Then if we find a contradiction, the proof is complete. From the assumption, we see that $n = 2 \cdot p$ for some positive integer p. If $p = 1$, then $n = 2$. If $p > 1$, then n is not prime because n is divisible by p and $p > 1$. In either case, we obtain a contradiction, so the initial implication is *true*.

Existence Proofs

Often we are confronted with a proposition of the form "$\exists x$ such that $P(x)$." This proposition guarantees the existence of at least one x for which the predicate $P(x)$ is true. In general, the proof of such a statement is *constructive*; that is, the proof is an algorithm for finding the object x. For example, let $P(x)$ be the predicate "x is an integer and $x^2 = 15{,}129$." We can decide whether or not there exists an x satisfying $P(x)$ by applying the well-known algorithm for extracting square roots. In this case, the correct value of x is 123. There are some existence proofs that are of necessity not constructive. They show that the required object must exist without necessarily showing how to find it.

Counterexamples

The last topic we will take up in relation to proof techniques has to do with demonstrating that a proposition of the form

$$\forall x\, P(x)$$

is *false*. Consider the following statement:

$$\forall \text{ positive integers } x \, \exists \text{ an integer } y \text{ such that } y^2 = x$$

We might try to prove this statement by any of the methods above, but as you might see already, these attempts would be futile. To conclude that the statement is *false*, we simply find one example of a positive integer x that fails to satisfy the equality for all y. In this example, $x = 3$ is sufficient. Thus, to show that a statement of the form

$$\forall x\, P(x)$$

is false, it suffices to find one **counterexample**, that is, one value of x that does not satisfy $P(x)$.

SELF TEST 3

1. Give a direct proof that for real numbers x and y, $|x| + |y| \geq |x + y|$.
2. Give a contrapositive proof that, for a real number x, if $|x| > 1$, then either $x > 1$ or $x < -1$.

Exercises 2.3

1. Give direct and contrapositive proofs of each of the following propositions.
   a. If n^2 is an even integer, then n is an even integer.
   b. If n and m are even integers, then $n + m$ is an even integer.
   c. If n and m are even integers, then $n - m$ is an even integer.
2. Give a proof by contradiction of each of the following propositions.
   a. If n and m are odd integers, then $n + m$ is an even integer.
   b. If n is an even integer and m is an odd integer, then $n + m$ is an odd integer.
3. Find a counterexample to show that the following proposition is false.

   If $x = p^2 + 1$, where p is a positive integer, then x is a prime number.

2.4   Mathematical Induction

Principle of Mathematical Induction

In this section, we will consider proofs based on the Principle of Mathematical Induction. Mathematical induction is useful for proving propositions that must be true for all integers or for a range of integers. Problems to which induction applies have the following form:

$$\text{Prove: } P(n) \text{ for all integers } n \geq k$$

The **Principle of Mathematical Induction** is stated as follows:

Let $P(n)$ be a proposition that is valid for $n \geq k$,
n, k integers.

If (1) $P(k)$ is *true*, and
   (2) $\forall n \geq k, P(n) \Rightarrow P(n + 1)$,
then $P(n)$ is *true* $\forall n \geq k$.

The statement $P(n)$ is called the **inductive hypothesis**, condition (1) is called the **base step**, and condition (2) is called the **induction step**. We have to be careful when using induction to check that both condition (1) and condition (2) are satisfied. To establish the implication in condition (2), we can use any of the methods for proofs discussed so far in this chapter. In general, an induction argument follows the following pattern: First we show that the proposition is *true* for the initial value, k. Then we assume that the proposition is *true* for $n \geq k$ and show that it is also true for $n + 1$. At first, when we say, "Assume that P(n) is true," and then derive P($n + 1$), it may seem that we are arguing in circles, but if you read the Principle of Mathematical Induction carefully, you will see that this is not the case.

EXAMPLE 2.11

Prove, by means of induction, that for all $n \geq 1$,

$$1 + 2 + 3 + \cdots + n = \frac{n(n + 1)}{2} \tag{2.1}$$

Solution

First, observe that the left side of the induction hypothesis calls for the sum of the first n integers; the ellipsis (. . .) indicates that the missing terms are to be supplied. The base step, P(1), is the statement

$$1 = \frac{1(1 + 1)}{2}$$

which is *true*. For the induction step, assume that Equation (2.1) holds for some n. To verify condition (2) of the Principle of Mathematical Induction we need to show that

$$1 + 2 + 3 + \cdots + n + (n + 1) = \frac{(n + 1)(n + 2)}{2} \tag{2.2}$$

But the left side of Equation (2.2) is

$$\frac{n(n + 1)}{2} + (n + 1) = \frac{n(n + 1) + 2(n + 1)}{2} \tag{2.3}$$

$$= \frac{(n + 1)(n + 2)}{2}$$

Thus we have verified conditions (1) and (2) of the Principle of Mathematical Induction, so we can conclude that Equation (2.1) is *true* for all $n \geq 1$.

In Example 2.11, we used the Principle of Mathematical Induction to establish an equality for a range of integers. We can also use induction to verify that from some point on (that is, for $n \geq k$) one expression is larger than another. Such inequalities are useful in comparing relative growth rates for integer expressions. For instance, it may be clear that at some point the square of an integer is larger than a constant multiple of it (that is, $n^2 > M \cdot n$ for a sufficiently large n.) You can use Example 2.12 as a model for building an inductive proof of the fact.

EXAMPLE 2.12

Show that for all $n \geq 5$,

$$2^n > n^2 \tag{2.4}$$

Solution

First observe that the base step, P(5), reduces to

$$2^5 > 5^2$$
$$32 > 25$$

which is certainly *true*.

Now, for the induction step, assume that Equation (2.4) is *true* for some $n \geq 5$. To obtain P($n + 1$), we replace the occurrence of n on the left side of the equation by ($n + 1$). This yields

$$\begin{aligned} 2^{n+1} &= 2 \cdot 2^n \\ &> 2 \cdot n^2 \\ &= n^2 + n^2 \end{aligned}$$

Since $n^2 > 2n + 1$ for $n \geq 5$, we have

$$2^{n+1} > n^2 + 2n + 1 = (n + 1)^2$$

for $n \geq 5$. Hence, we obtain the inequality

$$2^{n+1} > (n + 1)^2$$

as desired.

SELF TEST 4

1. Prove by induction that $n^3 + 2n$ is divisible by 3 for all $n \geq 1$.
2. Prove by induction that $n^2 > 2n + 1$ for $n \geq 3$.

Inductive Definition

Mathematical induction may also be used to define sequences of objects. An inductive definition follows the form of the Principle of Mathematical Induction. The first element of the sequence is defined, and then the nth element is defined in terms of preceding elements. In computer science, inductive definitions are called **recursive definitions**. A simple example of an inductive definition is the definition of the factorial of an integer. The factorial of an integer n is denoted $n!$. First, in the base step, we define $0!$ to equal 1. For the induction step, given $n!$ for $n \geq 0$, we define $(n + 1)! = (n!)(n + 1)$. By applying the Principle of Mathematical Induction, we can verify that $n!$ is defined for all $n \geq 0$. Applying our definition to $3!$, we have

$$
\begin{aligned}
3! &= 2! \cdot 3 \\
&= 1! \cdot 2 \cdot 3 \\
&= 0! \cdot 1 \cdot 2 \cdot 3 \\
&= 1 \cdot 1 \cdot 2 \cdot 3 \\
&= 6
\end{aligned}
$$

The definition of a factorial illustrates why the term *recursive* is applied to an inductive definition: The operation or computation being defined recurs in the definition.

In Example 2.11, we used $1 + 2 + \cdots + n$ to indicate the sum of the first n integers. We know that addition is actually defined as a binary operation because two numbers are added at a time. We can use an inductive definition to extend addition to a sum of an arbitrary sequence of numbers. We will use the Greek letter Σ (capital sigma) to denote the summation of a sequence of values. Let $x_1, x_2, \ldots,$ be real numbers. We define the sum of the first n elements of the sequence by

$$
\sum_{i=1}^{1} x_i = x_1
$$

and

$$
\sum_{i=1}^{k+1} x_i = \sum_{i=1}^{k} x_i + x_{i+1}
$$

For example,

$$\sum_{i=1}^{n} i^2$$

means the "sum of all values of i^2 as i takes on the values 1 to n." We can use summation notation to rewrite equation (2.1) as

$$\sum_{i=1}^{n} i = \frac{n(n + 1)}{2}$$

Other operations on real numbers, such as multiplication, can also be extended to an operation on an arbitrary sequence of numbers. We will use this idea later for other structures that have operations defined on them.

Notice how the definition of the general summation notation follows the form of the Principle of Mathematical Induction. That is, the base step is established, and then, given the term defined for n, the term for $n + 1$ is defined. This kind of inductive definition can also be used to formalize the definition of other operations on sequences of numbers. A **recurrence relation** is an equation that defines the ith value in a sequence of numbers in terms of the preceding $i - 1$ values. A factorial is an example of a sequence satisfying a recurrence relation, namely, $0! = 1$ and $n! = (n - 1)! \cdot n$. Observe that the initial values must be specified to get the recurrence relation started. In combinatorics, we will encounter several important recurrence relations. Some will have simple solutions, but others will be impossible to solve exactly. For now, we will be content with simple examples of recurrence relations.

EXAMPLE 2.13

Find the first six terms of the sequence satisfying the recurrence relation

$$x_1 = 2 \qquad x_2 = 1$$
$$x_{n+2} = 3x_n - 2x_{n+1}$$

for $n > 2$.

Solution

We are already given $x_1 = 2$ and $x_2 = 1$. Then

$$x_3 = 3x_1 - 2x_2 = 6 - 2 = 4$$
$$x_4 = 3x_2 - 2x_3 = 3 - 8 = -5$$
$$x_5 = 3x_3 - 2x_4 = 12 - (-10) = 22$$
$$x_6 = 3x_4 - 2x_5 = -15 - 44 = -59$$

Induction is an important proof technique, which can serve as a model for inductive (recursive) definitions. Later, we will use induction arguments in diverse situations. The following exercises give you some practice using induction. You might find it useful to return to this section for review as you progress through this book.

Exercises 2.4

1. Prove each of the following statements by induction.
   a. For all $n \geq 1$, $\sum_{i=1}^{n} i^2 = n(n + 1)(2n + 1)/6$.
   b. $(n^3 - n)$ is divisible by 3 for all $n \geq 0$.
   c. For each real number a different from 1 and for each integer $n \geq 0$,

$$\sum_{i=0}^{n} a^i = 1 + a + a^2 + a^3 + \cdots + a^n = \frac{a^{(n+1)} - 1}{a - 1}.$$

2. The Fibonacci sequence is the sequence of integers satisfying the recurrence relation

$$x_1 = 1 \quad \text{and} \quad x_2 = 1$$
$$x_{n+2} = x_n + x_{n+1}$$

   for $n \geq 1$. Compute x_3, x_4, and x_7.

3. Find the first six terms of the sequence satisfying the recurrence relation given below.

$$x_1 = 3, x_2 = 2, x_3 = 1$$
$$x_{n+3} = 2 \cdot x_n - x_{n+1} - x_{n+2}$$

4. Let M be a positive real number. Show that there exists k such that for $n \geq k$, $n^2 > M \cdot n + 1$. (*Hint:* First use an induction argument in the special case $M = 2$. Then generalize.)

*5. Let a be a positive real number. We define the *powers of a* inductively by

$$a^0 = 1$$

   and, for all $n \geq 0$,

$$a^{n+1} = a^n \cdot a$$

   Use an induction argument to show that for all n and m,

$$a^{n+m} = a^n \cdot a^m$$

2.5   Correctness of Algorithms (Optional) ━━━━━━

As we indicated in Chapter 1, there are two general issues about algorithms that we will consider in this book, (1) is the algorithm efficient? and (2) is the algorithm correct? The first question concerns storage requirements and execution time of an algorithm. We will discuss execution time in Chapter 4, but storage requirements need to be addressed from the perspective of data structures. The second question concerns proving that an algorithm does what it was designed to do. Proving that an algorithm is **correct**—that it does what it is supposed to do—requires proof methods similar to those we have just discussed. In this section, we will examine the form of such proofs. Verifying the correctness of an algorithm can be a difficult task; only relatively simple algorithms can be verified using the approach we will describe. However, understanding the techniques involved should help you design better algorithms.

Assertions

The approach we will use to prove correctness of algorithms is based on assertions about the variables used in the algorithm prior to, during, and after its execution. Such assertions are sometimes called **algorithm assertions**. Assertions take the form of predicates. We will use the notation $\{P\}A\{Q\}$ to denote "If execution of algorithm A starts with predicate P *true*, then it will terminate with predicate Q *true*." The predicate P is called the **precondition** for algorithm A; the predicate Q is called the **postcondition**. The statement $\{P\}A\{Q\}$ is itself a predicate. Proving the correctness of an algorithm amounts to proving that $\{P\}A\{Q\}$ is *true*.

───────────────────────────────

EXAMPLE 2.14

State the precondition and postcondition assertions for an algorithm that computes the sum of two numbers.

  Solution

The algorithm *Sum* is as follows:

Algorithm *Sum*
begin
    $z \leftarrow x + y$
end.

The precondition can be stated as

$\{x = x_1 \textbf{ and } y = y_1\}$

In the statement of the precondition, x and y are variables while x_1 and y_1 are arbitrary values of these variables. Thus the precondition states that x and y have arbitrary values. Then the postcondition is

$$\{z = x_1 + y_1\}$$

The algorithm can be written with its precondition and postcondition, inserted as comments.

Algorithm *Sum*
$\{x = x_1$ **and** $y = y_1\}$
begin
     $z \leftarrow x + y$
end.
$\{z = x_1 + y_1\}$

To prove the correctness of the algorithm in Example 2.14, we must prove that $\{P\}Sum\{Q\}$ is *true*. That is, we must prove that if "$x = x_1$ **and** $y = y_1$" is *true* and algorithm *Sum* terminates, then "$z = x_1 + y_1$" is *true*. Thus we need to be able to determine for every statement of the algorithm the effect of its execution on the variables. Let us consider various types of statements in turn.

Sequence Statements

To establish the correctness of a sequence or compound statement, we use the following inference rule, where A_1 and A_2 are algorithms:

> If $\{P\}A_1\{Q_1\}$ and $\{Q_1\}A_2\{Q\}$ are both *true*, then it follows that $\{P\}A_1; A_2\{Q\}$ is *true*.

Note that the postcondition for algorithm A_1 becomes the precondition for algorithm A_2. Thus if we wish to prove that some algorithm A is correct with respect to some initial conditions P, we can break A into small segments $A_1, A_2, \ldots, A_n$ and then obtain intermediate assertions $Q_1, Q_2, \ldots, Q_{n-1}$. By proving the n results

$$\{P\}A_1\{Q_1\}, \{Q_1\}A_2\{Q_2\}, \ldots, \{Q_{n-2}\}A_{n-1}\{Q_{n-1}\}, \{Q_{n-1}\}A_n\{Q\}$$

we will have established $\{P\}A\{Q\}$.

EXAMPLE 2.15

Prove that the following algorithm correctly interchanges the values of x and y.

Algorithm *Swap*
$\{x = x_1$ **and** $y = y_1\}$
begin
       temp ← x
       x ← y
       y ← *temp*
end.
$\{x = y_1$ **and** $y = x_1\}$

Solution

We insert intermediate assertions.

Algorithm *Swap*
$\{x = x_1$ **and** $y = y_1\}$
begin
       temp ← x
       $\{temp = x_1$ **and** $x = x_1$ **and** $y = y_1\}$
       x ← y
       $\{temp = x_1$ **and** $x = y_1$ **and** $y = y_1\}$
       y ← *temp*
end.
$\{x = y_1$ **and** $y = x_1\}$

Each assertion follows directly from the preceding assertion and the assignment statement. Hence, $\{x = x_1$ **and** $y = y_1\}$*Swap*$\{x = y_1$ **and** $y = x_1\}$ is correct.

Conditional Statements

Recall that when the conditional statement

if condition **then**
    Statement 1

is encountered, Statement 1 is executed only when the value of the condition is *true*. Thus there are two distinct paths through the **if–then** statement, depending on the truth value of condition. This leads to the following inference rule for **if–then** statements:

If {P **and** condition} Statement 1 {Q} is correct and the implication (P **and not** condition) $\Rightarrow$ Q is true, then it follows that

{P}
(**if** condition **then**
        Statement 1)
    {Q}

is correct.

The corresponding inference rule for

if condition **then**
        Statement 1
else
        Statement 2

is as follows:

If {P **and** condition} (Statement 1) {Q} and {P **and not** condition} (Statement 2) {Q} are both correct, then it follows that

{P}
(**if** condition **then**
        Statement 1
    else
        Statement 2)
    {Q}

is correct.

EXAMPLE 2.16

Prove that the algorithm *AbsoluteValue1* is correct.

Algorithm *AbsoluteValue1*
$\{x = x_1\}$
begin
        if $x < 0$ **then**
                $x \leftarrow -x$
end.
$\{x = |x_1|\}$

Solution

To verify the above algorithm, we need to show that

$\{(x = x_1) \textbf{ and } (x < 0)\}$
    $x \leftarrow -x$
$\{x = |x_1|\}$

is correct. The precondition $\{(x = x_1) \textbf{ and } (x < 0)\}$ and execution of the assignment statement permit us to conclude that $\{(x = x_1) \textbf{ and } (x > 0)\}$. It follows from the definition of absolute value that $x = |x_1|$.

We also need to show that $(x = x_1) \textbf{ and not } (x < 0) \Rightarrow (x = |x_1|)$. Since **not** $(x < 0)$ is $(x \geq 0)$ and since for $x \geq 0$, $x = |x|$, the implication is *true*. Therefore, the algorithm is correct.

EXAMPLE 2.17

Prove that the algorithm *AbsoluteValue2* is correct.

Algorithm *AbsoluteValue2*
$\{x = x_1\}$
begin
        if $x < 0$ **then**
                $y \leftarrow -x$
        else
                $y \leftarrow x$
end.
$\{y = |x_1|\}$

Solution

To verify the above algorithm we need to show that

$\{(x = x_1) \textbf{ and } (x < 0)\}$
        $y \leftarrow -x$
$\{y = |x_1|\}$

is correct. The argument is the same as for the first part of Example 2.16.
We also need to show that

$\{(x = x_1) \textbf{ and not } (x < 0)\}$
        $y \leftarrow x$
$\{y = |x_1|\}$

is correct. Here, as in Example 2.16, since for $x \geq 0$, $x = |x|$, the assignment statement will result in $y = |x_1|$. Therefore, we conclude that the algorithm is correct.

Iteration

The inference rule for iterative statements is more difficult than the others. Recall that when the iterative statement

while condition **do**
      Statement 1

is encountered, Statement 1 is executed as long as condition is *true*. When condition is **false**, execution proceeds with the statement following the **while . . . do** statement. The inference rule for the **while . . . do** statement requires that an assertion be true before the statement is executed and that it remain true after each iteration of the loop. This assertion is known as the **loop invariant**. If we can show that the loop invariant is *true* before the execution of the loop and that the invariant remains *true* after each execution of the loop, then to demonstrate the correctness of the loop, all we need to do is show that the loop condition is *false* when the loop terminates. More formally, if {Invariant **and** condition}(Statement){Invariant} is correct, then it follows that

{Invariant}
(**while** condition **do**
      Statement)
{Invariant **and not** condition}

is correct. Similar rules can be derived for the **for . . . do** and **repeat . . . until** statements. These are left as exercises.

EXAMPLE 2.18

Prove that the algorithm *Multiplication* is correct.

Algorithm *Multiplication*
$\{n \geq 0 \text{ and } x = x_1\}$
begin
      $i \leftarrow 0$
      $Product \leftarrow 0$
      while $i < n$ **do**
            begin
                  $Product \leftarrow Product + x$
                  $i \leftarrow i + 1$
            end
end.
$\{Product = n \cdot x_1\}$

Solution

To verify the above algorithm, we first need to produce the intermediate assertions, including the loop invariant assertion for the **while . . . do** loop.

$\{n \geq 0 \text{ and } x = x_1\}$
begin
　　　$i \leftarrow 0$
　　　$\{n \geq 0 \text{ and } x = x_1 \text{ and } i = 0\}$
　　　$Product \leftarrow 0$
　　　$\{n \geq 0 \text{ and } x = x_1 \text{ and } i = 0 \text{ and } Product = 0\}$
　　　$\{Product = i \cdot x \text{ and } i \leq n\}$
　　　while $i < n$ **do**
　　　　　$\{Product = i \cdot x \text{ and } i < n\}$
　　　　　begin
　　　　　　　$Product \leftarrow Product + x$
　　　　　　　$\{Product = (i + 1) \cdot x \text{ and } i < n\}$
　　　　　　　$i \leftarrow i + 1$
　　　　　　　$\{Product = i \cdot x \text{ and } i \leq n\}$
　　　　　end
end.
$\{Product = n \cdot x_1\}$

The assertion directly above the loop is the loop invariant assertion. The formulation of the proper loop invariant is often the most difficult step in proving the correctness of a loop.

　　The correctness of the algorithm up to the loop structure is easy to demonstrate. To verify the loop we need to show that

$\{Product = i \cdot x \text{ and } i \leq n\}$
(**while** $i < n$ **do**
　　　Statement)
$\{Product = n \cdot x_1\}$

is correct. Statement is the body of the loop. The algorithm segment

$\{Product = i \cdot x \text{ and } i < n\}$
($Product \leftarrow Product + x$)
$\{Product = (i + 1) \cdot x \text{ and } i < n\}$

is correct since the assignment statement adds x to *Product*. Similarly,

$\{Product = (i + 1) \cdot x \text{ and } i < n\}$
($i \leftarrow i + 1$)
$\{Product = i \cdot x \text{ and } i \leq n\}$

is also correct. Applying the iteration inference rule we conclude that

$\{Product = i \cdot x$ **and** $i \leq n\}$
(**while** $i < n$ **do**
         Statement)
$\{Product = n \cdot x$ **and** $i \geq n\}$

is correct. Since $(Product = n \cdot x$ **and** $i \geq n) \Rightarrow (Product = n \cdot x)$ is *true*, the correctness of the loop follows, as does the correctness of the entire algorithm.

Using the approach described in this section to verify algorithms is a difficult task. Complex algorithms will probably never be subjected to the level of rigor and detail we used here to verify simple algorithms, unless a major part of the work is done using a computer. Some progress has already been made in this direction. Other proof techniques, such as induction, may also be used with certain algorithms. In this text, we will use whatever approach seems most natural for the problem at hand.

Exercises 2.5

1. Prove the following algorithm is correct based on the stated preconditions and postconditions.

   Algorithm *Quadratic*
   $\{x$ is a real number$\}$
   begin
           $y \leftarrow ax$
           $y \leftarrow (y + b)x$
           $y \leftarrow y + c$
   end.
   $\{y = ax^2 + bx + c\}$

2. Verify the following algorithm.

   Algorithm *Largest*
   $\{a = a_1$ **and** $b = b_1\}$
   begin
           if $a > b$ **then**
                   $x \leftarrow a$
           else
                   $x \leftarrow b$
   end.
   $\{(x = a_1$ **and** $a_1 > b_1)$ **or** $(x = b_1$ **and** $b_1 \geq a_1)\}$

3. Verify the following algorithm.

Algorithm *Factorial*
$\{n = n_1\}$
begin
       $i \leftarrow 0$
       $Fact \leftarrow 1$
       while $i < n$ **do**
           begin
               $i \leftarrow i + 1$
               $Fact \leftarrow Fact \cdot i$
           end
end.
$\{Fact = n_1!\}$

*4. Prove the correctness of the algorithm in Example 2.18 using induction.

2.6 Basic Properties of Sets

Set theory forms the second half of the foundation of our study of discrete mathematics. In previous sections we have made references to properties of objects without formally discussing where these objects come from or how we can construct a unified view of classes of objects. The material in this section completes the picture.

A **set** is a collection of objects, which are called the **elements** of the set. Thus, sets are characterized by the concept of **belonging**; that is, a set consists of those objects that belong to it. We use the notation $a \in S$ to indicate that the object a is an element of the set S. We use the notation $a \notin S$ to indicate that the object a is not an element of the set S. In the remainder of the chapter, we will use uppercase letters for sets and lowercase letters for elements of sets.

Definitions of Sets

The first order of business is to answer the question: How can we describe a set? As usual, we are not satisfied with simple English descriptions such as, "S is the set of all the King's horses and all the King's men." There are two versions of formal set specifications we can use. The first is simple **enumeration**, or listing the elements of the set. For instance,

$$S = \{1, 2, 3, 4, 5, 6, 7, 8, 9, 10\}$$

defines the set consisting of the first ten positive integers. When a set is defined by enumeration, the order in which the elements are listed does not matter. Thus the set

S could also have been defined by

$$S = \{2, 3, 7, 1, 9, 4, 10, 5, 8, 6\}$$

Enumeration, however, is of limited use for defining large sets. In some cases, a natural ordering of set elements may make set definition easier. For instance, we can define sets consisting of consecutive elements from an ordered set, such as the integers or the letters of the English alphabet, using an ellipsis (. . .). Thus, the set S can also be defined by

$$S = \{1, \ldots, 10\}$$

The second method of set definition uses predicates. The form of this set specification is $S = \{x: P(x)\}$, which is read "S is the set of all elements x satisfying the predicate $P(x)$." The colon is sometimes replaced by a vertical bar. The set S above is defined in predicate form as

$$S = \{x: x \text{ is an integer and } 1 \leq x \leq 10\}$$

EXAMPLE 2.19

1. Enumerate the elements of the set $A = \{x: x$ is one of the first five letters in the lowercase alphabet$\}$.
2. Write a set specification in predicate form for the set of odd integers.

Solution

1. $A = \{a, b, c, d, e\} = \{a, \ldots, e\}$
2. There are many ways to do this. Here are two of them:

$$S = \{x: x \text{ is an odd integer}\}$$
$$S = \{x: \text{there is an integer } n \text{ such that } x = 2n + 1\}$$

Some Special Sets

We will now introduce some special sets and the notation we use for them. These sets occur frequently, and we will always use the same symbol to denote them.

- The **empty set** is the set that does not contain any elements. We denote the empty set by $\varnothing$ or $\{ \, \}$.
- **Z** denotes the set of integers.
- **R** denotes the set of real numbers.
- **Q** denotes the set of rational numbers.

- **N** denotes the set of *natural* numbers; that is,

$$\mathbf{N} = \{1, 2, 3, \ldots\}$$

- For each $n \in \mathbf{N}$, $S_n = \{1, \ldots, n\}$

EXAMPLE 2.20

Many modern programming languages require that all variables be declared to belong to a particular **data type**. Specifying a type is the same as specifying a set of values that variables of that type may take on.

For each pair of type and variable declarations below, write equivalent set specifications for the set of values the variable may take on.

1. **type** *footballyards* = integer range 1 . . . 100;
   variable *gain: footballyards;*

2. **type** *primarycolor* = (red, blue, yellow);
   variable *pallet: primarycolor;*

3. **type** *mixture* = **set of** *primarycolors*;
   variable *mix 1: mixture*

Solution

1. *gain* belongs to $\{x: x \in \mathbf{Z} \text{ and } 1 \le x \le 100\}$
2. *pallet* belongs to the set {red, blue, yellow}
3. The elements of type *mixture* are subsets of the set *primarycolor*. Thus *mix1* $\in$ {$\emptyset$, {red}, {blue}, {yellow}, {red, blue}, {red, yellow}, {blue, yellow}, {red, blue, yellow}}.

Operations on Sets

We continue now with some more definitions. Let S be a set. A **subset** of S is a set A such that if $x \in A$, then $x \in S$. We write $A \subseteq S$, or $S \supseteq A$. The two symbols are read, respectively, as "A is contained in S" and "S contains A." Two sets are **equal** if and only if each is a subset of the other.

There are several ways to combine sets to form new sets. The **union** of two sets A and B is the set

$$A \cup B = \{x: x \in A \text{ or } x \in B\}$$

The **intersection** of the sets A and B is the set

$$A \cap B = \{x : x \in A \text{ and } x \in B\}$$

The **complement** of set A relative to set B is the set

$$B - A = \{x : x \in B \text{ and } x \notin A\}$$

Sometimes in a given context we have a *universal* set such that all sets we consider are subsets of this universal set. Then we write $\sim A$ for the complement of A relative to the universal set. (The notation $\bar{A}$ is also used for the complement of A relative to the universal set.)

The **symmetric difference** of two sets A and B is the set

$$A \triangle B = \{x : (x \in A \text{ and } x \notin B) \text{ or } (x \in B \text{ and } x \notin A)\}$$

The union, intersection, and symmetric difference operations can be extended inductively to operations on arbitrary sequences of sets in the same way we extended addition earlier in this chapter. For instance, given sets $A_1, A_2, \ldots$, we define

$$\bigcup_{i=1}^{1} A_i = A_1$$

and

$$\bigcup_{i=1}^{n+1} A_i = \left(\bigcup_{i=1}^{n} A_i \right) \cup A_{n+1}$$

EXAMPLE 2.21

Let $A = \{1, 2, 4, 8, 16\}$, $B = \{2, 4, 6, 8, 10\}$, and $C = \{1, 3, 7, 15\}$.
   Find the following sets.

1. $A \cup B$ 5. $A \cup (B \cap C)$
2. $A \cap B$ 6. $(A \cup B) \cap (A \cup C)$
3. $A - C$ 7. $C - (B - A)$
4. $B \triangle C$ 8. $A \triangle (B \triangle C)$

   Solution

1. $A \cup B = \{x : x \in A \text{ or } x \in B\}$
   $\qquad\quad = \{1, 2, 4, 6, 8, 10, 16\}$
2. $A \cap B = \{x : x \in A \text{ and } x \in B\}$
   $\qquad\quad = \{2, 4, 8\}$

3. $A - C = \{x: x \in A \text{ and } x \notin C\}$
$\qquad = \{2, 4, 8, 16\}$

4. $B \triangle C = \{x: (x \in B \text{ and } x \notin C) \text{ or } (x \in C \text{ and } x \notin B)\}$
$\qquad = B \cup C \text{ (in this case)}$

5. $A \cup (B \cap C) = A \cup \varnothing$
$\qquad = A$

6. $(A \cup B) \cap (A \cup C) = \{1, 2, 4, 6, 8, 10, 16\} \cap \{1, 2, 3, 4, 7, 8, 15, 16\}$
$\qquad = \{1, 2, 4, 8, 16\}$
$\qquad = A$

7. $C - (B - A) = C - \{6, 10\}$
$\qquad = C$

8. $A \triangle (B \triangle C) = A \triangle \{1, 2, 3, 4, 6, 7, 8, 10, 15\}$
$\qquad = \{16\} \cup \{3, 6, 7, 10, 15\}$
$\qquad = \{3, 6, 7, 10, 15, 16\}$

We will often illustrate ideas about set operation by means of pictures called **Venn diagrams**. Although Venn diagrams are useful for illustrating sets and their properties, they can be misleading since an identity that appears to be true in one Venn diagram may not hold true in another.

In Venn diagrams, the universal set is represented by the enclosing rectangle, and the sets—usually named A, B, C—are represented as circles. The results of applying the given operations are shown as colored areas in Figure 2.2.

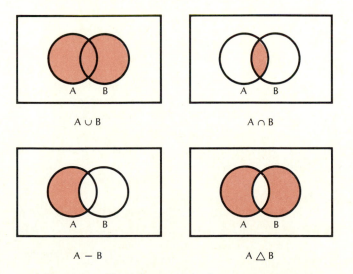

Figure 2.2. Venn diagrams for basic set operations

Properties and Identities

In the following discussion, $\mathcal{U}$ denotes the universal set containing the sets A, B, and C. There are numerous properties of sets that we can derive. Equalities that are true for all sets are called **identities**. Theorem 2.1 contains some useful properties of sets. Theorem 2.2 contains some set identities whose form is similar to some you have seen in algebra.

Theorem 2.1

1. $\emptyset \subseteq A$ for all sets A.
2. $A \subseteq A$ for all sets A.
3. If $A \subseteq B$ and $B \subseteq C$, then $A \subseteq C$.
4. $\sim(\sim A) = A$ for all sets A. (This is called the **involution** property.)
5. $A \cup \sim A = \mathcal{U}$ and $A \cap \sim A = \emptyset$ for all sets $A \subseteq \mathcal{U}$.
6. $A \cup \mathcal{U} = \mathcal{U}$ and $A \cap \mathcal{U} = A$ for all sets $A \subseteq \mathcal{U}$.
7. $A \cup \emptyset = A$ and $A \cap \emptyset = \emptyset$ for all sets A.
8. DeMorgan's Laws: For all sets A and B,

$$\sim(A \cup B) = \sim A \cap \sim B$$

and

$$\sim(A \cap B) = \sim A \cup \sim B$$

Proof

We will prove only (1) and (8), leaving the rest for the exercises.

(1) We need to show that the statement "If $x \in \emptyset$, then $x \in A$" is *true*. We use the vacuous case of implications discussed earlier. Since *no x* satisfies "$x \in \emptyset$," the statement must be *true* by definition of implication.

(8) Recall that two sets are equal if and only if each is a subset of the other. We apply this to the first of DeMorgan's laws.

Suppose that $x \in \sim(A \cup B)$. Then $x \notin (A \cup B)$ by definition of the complement. As a result, $x \notin A$ and $x \notin B$, so that $x \in \sim A \cap \sim B$.

Next suppose that $x \in \sim A \cap \sim B$. Then $x \in \sim A$ and $x \in \sim B$. Taking each in turn, we have $x \in \sim A \Rightarrow x \notin A$ and $x \in \sim B \Rightarrow x \notin B$. Hence, $x \notin (A \cup B)$ and, therefore, $x \in \sim(A \cup B)$.

The second of DeMorgan's laws can be proved similarly.

•   •   •

Theorem 2.2
Set Identities

Let A, B, C be sets.

Idempotent Law

1. a. $A \cup A = A$ b. $A \cap A = A$

Associative Law

2. a. $A \cup (B \cup C) = (A \cup B) \cup C$ b. $A \cap (B \cap C) = (A \cap B) \cap C$

Commutative Law

3. a. $A \cup B = B \cup A$ b. $A \cap B = B \cap A$

Distributive Law

4. a. $A \cup (B \cap C) = (A \cup B) \cap (A \cup C)$ b. $A \cap (B \cup C) = (A \cap B) \cup (A \cap C)$

•   •   •

We have omitted the proofs of these identities. You will be asked in the exercises to prove some of them and to draw Venn diagrams for the others.

The identities above appear in pairs in which each identity is obtained from the other by interchanging the roles of union and intersection. The names of the identities should be familiar to you from algebra. In algebra, however, they are applied to the operations of addition and multiplication on real numbers, rather than to union and intersection. Also, the symmetry between the two distributive identities is not valid in the case of real number arithmetic. The symmetry between the pairs of identities in Theorem 2.2 is the basis of an important principle called the Duality Principle.

Theorem 2.3
The Duality Principle

Any valid equality involving sets can be changed to an equally valid **dual** equality by applying the following rules:

Replace each $\cup$ by $\cap$ and each $\cap$ by $\cup$.
Replace each $\mathscr{U}$ by $\varnothing$ and $\varnothing$ by $\mathscr{U}$.

•   •   •

Applying the Duality Principle to the first of DeMorgan's Laws, we transform

$$\sim(A \cup B) = \sim A \cap \sim B$$

into

$$\sim(A \cap B) = \sim A \cup \sim B$$

which is the second of DeMorgan's Laws.

SELF TEST 5

1. Write a set specification in predicate form for each of the following.
   a. the set of nonnegative integers
   b. the set of prime numbers
2. Let $A = \{1, 2, 3, 4, 5\}$, $B = \{2, 4, 6\}$, and $C = \{1, 3, 5\}$. Find the following sets.
   a. $A \cup B$
   b. $A \cap B$
   c. $A - C$
3. Let A be an arbitrary set. Show that $\emptyset \cup A = A$ and $\emptyset \cap A = \emptyset$.

Exercises 2.6

1. a. Enumerate the elements of set A.

$$A = \{x : x \in \mathbf{Z} \text{ and } 10 \leq x \leq 17\}$$

   b. Write a set specification in predicate form for set A.

$$A = \{2, 5, 8, 11, \ldots\}$$

2. a. List all the subsets of the set $\{e, s, t\}$.
   b. Describe each of the following sets.

$$
\begin{array}{ll}
\mathbf{Z} - \mathbf{N} & \mathbf{Z} \cup \mathbf{Q} \\
\mathbf{N} - S_n & S_{17} - S_9 \\
\mathbf{R} - \mathbf{Q} & S_{10} \cap S_6
\end{array}
$$

3. For each type and variable declaration below, write an equivalent set specification for the set of values the variable may take on.

   a. **type** *clockface* = integer range 1 . . . 12;
     variable *hour*: *clockface*;

   b. **type** *worldpowers* = (USA, China, Russia);
     variable *bigone*: *worldpowers*;

   c. **type** *alliance* = **set of** *worldpowers*;
     variable *pact*: *alliance*;

4. Let $A = \{a, c, e, g, m\}$, $B = \{b, d, f, g, m\}$, and $C = \{a, b, h, p\}$. Find the following sets.

   a. $A \cup B$ e. $A \cup (B \cup C)$
   b. $A \cap B$ f. $(A \cup B) \cap (A \cup C)$
   c. $A - C$ g. $C - (B - A)$
   d. $B \triangle C$ h. $A \triangle (B \triangle C)$

5. Use DeMorgan's laws to simplify the following expressions.

   a. $\sim(\sim A \cup \sim B)$
   b. $\sim(A \cap \sim(B \cup C))$

6. a. Verify, by means of direct proofs, the associative set identities of Theorem 2.2.
   b. Draw Venn diagrams to illustrate the distributive set identities of Theorem 2.2.

2.7 More on Sets

In this section, we discuss the power and product sets, set partitions, some countability rules, and computer representation of sets. We will often ask in this section how many elements a given set has. We call the number of elements in a set the **cardinality** of the set and for set S denote it by $|S|$. If the cardinality of a set is an integer, we say that the set is **finite**; otherwise, we say the set is **infinite**. The sets S_n are finite (with n elements), and the sets **Z**, **Q**, **N**, and **R** are infinite. In discrete mathematics, we are mostly concerned with finite sets, or sets that are like the integers. The next theorem states a simple counting rule for the union of two sets. We omit the proof. The theorem has an extension, by induction, to the union of any finite collection of sets. As we build new kinds of sets from existing sets, we will give rules for determining their cardinality.

Theorem 2.4
The Principle of Inclusion and Exclusion

Let A and B be finite sets.

$$|A \cup B| = |A| + |B| - |A \cap B|$$

Power Sets

Let S be a set. The **power set** of S, denoted $\mathscr{P}(S)$, is the set of all subsets of S. For example, if S is the set $\{a, b, c\}$, then

$$\mathscr{P}(S) = \{\varnothing, \{a\}, \{b\}, \{c\}, \{a, b\}, \{a, c\}, \{b, c\}, \{a, b, c\}\}$$

Theorem 2.5 tells how many subsets there are of a set with cardinality n.

Theorem 2.5

If S is a set with $|S| = n$, n finite, then $|\mathscr{P}(S)| = 2^n$.

Proof

This is a good place to use an induction argument. Let $P(n)$ be the statement of the theorem, with $n \geq 0$.

For $n = 0$, $S = \{\ \}$. Hence $\mathscr{P}(S) = \{\ \{\ \}\ \}$ and $|\mathscr{P}(S)| = 1 = 2^0$. We see that $P(0)$ is *true*, so we have established the base step.

Next we assume that $P(n)$ is *true* and verify that $P(n + 1)$ is *true*. Let S be a set with $n + 1$ elements, say $S = \{a_1, a_2, \ldots, a_n, a_{n+1}\}$. Let $S' = \{a_1, a_2, \ldots, a_n\}$. Given a subset A of S, there are two possibilities: either $a_{n+1} \in A$ or $a_{n+1} \notin A$. In the first case, $A = A' \cup \{a_{n+1}\}$, where $A' \in \mathscr{P}(S')$. In the second case, A itself belongs to $\mathscr{P}(S')$. Thus $\mathscr{P}(S)$ consists of all the elements of $\mathscr{P}(S')$ and of all the sets formed by adding $\{a_{n+1}\}$ to each element of $\mathscr{P}(S')$. Since $|S'| = n$, $|\mathscr{P}(S')| = 2^n$, so we see that there are $2^n + 2^n = 2^{n+1}$ possible subsets of S.

Product Sets

Before we introduce the definition of a product set, we must define **ordered pair**. An ordered pair is a pair of objects (a, b) distinguishable by order. Ordered pairs are distinguished from sets with two elements by the property that two ordered pairs (a, b) and (c, d) are equal if and only if $a = c$ and $b = d$. In particular, $(a, b) = (b, a)$ if and only if $a = b$. Let A and B be sets. The **Cartesian product** of A and B, denoted $A \times B$, is the set of *ordered pairs* (a, b) such that $a \in A$ and $b \in B$. Thus,

$$A \times B = \{(a, b): a \in A \text{ and } b \in B\}$$

It follows that if either A or B is the empty set, then $A \times B$ is also the empty set.

EXAMPLE 2.22

Let $A = \{1, 2, 3\}$ and $B = \{a, b\}$. Find $A \times B$.

   Solution

$A \times B = \{(1, a), (1, b), (2, a), (2, b), (3, a), (3, b)\}$

We count the elements in this new kind of set. In Example 2.22 there are three elements in A, two elements in B, and six elements in $A \times B$. This is not surprising, given the rectangular layout of the product in three rows and two columns. We give the general formula in Theorem 2.6, without proof. The statement in the theorem is the reason for the term *product set*.

Theorem 2.6

Let A and B be sets.

$$\text{If } |A| = n, \text{ and } |B| = m, \text{ then } |A \times B| = nm.$$

$$\bullet \qquad \bullet \qquad \bullet$$

Let $A_1, A_2, \ldots, A_n$ be sets. The product of these sets is the set of *n-tuples* $(a_1, a_2, \ldots, a_n)$; that is,

$$\mathop{\times}_{i=1}^{n} A_i = \{(a_1, a_2, \ldots, a_n): a_i \in A_i, i = 1 \ldots n\}$$

If all A_i are equal to the same set A, we write A^n to denote the product set. In this case the *n*-tuples are often called *n-vectors*, or just *vectors*.

EXAMPLE 2.23

Let B $= \{0, 1\}$. According to the definition of product, for each $n \geq 1$,

$$B^n = \{(b_1, \ldots, b_n): b_i = 0 \text{ or } 1, 1 \leq i \leq n\}$$

List the elements of B^3.

Solution

For $n = 3$, we get the set

$$B^3 = \{(0, 0, 0), (0, 0, 1), (0, 1, 0), (0, 1, 1), (1, 0, 0), (1, 0, 1), (1, 1, 0), (1, 1, 1)\}$$

In Example 2.23, we found that $|B^3| = 8$, just as the cardinality of the power set of a set with three elements is 8. This is not a coincidence. B^n consists of vectors of 0's and 1's, which are often called *bit strings* in computer science. In the next example, we show how to represent subsets of a given set as bit strings.

EXAMPLE 2.24

1. Let S be a finite set with n elements, $n \geq 1$. Show that a subset A of S can be represented as an element of the set B^n. The sets B^n were defined in Example 2.23.
2. Let S $= \{a, b, c, d, e\}$, and let A $= \{a, c, d\}$ be a subset of S. Find the elements of B^5 that represent A and $\sim$A. What subset of S does the vector $(0, 1, 1, 0, 1)$ represent?

Solution

1. Let S $= \{s_1, s_2, \ldots, s_n\}$, and let A $\subseteq$ S. Define an element $b = (b_1, b_2, \ldots, b_n) \in B^n$ by setting $b_i = 1$ if $s_i \in$ A and $b_i = 0$ if $s_i \notin$ A. For instance, the set S itself will be associated with a vector consisting entirely of 1's. The empty set will be associated with a vector consisting entirely of 0's. The n-tuple b is called the **characteristic vector** of the set A.
2. We consider the elements a, b, c, d, e to be numbered from 1 to 5. Since elements 1, 3, and 4 belong to A, the vector b must have a 1 in its first, third, and fourth positions. Thus $b = (1, 0, 1, 1, 0)$. Observe that the complement of A will have the representation $c = (0, 1, 0, 0, 1)$.

Finally the vector $(0, 1, 1, 0, 1)$ contains a 1 in its second, third, and fifth positions, so it corresponds to the subset $\{b, c, e\}$.

Theorem 2.7 summarizes our current knowledge of counting rules. We omit the proof.

Theorem 2.7

Let A, B, S be finite sets.

1. The Principle of Inclusion and Exclusion

$$|A \cup B| = |A| + |B| - |A \cap B|$$

2. $|A \times B| = |A| \cdot |B|$
3. $|\mathscr{P}(S)| = 2^{|S|}$

•   •   •

Partitions of a Set

The last topic we will consider in this chapter concerns partitions of a set. Let S be a set. A **partition** of S is a set

$$\pi = \{A_1, A_2, \ldots, A_i, \ldots, A_k\}$$

where

1. each A_i is a nonempty subset of S, $i = 1, \ldots, k$,
2. the A_i *cover* S, in that

$$S = A_1 \cup A_2 \cup \ldots \cup A_k$$

3. the A_i are *mutually disjoint*, in that

$$\text{If } i \neq j, \text{ then } A_i \cap A_j = \varnothing$$

We call the sets A_i the **blocks** of a partition. Figures 2.3(a) and (b) show some intuitive examples of partitions. For a more explicit example, let A_1 be the set of even integers, and let A_2 be the set of odd integers. Then $\pi = \{A_1, A_2\}$ is a partition of **Z**.

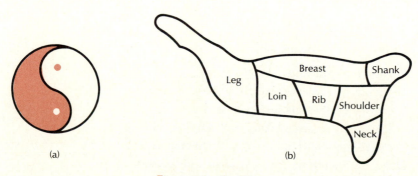

(a)   (b)

Figure 2.3. Partitions

EXAMPLE 2.25

Let $S = S_9$. Let $A_1 = \{1, 5\}$, $A_2 = \{2\}$, $A_3 = \{7, 8, 9\}$, and $A_4 = \{3, 4, 6\}$. Show that $\pi = \{A_1, \ldots, A_4\}$ is a partition of S.

Solution

Each of the sets is nonempty and the sets cover S. By inspection, they are mutually disjoint. Hence, π is a partition.

SELF TEST 6

1. Let $A = \{1, 2, 3\}$ and $B = \{x, y, z\}$. How many elements belong to the Cartesian product of A and B? List them.

2. Let A be the set of distinct letters in the word *Mississippi*. List all the elements of $\mathscr{P}(A)$.

3. A survey of students taking Discrete Mathematics found that 59 had never heard of the Principle of Inclusion and Exclusion and 87 had never heard of mathematical induction. Moreover, 12 students had never heard of both the Principle of Inclusion and Exclusion and mathematical induction. How many had never heard of at least one of the topics?

Exercises 2.7

1. Let $S = S_4$. List the elements of $\mathscr{P}(S)$.

2. Let $A = \{1, 3, 5, 7\}$ and $B = \{x, y, z\}$. How many elements are there in the product set $A \times B$? List the elements of $A \times B$.

3. a. Extend the Principle of Inclusion and Exclusion to count the union of three sets. (Venn diagrams may be helpful.)
   b. Consider the following situation:

   There are 25 students taking English.
   There are 27 students taking discrete mathematics.
   There are 12 students taking Russian I.
   There are 20 students taking both English and discrete mathematics.
   There are no students taking all three courses.
   There are 5 students taking English and Russian I.
   There are 3 students taking Russian I and discrete mathematics.

How many students are taking at least one of the three courses? How many of the students involved are taking only Russian I?

4. Let $S = \{x, y, u, v, w, z\}$. Let $A = \{x, y, z\}$.

   a. Represent A as an element of B^6.

   b. What subset of S does the vector $(0, 1, 0, 0, 1, 1)$ represent?

   c. Compute $|\mathscr{P}(S)|$ and $|B^6|$. Do the results surprise you?

5. Find all partitions of the set $\{\alpha, \beta, \delta\}$.

6. If the cardinality of A is $|A|$ and the cardinality of B is $|B|$, find the cardinality of $\mathscr{P}(A \times B)$.

A Look at Knowledge-Based Systems

Researchers in the field of artificial intelligence (AI) within computer science have long held the hope that computers could be made to behave more like people. Over the years, in work on robotics, game playing, theorem proving, and analysis of natural languages, researchers in AI have made significant progress in this direction. Some of the efforts have been devoted to trying to make machines perform reasoning through applications of formal logic. While this approach is only one of many (and has its critics in the AI community), it has been successfully applied.

In more recent applications, efforts have been made to develop machines that can supplement the work of experts in specialized fields. To date, systems that aid in medical diagnosis, configuration of computer systems, circuit analysis, speech understanding, chemistry, and mathematics have been implemented successfully. The common thread in these **expert systems** is the capture of the knowledge and experience of experts in a **knowledge base** consisting of a set of facts and a set of inference rules. On top of the knowledge base is an **inference engine**. The inference engine gives the expert system its problem-solving capability. The inference engine, for instance, can generate answers to questions even if the answer was not directly built in to the knowledge base when it was constructed. Figure 2.4 shows a conceptual view of the structure of an expert system. Once all the components are in place, a user can interact with the system as if it were the expert.

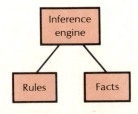

Figure 2.4. Structure of an expert system

Let us consider an example of the use of an expert system by describing a set of facts and rules that could be used in a simple expert system for a family.

An Example: Basic Facts About the Family

The basic facts in our system will be predicates of the form "X is the father of Y" and "A is a female." Actual expert systems and so-called *logic languages*, such as Prolog and LOGLisp, have a more cryptic representation of the predicates. In Prolog, for instance, the natural English predicate "John is the father of Jane" would be written as "father(john, jane)." Here are the basic facts represented in our system:

John is the father of Andy.	Mary is the mother of Andy.
John is the father of Ann.	Mary is the mother of Ann.
Bill is the brother of John.	Linda is the mother of Frank.
John is the husband of Mary.	Linda is the wife of Bill.
John is a male.	Andy is a male.
Bill is a male.	Frank is a male.
Mary is a female.	Ann is a female.
Linda is a female.	
Andy is older than Ann.	Ann is older than Frank.
Bill is older than Linda.	Mary is older than John.

Inference Rules

The information given above is by no means complete. We can also give rules that allow new facts to be confirmed by the system. These are if–then (or situation–action) rules. Rules make use of variables to allow generality, and the rules can handle more complicated situations with the logical operations **and**, **or**, and **not**. For our simple case, we have the following rules:

1. If X is the husband of Y, then Y is the wife of X.
2. If X is the wife of Y, then Y is the husband of X.
3. If A is the father of B, or A is the mother of B, and B is a male, then B is the son of A.
4. If A is the father of B, or A is the mother of B, then A is a parent of B.
5. If X is the son of Y, and Y is the brother of Z, then X is the nephew of Z.
6. If U is the nephew of W, and W is a male, then W is the uncle of U.

Queries

Given the basic facts and the inference rules, the expert system can respond to questions or problems presented to it. It always has the option of responding "I don't know" to any query it cannot otherwise satisfy. Here are some queries for our family system:

Is Bill the husband of Linda?

Who is the wife of John?

Who are the parents of Frank?

Who is the uncle of Andy?

List all the uncles.

Is John older than Linda?

Who is the mother of John?

To respond to a query, the inference engine must search through the collection of facts and rules until it finds an answer or concludes that it does not have enough information to answer the question. Here we examine one way the system can respond to the question, "Is Bill the husband of Linda?" First, it could look for the facts given about Linda and Bill:

Bill is the brother of John.   Linda is the mother of Frank.

Linda is the wife of Bill.   Bill is a male.

Linda is a female.   Bill is older than Linda.

None of these satisfies the query directly, so the system must examine inferences. There is only one inference that has the conclusion ". . . is the husband of . . . ," namely,

If X is the wife of Y, then Y is the husband of X.

Since the fact "Linda is the wife of Bill" matches the pattern of the *if* part of the rule, the conclusion "Bill is the husband of Linda" holds. The system will respond to the question with "Yes."

Answering this query was not too difficult, but satisfying the query "Who is the uncle of Andy?" will require more searching of the facts and rules to find matching patterns. The complexity of the search and the management of the interrelationships among derived facts is the real challenge in designing an effective expert system.

Exercises 2.8

Test your understanding of the concepts presented here by completing these exercises.

1. Follow the rules and determine the answers to the rest of the queries that begin on page 69.
2. Add a rule, or a set of rules, that concludes "X is the cousin of Y."

SUMMARY

1. **Logic** consists of a set of rules for drawing inferences.

2. In an axiomatic system, we assume that certain key statements are true. These statements are called **axioms**.

3. A **proposition** is a statement that has a truth value; that is, the statement is either *true* or *false*.

4. Propositions can be built from other propositions through the use of **logical operations**. The logical operations are **and**, **or**, **if . . . then**, and **not**.

5. Two propositions are **logically equivalent** if they have the same truth value.

6. A **predicate**, P(x), is a statement that concerns properties of the object x. *For all* ($\forall$) and *there exists* ($\exists$) are **predicate quantifiers**.

7. Proving an implication P $\rightarrow$ Q is true by assuming P true and showing that Q must be true is called a **direct argument**.

8. Proof by **contraposition** is based on the fact that **not** Q $\rightarrow$ **not** P is logically equivalent to P $\rightarrow$ Q.

9. A proof of an implication that shows that the assumption "Q is false" together with the premise "P is true" leads to a contradiction is called a proof by **contradiction**.

10. To prove a proposition of the form "$\exists x$ such that P(x)," we give an algorithm for actually exhibiting the object x. This is called a **constructive proof**.

11. To show that a statement of the form "$\forall x$ P(x)" is false, it suffices to find one counterexample; that is, one value of x that does not satisfy P(x).

12. **Mathematical induction** is useful for proving propositions that must be true for all integers or for a range of integers.

13. The **Principle of Mathematical Induction** is:

> Let P(n) be a proposition that is valid for $n \geq k$,
> n, k integers.
>
> If   (1) P(k) is *true*
> and (2) $\forall n \geq k$, P(n) $\rightarrow$ P(n + 1)
> then P(n) is *true* $\forall n \geq k$.

14. A **recursive definition** follows the form of the Principle of Induction. The first element of the sequence is defined, and then the nth element is defined in terms of preceding elements.

15. A **recurrence relation** is an equation that defines the ith value in a sequence of numbers in terms of the preceding $i - 1$ values.

16. The notation {P}A{Q} means "If execution of algorithm A starts with predicate P true, then it will terminate with predicate Q true." The predicate P is called the precondition for algorithm A; the predicate Q is called the postcondition.

Proving the correctness of an algorithm amounts to proving that $\{P\}A\{Q\}$ is true.

17. A **set** is a collection of objects. The objects in a set are called the **elements** of the set.

18. Let S be a set. A **subset** of S is a set A such that if $x \in A$ then $x \in S$.

19. Two sets are **equal** if and only if each is a subset of the other.

20. The **union** of two sets A and B is the set

$$A \cup B = \{x: x \in A \text{ or } x \in B\}$$

21. The **intersection** of the sets A and B is the set

$$A \cap B = \{x: x \in A \text{ and } x \in B\}$$

22. The **complement** of set A relative to set B is the set

$$B - A = \{x: x \in B \text{ and } x \notin A\}$$

23. **Theorem:**
   a. $\emptyset \subseteq A$ for all sets A.
   b. $A \subseteq A$ for all sets A.
   c. If $A \subseteq B$ and $B \subseteq C$ then $A \subseteq C$.
   d. $\sim(\sim A) = A$ for all sets A.
   e. $A \cup \sim A = \mathcal{U}$ and $A \cap \sim A = \emptyset$ for all sets $A \subseteq \mathcal{U}$
   f. $A \cup \mathcal{U} = \mathcal{U}$ and $A \cap \mathcal{U} = A$ for all sets $A \subseteq \mathcal{U}$
   g. $A \cup \emptyset = A$ and $A \cap \emptyset = \emptyset$ for all sets A
   h. DeMorgan's Laws: for all sets A and B,
      i. $\sim(A \cup B) = \sim A \cap \sim B$, and
      ii. $\sim(A \cap B) = \sim A \cup \sim B$

24. **Theorem:** Let A, B, C be sets.

$$A \cup A = A \qquad\qquad A \cap A = A$$
$$A \cup (B \cup C) = (A \cup B) \cup C \qquad A \cap (B \cap C) = (A \cap B) \cap C$$
$$A \cup B = B \cup A \qquad\qquad A \cap B = B \cap A$$
$$A \cup (B \cap C) = (A \cup B) \cap (A \cup C) \qquad A \cap (B \cup C) = (A \cap B) \cup (A \cap C)$$

25. **Theorem: The Duality Principle.** Any valid equality involving sets can be replaced by an equally valid dual equality by applying the following rules:

   Replace each $\cup$ by $\cap$ and each $\cap$ by $\cup$.
   Replace each $\mathcal{U}$ by $\emptyset$ and $\emptyset$ by $\mathcal{U}$.

26. The number of elements in a set S is called the **cardinality** ($|S|$) of the set S.

27. Let S be a set. The **power set** of S, denoted $\mathscr{P}(S)$, is the set of all subsets of S.

28. An **ordered pair** of objects is a pair of objects (a, b) distinguishable by order.

29. Let A and B be sets. The **Cartesian product** of A and B, denoted $A \times B$, is the set of ordered pairs (a, b) such that $a \in A$ and $b \in B$.

30. **Theorem**: Let A, B, S be finite sets.
    a. $|A \cup B| = |A| + |B| - |A \cap B|$
    b. $|A \times B| = |A| \cdot |B|$
    c. $|\mathscr{P}(S)| = 2^{|S|}$

31. Let S be a set. A **partition** of S is a set

$$\pi = \{A_1, A_2, \ldots, A_i, \ldots, A_k\},$$

where

    a. each A_i is a non-empty subset of S, $i = 1, \ldots, k$
    b. $S = A_1 \cup A_2 \cup \ldots \cup A_k$
    c. If $i \neq j$ then $A_i \cap A_j = \varnothing$

REVIEW PROBLEMS

1. Let P be the proposition "It is dark outside" and let Q be the proposition "The stars are out." Give simple English sentences for each of the following statements.
    a. **not** P
    b. P **and** Q
    c. P **or** Q
    d. **not** P **and not** Q
    e. **not not** Q
    f. Q **or not** P

2. Let P be "Logic is easy" and let Q be "Logic is fun." Write each of the following sentences in symbolic form using P and Q.
    a. Logic is easy and fun.
    b. Logic is fun but not easy.
    c. Logic is neither easy nor fun.
    d. Logic is not fun or it is easy.
    e. It is not true that logic is easy and fun.
    f. It is not true that logic is easy or not fun.

3. Construct a truth table for (P **or** Q) **and** R.

4. Construct a truth table for **not** (**not** P **and** Q).

5. Construct a truth table for (**not** P **or** Q) $\Rightarrow$ P.

6. Show that $P \Rightarrow Q$ is logically equivalent to **not** P **or** Q.

7. Show that (P **and** Q) $\Rightarrow$ R is logically equivalent to $P \Rightarrow (Q \Rightarrow R)$.

8. A compound proposition that is always *true* regardless of the elementary propositions within it is called a **tautology**. This means that the truth table for the proposition contains only T's in its results column. Using truth tables, verify that each of the following propositions are tautologies.
    a. $Q \Rightarrow (P \Rightarrow Q)$
    b. **not** $P \Rightarrow (P \Rightarrow Q)$
    c. (P **and** Q) $\Rightarrow$ (P **or** Q)
    d. $P \Rightarrow$ (**not** $P \Rightarrow Q$)

9. A compound proposition that is always *false* regardless of the elementary propositions within it is called a **contradiction**. Using truth tables, verify that each of the following propositions are contradictions.

   a. P **and not** P
   b. (P **and** Q) **and not** (P **or** Q)
   c. **not** (P **or not** P)

10. Express **and** in terms of **or** and **not**.

11. Express **or** in terms of **and** and **not**.

12. Use DeMorgan's Laws to simplify each of the following.

   a. **not** (P **and not** Q)
   b. **not** (**not** P **and not** Q)
   c. **not** (**not** P **or** Q)

13. Consider the new propositional connective **xor**, called **exclusive disjunction**. P **xor** Q means "P **or** Q but **not** both."

   a. Construct a truth table for **xor**.
   b. Express **xor** in terms of **or**, **and**, and **not**.

14. The propositional connective **nand** is defined by the following truth table:

P	Q	P **nand** Q
T	T	F
T	F	T
F	T	T
F	F	T

   a. Show that P **nand** P is logically equivalent to **not** P.
   b. Show that (P **nand** P) **nand** (Q **nand** Q) is logically equivalent to P **or** Q.
   c. Using only the **nand** operator, derive an expression equivalent to P **and** Q.

15. Using x to stand for "cat" and $P(x)$ to stand for the predicate "x has whiskers," write each of the following statements in symbolic form.

   a. All cats have whiskers.
   b. There exists a cat without whiskers.
   c. There does not exist a cat without whiskers.
   d. No cat has no whiskers.

16. Using x to stand for "man" and $P(x)$ to stand for "x is a fool," express each of the following statements in symbolic form. Then express the negation of the statements in symbolic form.

   a. All men are fools.   c. All men are not fools.
   b. Every man is not a fool.   d. No men are fools.

17. Use a direct argument to prove each of the following statements.

   a. If an integer is divisible by 15, then it is divisible by 3.

b. The product of two odd integers is odd.

c. If $x > y$, then $x - y$ is positive.

18. Use a proof by contraposition to establish each of the statements in problem 17.

19. Using proof by contradiction, prove each of the following statements.

a. If $yx = y$ and $y \neq 0$, then $x = 1$.

b. The sum of an odd integer and an even integer is odd.

20. Prove: $1 + 4 + 7 + \cdots + (3n - 2) = (n(3n - 1))/2$

21. Expand each of the following summations, showing at least four distinct terms.

a. $\displaystyle\sum_{x=1}^{n} (1/x)$

b. $\displaystyle\sum_{i=1}^{n} (x_i - \mu)^2$

22. Prove: $\displaystyle\sum_{i=1}^{n} 1/((2i - 1)(2i + 1)) = n/(2n + 1)$

23. Give an inductive definition of a^b (exponentiation) using only multiplication and addition. Assume $a \in \mathbf{R}$, $b \in \mathbf{Z}$, and $b \geq 0$.

24. Using induction, prove $a^m a^n = a^{m+n}$, where $a \in \mathbf{R}$ and $m, n \in \mathbf{N}$.

25. Verify the following algorithm.

$\{x = x_1\}$
begin
      $y \leftarrow ax$
      $y \leftarrow y + b$
end.
$\{y = ax_1 + b\}$

26. Prove that the following algorithm is correct.

$\{n \geq 0 \textbf{ and } x = x_1\}$
begin
      $i \leftarrow 0$
      $Power \leftarrow 1$
      while $i \leq n$ **do**
            begin
                 $Power \leftarrow Power \cdot x$
                 $i \leftarrow i + 1$
            end
end.
$\{Power = x_1{}^n\}$

27. Which of the following sets are equal?

$$\{\alpha, \beta\}, \{\alpha, \gamma\}, \{\beta, \alpha\}, \{\alpha, \beta, \alpha\}, \{\gamma, \alpha, \gamma\}$$

28. List the elements of the following sets if the universal set $\mathscr{U} = \{a, b, c, \ldots, z\}$.
    a. $A_1 = \{x : x \text{ is a consonant}\}$
    b. $A_2 = \{x : x \text{ follows the letter } r \text{ in the alphabet}\}$
    c. $A_3 = \{x : x \text{ is a letter in the word } discrete\}$
    d. $A_4 = \sim A_1$

29. Let $A = \{1, 2, 3, 4, 5, 6\}$, $B = \{x : x = 2n, 1 \leq n \leq 3\}$, $C = \{3, 5, 10, 17, 26\}$. Find each of the following.
    a. $A \cap B$ e. $A \cup C$
    b. $A \cap C$ f. $A \cup B \cup C$
    c. $A \cap B \cap C$ g. $A - (B - A)$
    d. $A \cup B$

30. Prove that $A \bigtriangleup B = B \bigtriangleup A$ for any sets A and B.

31. Prove that $A = \{1, 2, 4, 6\}$ is not a subset of $B = \{x : x = 2p + 1, p \in \mathbf{N}\}$.

32. Prove that $(\varnothing \cup A) \cap (B \cup A) = A$ for any sets A and B.

33. Write the equality dual to $(A \cup B) \cap (A \cup \sim B) = A \cup \varnothing$.

34. Calculate $\mathscr{P}(\{\alpha, \beta\})$ and $\mathscr{P}((\mathscr{P}\{\alpha, \beta\}))$.

35. Show that for finite sets A, B, C,

$$|A \cup B \cup C| = |A| + |B| + |C| - |A \cap B| - |A \cap C| - |B \cap C| + |A \cap B \cap C|$$

36. Make a sequence of Venn diagrams to illustrate the following identity.

$$A \cup (B \cap C) = (A \cup B) \cap (A \cup C)$$

37. Suppose $A = \{1, 2, 3, 4\}$ and $B = \{5, 6, 7\}$. List all sets consisting of one element of A and one element of B. List the members of $A \times B$.

38. If A and B are sets and $|A \times B| = 8$, what are the possible values of $|A|$ and $|B|$?

39. Give examples of sets A and B such that $|A \cup B| < |A| + |B|$. Is it possible to find sets A and B such that $|A| + |B| < |A \cup B|$? Explain.

40. A certain state requires license plates to have exactly six characters. If the first three must be letters of the alphabet and the second three must be digits, how many distinct license plates can there be?

41. A pizza parlor offers the following toppings for their pizzas: sausage, pepperoni, mushrooms, green peppers, and onions. How many different pizzas can one make if the toppings can be added in any combination?

42. Define the sets R_i by $R_i = \{x : x \in S_{20} \text{ and } x \text{ has remainder } i \text{ upon division by } 5\}$, where $i = 0, 1, 2, 3, 4$.
    a. List all the elements of R_0.
    b. Compute the union of all the R_i.
    c. Compute the intersection of all R_i and R_j, $i \neq j$.
    d. Is the collection of all R_i a partition of S_{20}? Why or why not?

◼ Programming Problems

43. Design an algorithm to generate the power set of $\{0, 1, 2, 3, \ldots, n\}$ for any positive integer n.

44. We showed how to represent subsets of a given set with bit strings. For example, the subset $\{1, 5, 6, 8, 9\}$ of S_9 can be represented by the following bit string.

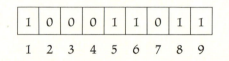

1	0	0	0	1	1	0	1	1
1	2	3	4	5	6	7	8	9

   a. Design an algorithm that accepts two subsets A and B of S_n, determines their bit string representations, and uses them to compute $A \cup B$.
   b. In a manner similar to part (a), design algorithms that compute $A \cap B$ and $A \times B$.
   c. Design an algorithm that uses the bit string representation of subsets of S_n to determine, for a given set $A \subseteq S_n$ and a given integer $i \in S_n$, whether $i \in A$.
   d. Design an algorithm which uses the bit string representation of a set to compute the cardinality of a set $A \subseteq S_n$.

SELF TEST ANSWERS

Self Test 1

1. a. P **and** R is *false*, since R is *false*.
   b. P **or** S is *true*, since P is *true*.
   c. P $\Rightarrow$ Q is *true*, since P and Q are *true*.
   d. R $\Rightarrow$ (**not** S is *true*, since R is *false*.

2.

P	Q	P **or** (**not** Q)
T	T	T
T	F	T
F	T	F
F	F	T

   When P is *true* and Q is *false*, the expression is *true*.

3. a. The sunset is red **and** tomorrow will be clear.
   b. The sunset is red **or** tomorrow will be clear.
   c. If the sunset is red, then tomorrow will be clear.

4. Let P = "The cat is sleeping" and Q = "The dog is barking." Then (a) becomes **not** (P **and** Q) and (b) becomes (**not** P **or not** Q). These are equivalent by DeMorgan's laws.

Self Test 2

1. a. The statement is *false*. (Consider, for example, $x = 1$.)
   b. The statement is *true* for $x = 2$.
2. The negation is the statement: "There exists an x such that for all y, $y^2 \leq x$."

Self Test 3

1. Use the fact that for any real number z, $z^2 = |z|^2$. For any real numbers x and y, $|x + y|^2 = (x + y)^2 = x^2 + 2xy + y^2$. But $2xy \leq 2|x| |y|$, so we get $|x + y|^2 \leq (|x| + |y|)^2$, and the inequality follows.
2. The contrapositive of the expression is as follows:

$$\text{For any real number } x, \text{ if } -1 \leq x \leq 1, \text{ then } |x| \leq 1.$$

To prove this statement observe that when $0 \leq x \leq 1$, $|x| = x \leq 1$. Moreover, when $-1 \leq x \leq 1$, $|x| = -x \leq 1$, as well.

Self Test 4

1. Let $P(n)$ be the statement "$n^3 + 2n$ is divisible by 3" for $n \geq 1$. $P(1)$ is *true*, since $1^3 + 2 = 3$ is divisible by 3. Assume $P(n)$ is true for some $n \geq 1$. Then

$$(n + 1)^3 + 2(n + 1) = n^3 + 3n^2 + 3n + 1 + 2n + 2$$
$$= (n^3 + 2n) + 3n^2 + 3n + 3$$

The first summand is divisible by 3, by the induction hypothesis. The second summand is divisible by 3 because 3 is a factor. Thus the entire expression is divisible by 3.
2. Let $P(n) = n^2 > 2n + 1$, $n \geq 3$. $P(3)$ is *true* since $3^2 = 9 > 2 \cdot 3 + 1 = 7$. Assume $P(n)$ is true for some $n \geq 3$. Then

$$(n + 1)^2 = n^2 + 2n + 1 > (2n + 1) + (2n + 1) > 2(n + 1)$$

Self Test 5

1. a. $\{x : x \text{ is an integer and } x \geq 0\}$
   b. $\{x : x \text{ is an integer divisible only by itself and 1}\}$
2. a. $A \cup B = \{1, 2, 3, 4, 5, 6\}$
   b. $A \cap B = \{2, 4\}$
   c. $A - C = \{2, 4\}$

3. $\varnothing \cup A = \{x: x \in \varnothing$ **or** $x \in A\} = \{x: x \in A\} = A$
$\varnothing \cap A = \{x: x \in \varnothing$ **and** $x \in A\} = \varnothing$

Self Test 6

1. $A \times B$ has nine elements.

$$A \times B = \{(1, x), (1, y), (1, z), (2, x), (2, y), (2, z), (3, x), (3, y), (3, z)\}$$

2. Since $A = \{i, m, p, s\}$, $\mathscr{P}(A)$ has 2^4, or 16 elements.

$$\mathscr{P}(A) = \{\varnothing, \{i\}, \{m\}, \{p\}, \{s\}, \{i, m\}, \{i, p\}, \{i, s\}, \{m, p\}, \{m, s\},$$
$$\{p, s\}, \{i, m, p\}, \{i, m, s\}, \{i, p, s\}, \{m, p, s\}, \{i, m, p, s\}\}.$$

3. Let A be the set of students who haven't heard of the Principle of Inclusion and Exclusion. Let B be the set of students who haven't heard of mathematical induction. Then $A \cap B$ is the set of students who haven't heard of both. Moreover, the set of students who haven't heard of at least one of the topics is $A \cup B$. By the Principle of Inclusion and Exclusion,

$$|A \cup B| = |A| + |B| - |A \cap B|$$
$$= 59 + 87 - 12$$
$$= 134$$

3 Relations and Functions

The idea of associating two or more related objects is universal. We grow up in the midst of people called relatives, and throughout life we experience a wide variety of personal relationships. Moreover, the process of abstraction, in which even small children engage, requires us to make connections between apparently different kinds of objects. Typical comparisons using such words as *bigger*, *better*, and *faster* are assertions about relationships between two objects. This chapter contains a formal treatment of relations in general and of special relations called functions.

3.1 Relations

In our study of relations, it will be important to distinguish relations between like objects—objects belonging to the same set—and relations between objects belonging to different sets. In both cases relations are built out of ordered pairs.

We begin our study of relations by describing the relationship between students and courses they take. Let S be the set of students registered at VLSI Tech (Very Large State Institute of Technology). Let C be the set of courses offered at VLSI Tech. For each student, there is a set of courses for which the student is registered. Similarly, for each course, there is a set of students registered in the course. We can, with the Registrar's help, construct a large set of ordered pairs (s, c) from $S \times C$ containing all associations between students, s, and courses, c. For each student there will be several such pairs. For most courses there will be at least one ordered pair in the collection, but for some there may be none. This collection of ordered pairs captures the relation between students and courses.

Rather than looking at pairs in $S \times C$, we could look at ordered pairs in $S \times \mathscr{P}(C)$. That is, for each student s, we could find the *set* of courses, say A, in which the student is enrolled to form the ordered pair (s, A). There will be one and only one pair (s, A) for each student. However, for certain sets $A \in \mathscr{P}(C)$, there will be no corresponding ordered pair in $S \times \mathscr{P}(C)$.

Binary Relations

A **binary relation** between sets A and B is a subset R of $A \times B$. That is, a binary relation is a collection of ordered pairs from $A \times B$. In the example about students and courses discussed above, we constructed sets of ordered pairs relating either students and courses or students and sets of courses. Thus, both sets of ordered pairs are relations.

If A and B are equal, we refer to the relation as a **relation on** the set A. Since a relation R is a subset of $A \times B$, any relation R has a complementary relation $\bar{R}$, which is the complement of the set R relative to $A \times B$.

EXAMPLE 3.1

Let $A = \{1, 2, 3\}$ and $B = \{a, b\}$. Define some relations between A and B.

Solution

Here are some relations between A and B:

$$R_1 = \{(1, a), (2, b), (3, a), (1, b)\}$$
$$R_2 = \{(3, b)\}$$
$$R_3 = A \times B$$
$$R_4 = \varnothing$$

EXAMPLE 3.2

Let $A = \{1, 2, 3\}$. Let R be the relation on A consisting of ordered pairs (a, b) such that $a \geq b$. List the elements of R.

Solution

$$R = \{(3, 3), (3, 2), (3, 1), (2, 2), (2, 1), (1, 1)\}$$

EXAMPLE 3.3

If $|A| = 4$ and $|B| = 3$, how many relations between A and B are there?

Solution

Since there are 12 elements of $A \times B$, there are 2^{12} subsets of $A \times B$, which is the same as the number of relations between A and B.

Graphical Representation of Relations

One way to describe a relation is to give it an English description, such as "greater than." Another way is to list the set of ordered pairs belonging to the relation. A

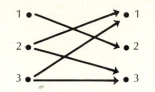

Figure 3.1. Graphical representation of a relation

third way is to draw a graph in which arrows between symbols correspond to the ordered pairs. The three representations are equivalent, and one can choose to use the representation that is most appropriate in a given situation.

Let $A = \{1, 2, 3\}$, and let R be the relation on A defined as $R = \{(1, 2), (2, 1), (2, 3), (3, 1), (3, 3)\}$. A graphical representation of R is shown in Figure 3.1. The elements of A are listed twice, and arrows represent the ordered pairs. That is, there is an arrow from a to b if and only if $(a, b) \in R$, for any a and b in A.

EXAMPLE 3.4

Draw graphical representations of the relations R_1, R_2, R_3, and R_4 from Example 3.1.

Solution

The graphical representations are pictured in Figure 3.2.

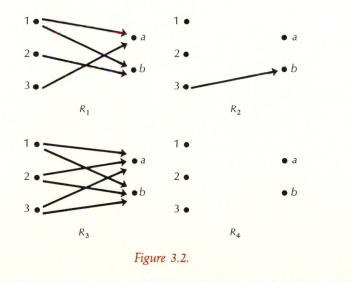

Figure 3.2.

EXAMPLE 3.5

List the ordered pairs belonging to the relation shown in Figure 3.3.

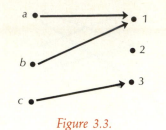

Figure 3.3.

Solution

The relation defined in Figure 3.3 is given by the set R below.

$$R = \{(a, 1), (b, 1), (c, 3)\}$$

If R is a relation on a set A, the graphical representation can take quite a different form. Rather than show two copies of A with arrows between them, we can show one copy of A with arrows between related points. Figure 3.4 shows such a representation for the relation represented in Figure 3.1, $R = \{(1, 2), (2, 1), (2, 3),$ $(3, 1), (3, 3)\}$. Such a representation is called a **directed graph**. Directed graphs will be studied in greater detail in Chapter 6. We will use them here only as a means of representing relations.

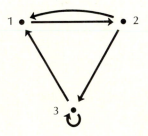

Figure 3.4. Directed graph representation of a relation on a set

---

EXAMPLE 3.6

Draw the directed graph representation of the relation "less than or equal to" on the set S_3.

Solution

The solution is shown in Figure 3.5. Since $1 \leq 1$, there is an arrow that starts at 1 and ends at 1. This is called a self-loop, or loop.

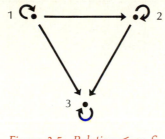

Figure 3.5. Relation $\leq$ on S_3

It is somewhat cumbersome to have to write such expressions as $(x, y) \in R$. In many instances, it will be more natural to use the notation xRy, which is read "x is R related to y," or "xRy." This form is used for common relations such as equality, where we write $x = y$ to mean "$(x, y) \in \{(a, b): a = b\}$."

Matrix Representation of Relations

There is one more commonly used representation of relations that is more convenient for computations. Let A be a set with n elements, and let B be a set with m elements.

$$A = \{a_1, a_2, \ldots, a_n\}$$
$$B = \{b_1, b_2, \ldots, b_m\}$$

Let R be a relation between A and B. Define the $n \times m$ matrix M by

$$M(i, j) = \begin{cases} false \text{ if } (a_i, b_j) \notin R \\ true \text{ if } (a_i, b_j) \in R \end{cases}$$

for $i = 1, \ldots, n$ and $j = 1, \ldots, m$. M is called the *logical matrix* for R. Recall that subsets of a given set can be represented as a vector of 0's and 1's, where a 1 in a particular position indicates that an element is a member of the subset. M is serving the same purpose here, since a relation is a subset of the product set $A \times B$. The difference is that rather than use 0 and 1, we use *false* and *true*. Also, rather than use a vector with a single index to represent the product set, we use the more natural matrix form. Given a relation described by some other method, it is possible to construct the matrix form. Also, a matrix with *true* and *false* entries can be interpreted as a relation between two sets. The matrix below represents the relation $R = \{(1, 2), (2, 1), (2, 3), (3, 1), (3, 3)\}$. Note that we use T for *true* and F for *false*.

$$\begin{bmatrix} F & T & F \\ T & F & T \\ T & F & T \end{bmatrix}$$

EXAMPLE 3.7

1. Let A be the set S_3, and let B be the set S_2. Let R be the relation between A and B.

$$R = \{(1,\ 1),\ (1,\ 2),\ (2,\ 1),\ (3,\ 2)\}$$

Write the matrix representing R.

2. Describe the relations corresponding to the two matrices given below.

$$M_1 = \begin{bmatrix} T & F & T & T \\ F & T & F & T \\ T & F & F & T \end{bmatrix} \qquad M_2 = \begin{bmatrix} F & T & F \\ T & F & T \\ T & T & T \end{bmatrix}$$

Solution

1. Since $|A| = 3$ and $|B| = 2$, the matrix representing R must have three rows and two columns. In row 2, column 2, for instance, the entry will be *false*, since $(2, 2)$ is not an element of R. The complete matrix representation is the matrix M below.

$$M = \begin{bmatrix} T & T \\ T & F \\ F & T \end{bmatrix}$$

2. Consider the matrix M_1 first. There are three rows and four columns, so M_1 can be interpreted as representing a relation R_1 between S_3 and S_4. Clearly, $1R_1 1$, $2R_1 2$, and so on. The graph of R_1 is shown in Figure 3.6(a). Next consider the matrix M_2, which has three rows and three columns. M_2 can be interpreted as representing a relation R_2 on the set S_3. The graph of M_2 is shown in Figure 3.6(b).

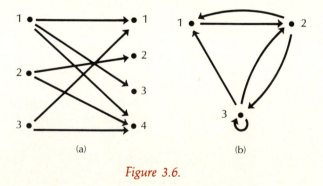

(a)   (b)

Figure 3.6.

SELF TEST 1

1. Let $A = \{1, 2, 3, 4\}$, and let R be the relation on A given by $R = \{(1, 2), (1, 3), (2, 4)\}$. Show the following representations of R.
   a. the graph representation
   b. the directed graph representation
   c. the matrix representation

2. Let $A = \{1, 2, 3, 4, 5, 6\}$, and let R be the relation on A given by $R = \{(x, y): x < y$ **or** x is prime$\}$. Give the matrix representation of R.

3. Write the relation shown in Figure 3.7 as a set of ordered pairs.

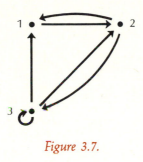

Figure 3.7.

Exercises 3.1

1. Let $A = \{1, 3, 5, 9\}$, and let $B = \{u, v, w\}$. Represent the following relations in graphical form.
   a. $R_1 = \{(1, v), (1, w), (5, u), (9, v), (9, u)\}$
   b. $R_2 = \{(1, v), (3, v), (5, v), (9, v)\}$
   c. $R_3 = \{(5, u), (5, v), (5, w)\}$

2. Let Q be the relation on S_4 given by uQv if $u \neq v$. Represent Q in each of the following ways.
   a. as a set of ordered pairs
   b. in graphical form
   c. in matrix form

3. Let A be a set. Describe "is an element of" as a relation between A and $\mathcal{P}(A)$.

4. For each of the relations represented in Figure 3.8, give the corresponding set of ordered pairs and the matrix representation.

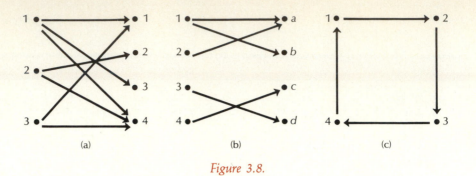

(a)　　　　　　　(b)　　　　　　　(c)

Figure 3.8.

5. For each of the following relation matrices, list the set of ordered pairs that belong to the relation and draw the graph form.

a.　　1　2　3　4　　　　b.　　　1　2　　　　c.　　　1　2　3

　　1 [T　F　F　T]　　　　1 $\begin{bmatrix} F & T \\ F & T \end{bmatrix}$　　　1 $\begin{bmatrix} T & F & T \\ T & F & F \\ F & T & T \end{bmatrix}$
　　　　　　　　　　　　2　　　　　　　　　2
　　　　　　　　　　　　　　　　　　　　　3

6. Let A and B be sets such that $|A| = m$ and $|B| = n$. How many relations between A and B are there?

3.2 Properties of Relations

Relations on a set can be classified according to certain properties. Let R be a relation on a set A.

- We say that R is **reflexive** if for all $x \in A$, xRx.
- We say that R is **symmetric** if for all $x, y \in A$, $xRy \Rightarrow yRx$.
- We say that R is **transitive** if for all $x, y, z \in A$, xRy and $yRz \Rightarrow xRz$.

Reflexive Relations.　The directed graph of every reflexive relation includes an arrow from every point to the point itself. Examples of simple reflexive relations are illustrated in Figure 3.9.

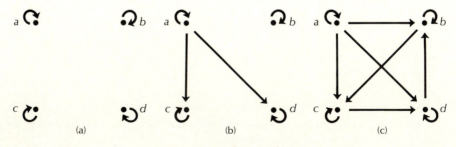

(a)　　　　　　　(b)　　　　　　　(c)

Figure 3.9. Reflexive relations

Here are some familiar examples of reflexive relations:

- subset of ($\subseteq$), on a power set (Every set is its own subset.)
- having the final examination scheduled at the same time, for a set of courses (Every course has its final examination scheduled at the same time as its final examination.)
- being divisible by a positive integer, for the set of positive integers (Every positive integer is divisible by itself.)

Symmetric Relations.   The reason for the name of the symmetric property can be seen in the matrices shown below. Notice that each of the matrix representations for the symmetric relations are symmetric with respect to the main diagonal. Directed graph representations of symmetric relations are also readily recognized because for every arrow from a to b there must also be an arrow from b to a.

$$
\begin{array}{c c c c c}
 & a & b & c & d \\
a & F & F & F & F \\
b & F & F & T & F \\
c & F & T & F & F \\
d & F & F & F & F
\end{array}
\qquad
\begin{array}{c c c c c}
 & a & b & c & d \\
a & F & T & F & T \\
b & T & F & T & F \\
c & F & T & F & F \\
d & T & F & F & F
\end{array}
\qquad
\begin{array}{c c c c c}
 & a & b & c & d \\
a & F & F & F & F \\
b & F & F & F & F \\
c & F & F & F & F \\
d & F & F & F & F
\end{array}
$$

         (a)   (b)   (c)

Here are some examples of symmetric relations:

- sitting next to each other in class (If I sit next to you, then you sit next to me.)
- being integers of the same parity (If x has the same parity—is even or odd—as y, then y has the same parity as x.)
- being equal sets (If set A equals set B, then set B equals set A.)

Transitive Relations.   Figure 3.10 illustrates transitive relations. Note that it is not easy to determine if a relation is transitive from either the matrix or the graph representation. Here are some examples of transitive relations:

- subset of ($\subseteq$), on a power set (If A is a subset of B and B is a subset of C, then A is a subset of C.)
- being a descendent of (If A is a descendent of B and B is a descendent of C, then A is a descendent of C.)

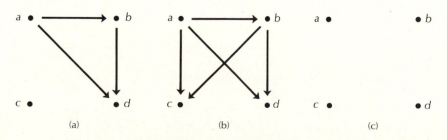

        (a)   (b)   (c)

Figure 3.10. Transitive relations

EXAMPLE 3.8

Determine if each of the following relations is reflexive, symmetric, or transitive.

1. Equality $(=)$, on a nonempty set A
2. "Less than" $(<)$, on $\mathbf{Z}$
3. The relation $R = \{(1, 1), (2, 3), (3, 1)\}$, on the set S_3
4. The empty relation $\varnothing$, on S_3
5. The relation $R = \{(a, b), (c, d)\}$, on $\{a, b, c, d\}$

Solution

1. Let R be the equality relation on the set A. Then R is reflexive, symmetric, and transitive. Consider reflexivity first. If $x \in A$, then $x = x$, so xRx. R is also symmetric because $x = y \Rightarrow y = x$, for all $x, y \in A$. Finally, transitivity follows from the fact that things equal to the same thing are equal to each other.
2. Let R be the "less than" relation on the set $\mathbf{Z}$. Then R is transitive, but not reflexive or symmetric.
3. Since $(2, 2) \notin R$, R is not reflexive. Since $(2, 3) \in R$, but $(3, 2) \notin R$, R is not symmetric. Finally, $(2, 3) \in R$ and $(3, 1) \in R$, but $(2, 1) \notin R$, so R is not transitive. Hence R has none of the three properties.
4. Let $\varnothing$ be the empty relation on the set S_3. At first glance, it might seem that $\varnothing$ is not reflexive, symmetric, or transitive. But let's look at the condition for symmetry more closely. It can be restated as the following implication:

$$\text{If } (x, y) \in \varnothing, \text{ then } (y, x) \in \varnothing$$

Since $(x, y) \in \varnothing$ is never true, the implication is vacuously true. By similar reasoning, the empty relation is also transitive. However, the empty relation is not reflexive, because the condition for reflexivity is the implication

$$\text{If } x \in S_3, \text{ then } (x, x) \in \varnothing$$

and, for instance, $1 \in S_3$, but $(1, 1) \notin \varnothing$.
5. Clearly R is neither reflexive nor symmetric. R is transitive, however, since there are *no x, y, z* $\in \{a, b, c, d\}$ such that xRy and yRz. Thus the condition for transitivity is vacuously satisfied.

Equivalence Relations and Partitions

Relations that are reflexive, symmetric, and transitive are called **equivalence relations**. An equivalence relation is an abstraction of equality. When we think of objects being equivalent in some way, we treat them as nondistinguishable. They share a collection of properties that make them equivalent. An equivalence relation on a set A leads to a natural partitioning of A into groups of like objects. We can then think of the elements of the same block of the partition as being equivalent.

Let S be a nonempty set. Let R be an equivalence relation on S. For each $x \in S$, we define the **equivalence class** of x relative to R, denoted $[x]$, by

$$[x] = \{y \in S : xRy\}$$

EXAMPLE 3.9

Let $S = S_6$. For $x, y \in S$, let xRy if $x - y$ is divisible by 2. Show that R is an equivalence relation. Find the equivalence classes relative to R.

Solution

Clearly R is reflexive. If xRy, then $x - y$ is divisible by 2, so $y - x$ is also divisible by 2. Hence, R is symmetric. For transitivity, suppose that xRy and yRz. Then there exists integers k and m such that $x - y = 2k$ and $y - z = 2m$. Finally,

$$x - z = (x - y) + (y - z) = 2k + 2m = 2(k + m)$$

Thus R is transitive, and it is an equivalence relation on S. Now let's find the equivalence classes relative to R.

$$[1] = \{x \in S_6 : x - 1 \text{ is divisible by 2}\}$$
$$= \{1, 3, 5\}$$
$$[2] = \{x \in S_6 : x - 2 \text{ is divisible by 2}\}$$
$$= \{2, 4, 6\}$$

If we were to continue, we would find that $[1] = [3] = [5]$, and $[2] = [4] = [6]$, so there are only two distinct equivalence classes for this equivalence relation. The sets $[1]$ and $[2]$ are nonempty and disjoint, and $[1] \cup [2] = S_6$. This means that $\{[1], [2]\}$ is a partition of S_6. Theorem 3.1 shows the connection between equivalence relations and partitions.

Theorem 3.1

Let S be a nonempty set. Let R be an equivalence relation on S. Then $\{[x] : x \in S\}$ is a partition of S. Conversely, given a partition π of S, we can define the relation R_π on S.

$$R_\pi = \{(x, y) : [x] = [y]\}$$

Then R_π is an equivalence relation on S.

Proof

We will prove one direction of the theorem and leave the other as an exercise. Let R be an equivalence relation on a nonempty set S. In order to show that the set of equivalence classes determined by R forms a partition of S, we need to show three things:

1. For all $x \in S$, $[x] \neq \emptyset$
2. If $[x] \neq [y]$, then $[x] \cap [y] = \emptyset$
3. $\bigcup \{[x]: x \in S\} = S$

First we observe that $x \in [x]$ for all $x \in S$, since R is reflexive. This also implies that each $[x] \neq \emptyset$, and that $S = \bigcup \{[x]: x \in S\}$. Thus we have shown the first and third things to be true. For the second, suppose that $x, y \in S$ and $[x] \cap [y] \neq \emptyset$. Then there exists $z \in S$ such that $z \in [x] \cap [y]$. Then xRz and yRz. Using symmetry, we get xRz and zRy, so by transitivity xRy. Now we can show that $[x] = [y]$.

Suppose that $t \in [x]$. Then tRx. Combining tRx with xRy, we see that tRy. Hence $t \in [x] \Rightarrow t \in [y]$, or $[x] \subseteq [y]$. Similar reasoning shows that $[x] \supseteq [y]$, so $[x] = [y]$.

We have proved that if R is an equivalence relation on a set S, then the set of equivalence classes of R is a partition of S.

EXAMPLE 3.10

Let $S = \{a, b, c, d, e, f\}$. Let π be the partition of S.

$$\pi = \{\{a, b\}, \{c, e, f\}, \{d\}\}$$

Find the equivalence relation R_π whose existence is guaranteed by Theorem 3.1.

Solution

According to Theorem 3.1, if $x, y \in S$, then $xR_\pi y$ if and only if x and y belong to the same block. Applying this to the block $\{a, b\}$, there are four relationships: $aR_\pi a$, $aR_\pi b$, $bR_\pi a$, $bR_\pi b$. A complete list of the ordered pairs in R_π is given below.

$\{(a, a), (a, b), (b, a), (b, b),$
$(c, c), (c, e), (c, f), (e, c), (e, e), (e, f), (f, c), (f, e), (f, f),$
$(d, d)\}$

EXAMPLE 3.11

Let A be the set of binary bit strings of length 4. Define the relation R on A as follows: aRb, $a, b \in A$, if and only if a and b have the same number of 0's. Is R an equivalence relation?

Solution

Clearly R is reflexive, since any bit string has the same number of 0's as itself. Just as clearly, R is symmetric and transitive, since if a has the same number of 0's as b, then b has the same number of 0's as a, and since if a has the same number of 0's as b and b has the same number of 0's as c, then a has the same number of 0's as c. There are five equivalence classes: the bit strings with no 0's, one 0, two 0's, three 0's, and four 0's.

Order Relations

There are two other important properties of relations. We say that a relation R on a set S is **irreflexive** if $x\overline{R}x$ for all $x \in S$. We say that R is **antisymmetric** if for all x, $y \in S$, xRy and yRx implies that $x = y$. Do not confuse the irreflexive property with the absence of the reflexive property. There are relations that are neither reflexive nor irreflexive. Similarly, there are relations that are neither symmetric nor antisymmetric.

EXAMPLE 3.12

1. Show that the relation "greater than or equal to" on $\mathbf{Z}$ is antisymmetric.
2. Show that the relation on the set $\{a, b, c\}$ given by the set of ordered pairs $\{(a, b), (b, c), (c, a)\}$ is irreflexive.

Solution

1. The relation "greater than or equal to" on the set of integers is antisymmetric, because if $x, y \in \mathbf{Z}$, then $(x \geq y \text{ and } y \geq x) \Rightarrow (x = y)$.
2. The relation on the set $\{a, b, c\}$ given by the set of ordered pairs $\{(a, b), (b, c), (c, a)\}$ is irreflexive, because it does not contain any of the ordered pairs (a, a), (b, b), (c, c).

Just as we used the reflexive, symmetric, and transitive properties to characterize equivalence relations, we often wish to characterize precedence, such as when comparing the sizes of integers or determining which set is a subset of another. Examining the relation "is a subset of," we find that it is reflexive, transitive, and antisymmetric.

A relation on a set A which is reflexive, antisymmetric, and transitive is called a **partial ordering** on the set. The reason for the word *partial* is that not every pair of elements of A must be related. Consider Figure 3.11. We have $C \subseteq A$, but $B \subseteq C$ and $C \subseteq B$ are both *false*. In such cases, B and C are said to be incomparable.

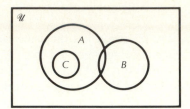

Figure 3.11. Subset as a partial ordering

Probably the most familiar example of a partial ordering is the "less than or equal to" relation on the set of real numbers. A set A with a partial ordering is called a **partially ordered set**, or **poset**.

The essence of a relation which orders a set is the transitive property; a relation that is transitive is simply called an **ordering**. Orderings are important in computer science in sorting applications, but we need to have more than just the transitivity property to make sorting possible. Sorting is generally carried out on sets that are *linearly ordered*.

Let A be a set and let R be a relation on A. R is a **linear order** on A if and only if R has the properties of (1) antisymmetry, (2) transitivity, and (3) trichotomy (for any $x, y \in A$, either xRy, yRx, or $x = y$). The relations "less than or equal to" and "less than" on the set $\mathbf{R}$ are linear orders.

EXAMPLE 3.13

Show that the relation R, defined on the set A of alphabetic characters as $a_i R a_j$, a_i, $a_j \in A$, if and only if a_i precedes a_j alphabetically, is a linear order.

Solution

We need to show that R has antisymmetry, transitivity, and trichotomy. Antisymmetry follows from the fact that if an alphabetic character a_i precedes an alphabetic character a_j alphabetically and a_j alphabetically precedes a_i, then $a_i = a_j$. Transitivity is just as easy to demonstrate, and trichotomy follows from the fact that if two alphabetic characters are distinct, one must precede the other in the alphabet.

Products of linearly ordered sets can be given a linear order called **lexicographic order**. Let A be a set of sequences of the form $a = (a_1, a_2, \dots)$, and let a linear ordering R_i be defined for each position i of the sequence. Then the sequence $x = (x_1, x_2, \dots)$ precedes the sequence $y = (y_1, y_2, \dots)$ in lexicographic order if and only if for the smallest integer k such that $x_k \neq y_k$, $x_k R_k y_k$.

Lexicographic order is used to order such objects as entries in an index as well as character strings in a programming language. The string 'Smith' precedes the string

'Smythe' in lexicographic order because the first position in which the strings differ is the third position and 'i' precedes 'y' in alphabetical order.

Transitive Closure

We will now examine one approach for determining if a relation is transitive. If the relation is not transitive, we will show how to extend it so that it is a transitive relation. For example, let S be the set of all people, and let R be the relation "is an acquaintance of." R is not transitive, but there is a natural way to extend the relation so that it forms the basis of a transitive relation. We say that two people are acquainted in an extended sense if they have a mutual acquaintance. Then we extend this by allowing chains of two intermediate acquaintances, then three, and so on. When we're done, the completely extended relation is surely transitive. We formalize this process as follows:

Let R be a relation on a set S. Define the relation R^* on S as follows: for all $x, y \in S$, xR^*y if and only if there exists a sequence $x_1, x_2, \ldots, x_k \in S$, $k \geq 2$, such that $x = x_1$, $y = x_k$, and for all $i = 1, \ldots, k - 1$, x_iRx_{i+1}. R^* is called the **transitive closure** of the relation R. Figure 3.12 illustrates extending the relation R by using intermediate chains of relations. Observe that, as sets, $R \subseteq R^*$. Moreover, R is transitive if and only if $R = R^*$.

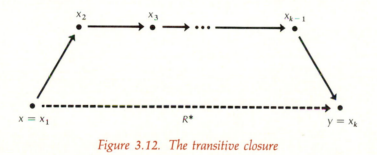

Figure 3.12. The transitive closure

EXAMPLE 3.14

Let R be the relation $\{(a, b), (b, c), (c, a)\}$ on the set $S = \{a, b, c\}$. Find the transitive closure of R.

Solution

First of all, $(a, b), (b, c), (c, a) \in R^*$ since they are in R. Since aRb and bRc, we conclude that aR^*c. Similarly, bRc and cRa implies that bR^*a. Continuing in this fashion, we see that the transitive closure of R is

$$\{(a, b), (b, c), (c, a), (a, c), (b, a), (c, b), (a, a), (b, b), (c, c)\}$$

SELF TEST 2

1. Consider the set $A = \{a, b, c\}$ and the relations on A represented by the directed graphs of Figure 3.13. Determine which are reflexive, irreflexive, symmetric, antisymmetric, and transitive.

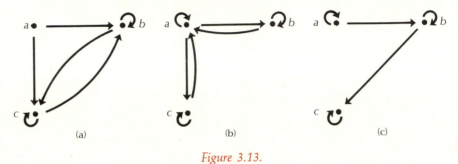

(a)   (b)   (c)

Figure 3.13.

2. Show that the relation $R = \{(a, a), (b, b), (c, c), (a, b), (b, a)\}$ on $A = \{a, b, c\}$ is an equivalence relation.
3. Describe the equivalence classes for the relation of exercise 2.
4. Graph the partial ordering "is a subset of" on the power set $\mathscr{P}(A)$, where $A = \{a, b\}$.
5. Find the transitive closure of the relation

$$R = \{(b, a), (b, c), (c, a), (a, c)\}$$

defined on $A = \{a, b, c\}$.

Exercises 3.2

1. Find examples of reflexive relations that are not symmetric or transitive and examples of relations that are transitive and symmetric, but not reflexive, and so on.
2. What is wrong with the following "proof" that the reflexive property follows from the symmetric and transitive properties?

   Assume relation R is symmetric and transitive. Then $xRy \Rightarrow yRx$ by symmetry. Since xRy and yRx, then by transitivity we have xRx. Therefore, R is reflexive.

3. Determine if each of the following relations is reflexive, symmetric, transitive, irreflexive, or antisymmetric.
   a. "is the brother of," on the set of males

b. "has the same birthday as," on the set of all people

c. the relation R, on the set S_4

$$S_4 = \begin{bmatrix} T & F & T & T \\ F & T & F & T \\ T & T & T & F \\ F & F & F & T \end{bmatrix}$$

d. "is a multiple of," on the set of positive integers

4. a. Give an example of a relation that is neither reflexive nor irreflexive.

   b. Give an example of a relation that is neither symmetric nor antisymmetric.

5. Show that a relation R is irreflexive if and only if $\bar{R}$ is reflexive.

6. Find the transitive closure of each of the relations in exercise 3.

7. Determine which of the relations in exercise 3 are partial orderings and which are linear orderings.

8. Find the transitive closure of the relation shown in Figure 3.14.

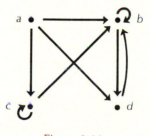

Figure 3.14.

9. Let $S = \{A_1, A_2, \ldots, A_n\}$ be a set of algorithms. Define the relation R on S as A_iRA_j if and only if A_i calls A_j. In particular, let $S = \{A, B, C, D, E\}$ be a set of algorithms, and suppose the relation "calls" is defined as follows:

A calls B

A calls C

B calls C

C calls E

D calls C

E calls B

a. Compute the transitive closure of "calls."

b. An algorithm A_i is recursive if A_iRA_i is in the transitive closure of "calls." Does S contain any recursive algorithms?

3.3   Composition of Relations

We will now examine several ways of creating new relations from existing ones. Let R be a relation between sets A and B, and let S be a relation between B and C. The **composition** of R and S is the relation between A and C, denoted $S \circ R$, given by

$$S \circ R = \{(x, z): x \in A, z \in C, \text{ and} \\ \text{there exists } y \in B \text{ such that } xRy \text{ and } ySz\}$$

Figure 3.15 shows the composite relation $S \circ R$. One can view the composite relation as a means of linking elements of A to elements of C by using elements of B as intermediate points.

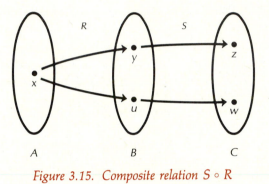

Figure 3.15. Composite relation $S \circ R$

EXAMPLE 3.15

Let R be the relation between S_3 and S_4, and let S be the relation between S_4 and S_2 as shown in Figure 3.16. Find the composite relation $S \circ R$ between S_3 and S_2.

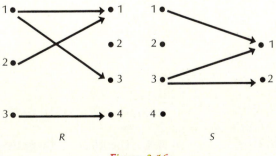

Figure 3.16.

Solution

The composite relation can be obtained by finding all arrows from points in S_3 that end at points in S_4 from which other arrows leave to reach points in S_2. For instance $(1, 1)$ and $(1, 2) \in S \circ R$ since there is an arrow from 1 to 3 in R (that is, $(1, 3) \in R$), and there are arrows from 3 to 1 and from 3 to 2 in S (that is, $(3, 1)$ and $(3, 2) \in S$). The complete solution is shown in Figure 3.17.

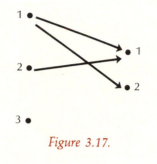

Figure 3.17.

Logical Matrix Product

We would like to define a matrix operation that combines the matrices representing two relations to obtain the matrix representing the composite relation. Let A be the set $\{a_1, a_2, \ldots, a_m\}$, let B be the set $\{b_1, b_2, \ldots, b_n\}$, and let C be the set $\{c_1, c_2, \ldots, c_p\}$. Let R be a relation between A and B, and let S be a relation between B and C. Then R has an $m \times n$ matrix representation M_1, and S has an $n \times p$ matrix representation M_2. The composite relation $S \circ R$ has an $m \times p$ matrix representation M_3. The operation we seek, which we will denote by the symbol $\cdot$, allows us to write

$$M_3 = M_1 \cdot M_2$$

We can find the required operation by considering Figure 3.18, which shows the relations R and S. For each $i \in \{1, \ldots, m\}$ and $j \in \{1, \ldots, p\}$, $M_3(i, j) = true$ if and only if $a_i \, S \circ R \, a_j$, which in turn holds if and only if there exists $k \in \{1, \ldots, n\}$ such that $a_i R b_k$ and $b_k S c_j$. Using the definitions of the logical operations **and** and **or**, we obtain the formula given below.

$$M_3(i, j) = [M_1(i, 1) \textbf{ and } M_2(1, j)] \textbf{ or }$$
$$[M_1(i, 2) \textbf{ and } M_2(2, j)] \textbf{ or } \ldots$$
$$[M_1(i, n) \textbf{ and } M_2(n, j)]$$

We call the derived matrix, M_3, the **logical (Boolean) matrix product** of the matrices M_1 and M_2.

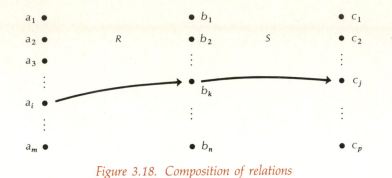

Figure 3.18. Composition of relations

EXAMPLE 3.16

Let M_1 and M_2 be the relation matrices.

$$M_1 = \begin{bmatrix} T & T & F \\ F & T & F \end{bmatrix} \qquad M_2 = \begin{bmatrix} T & F & T & T \\ T & F & F & F \\ F & T & F & T \end{bmatrix}$$

1. Compute $M_3 = M_1 \cdot M_2$.
2. Interpret the product as the composition of relations.

Solution

1. M_1 is a 2×3 matrix and M_2 is a 3×4 matrix, so the product M_3 must be a 2×4 matrix. Consider the entry $M_3(2, 3)$. By definition

$$M_3(2, 3) = [M_1(2, 1) \text{ and } M_2(1, 3)] \text{ or}$$
$$[M_1(2, 2) \text{ and } M_2(2, 3)] \text{ or}$$
$$[M_1(2, 3) \text{ and } M_2(3, 3)]$$

This amounts to placing the second row of M_1 (F   T   F) next to the third column of M_2 (T   F   F) and looking for matching T's. Since none match in this case, $M_3(2, 3) = F$. The complete result is

$$\begin{bmatrix} T & F & T & T \\ T & F & F & F \end{bmatrix}$$

2. R and S can be interpreted as relations between S_2 and S_3 and between S_3 and S_4, respectively. These are shown in Figure 3.19. Consider the entry corresponding to $(2, 3)$ in the composite relation. There is an arrow from 2 to 2 on the R side, but there is not an arrow from 2 to 3 on the S side. Hence, there should not be an arrow from 2 to 3 in the composite relation $S \circ R$. This is precisely the same

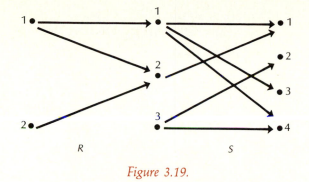

Figure 3.19.

as saying that $M_3(2, 3)$ should be *false*, as we computed in the first part of this example. The graph of $S \circ R$ is shown in Figure 3.20.

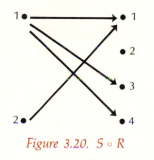

Figure 3.20. $S \circ R$

The Identity Relation

A relation that can be defined on any set A is the **identity** relation, denoted i_A.

$$i_A = \{(x, x) : x \in A\}$$

Thus for all $x, y \in A$, $x i_A y$ if and only if $x = y$. Identity relations have the important property given in Theorem 3.2.

Theorem 3.2

If R is a relation between sets A and B, then

$$R \circ i_A = i_B \circ R = R$$

Proof

It is enough to apply the definition of composition directly.

$$R \circ i_A = \{(x, z): x \in A, z \in B \text{ such that}$$
$$\qquad \text{there exists } y \in A \text{ such that } xi_Ay \text{ and } yRz\}$$
$$= \{(x, z): xRz\}$$
$$= R$$

Since xi_Ay if and only if $x = y$, we have $R \circ i_A = \{(x, z): xRz\} = R$. The proof that $i_B \circ R = R$ is similar.

Inverse Relations

Every relation R between sets A and B is a subset of $A \times B$. We can reverse the roles of A and B to obtain a relation between B and A called the **inverse relation** of R. The inverse relation of R, denoted R^{-1}, is the relation between B and A given by

$$R^{-1} = \{(y, x): (x, y) \in R\}$$

EXAMPLE 3.17

Find the inverse of the relation given in Figure 3.21.

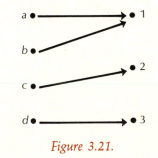

Figure 3.21.

Solution

Since xRy is identical to $yR^{-1}x$ for all $x \in A$ and $y \in B$, drawing R^{-1} is simply a matter of reversing each of the arrows in the figure. The result is shown in Figure 3.22.

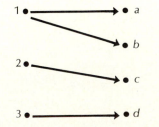

Figure 3.22. The inverse of the relation in Figure 3.23

SELF TEST 3

1. Let R be the following relation between S_3 and S_5:

$$R = \{(1, 1), (1, 3), (2, 3), (2, 4), (3, 1), (3, 4), (3, 5)\}$$

Let S be the following relation between S_5 and S_2:

$$S = \{(1, 1), (1, 2), (2, 1), (4, 2), (5, 1), (5, 2)\}$$

   a. Find the composite relation $S \circ R$ between S_3 and S_2 directly from the definition.
   b. Find $S \circ R$ using the logical matrix product.
2. Let R be the following relation between S_3 and S_4:

$$R = \{(1, 1), (1, 3), (2, 3), (2, 4), (3, 1), (3, 4)\}$$

   Find R^{-1}, the inverse of R.
3. Let $M_1 = \begin{bmatrix} F & T \\ T & F \\ F & T \end{bmatrix}$, and let $M_2 = \begin{bmatrix} T & F & F \\ T & T & T \end{bmatrix}$.

   a. Compute $M_1 \cdot M_2$.
   b. Compute $M_2 \cdot M_1$.

Exercises 3.3

1. Let R and S be the following relations on $A = \{1, 2, 3, 4\}$:

$$R = \{(1, 1), (1, 3), (3, 2), (3, 4), (4, 2)\}$$
$$S = \{(2, 1), (3, 3), (3, 4), (4, 1)\}$$

   Find each of the following composite relations on A.
   a. $R \circ S$ c. $R \circ R$
   b. $S \circ R$ d. $S \circ S$
2. Prove that $T \circ (S \circ R) = (T \circ S) \circ R$, where R is a relation between A and B, S is a relation between B and C, and T is a relation between C and D.
3. Let $A = \{1, 2, 3, 4, 6\}$, and let R be the relation on A defined by "x divides y," $x, y \in A$. Find R^{-1}, the inverse of R.
4. Prove that if R is either the empty relation or the universal relation on a set A, then $R \circ R = R$.

5. Show that a relation R on A is reflexive if and only if $i_A \subseteq R$.

6. Show that a relation R on a set A is symmetric if and only if $R = R^{-1}$.

7. Show that a relation R on a set A is transitive if and only if $R \circ R \subseteq R$.

3.4   Functions

Relations are useful for modeling associations between pairs of elements, but there are situations where more special conditions are needed. For instance, it is often important that a person be identified by a unique name. However, the association between names and people is not unique—look in your telephone book, wherever you are, and you will find several entries with the same name. Thus other systems are used for assigning identification to individuals. For each person in the United States, it is important that he or she has only one Social Security number. This kind of mapping between objects is the subject of this section.

A **function** from set A to set B is a relation F between A and B satisfying two conditions:

1. For all $x \in A$, there exists $y \in B$ such that $(x, y) \in F$.

2. For all $x \in A$, y, $z \in B$, $(x, y) \in F$ **and** $(x, z) \in F$ implies $y = z$.

EXAMPLE 3.18

Which of the relations depicted in Figure 3.23 are functions?

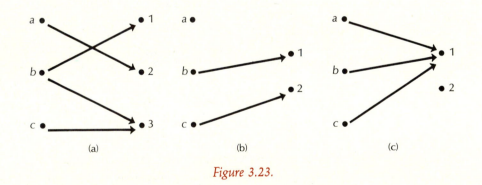

(a)   (b)   (c)

Figure 3.23.

Solution

The first condition says that there must be at least one arrow leaving each point in A. The second condition says that there can be no more than one arrow leaving each point in A. We see that the relation in (a) satisfies the first condition, but not the second condition, so it is not a function. Similarly, the relation shown in (b) satisfies the second condition, but not the first condition, so it is not a function. The relation shown in (c) is a function from $\{a, b, c\}$ to $\{1, 2\}$.

Usually we denote relations with capital letters. With functions the convention is to use lowercase letters. Let f be a function from A to B. Because each element x of A appears in one and only one pair $(x, y) \in f$, it is possible to write $y = f(x)$ whenever $x \in A$. This notation suggests **mapping** the element x to the element y. Functions are often referred to as mappings or transformations. The unique element $y = f(x)$ of B assigned to $x \in A$ by f is called the **image** of x under f. We write $f: A \to B$ to indicate that f is a function from A to B. The set A is called the **domain** of f, and the set B is called the **codomain** of f. The **range** of f, denoted by $f[A]$, is the set of all images; that is,

$$f[A] = \{f(x): x \in A\}$$

Two functions f and g are said to be **equal** if they have the same domain and codomain, and for all x in the domain $f(x) = g(x)$.

More generally, if C is a subset of A, the **image of C by f**, denoted $f[C]$, is the set

$$f[C] = \{y \in B: \text{there exists an } x \in C \text{ such that } f(x) = y\}$$

EXAMPLE 3.19

Determine whether each of the following relations is a function. For those that are functions, give the domain and range of the function.

1. $\{(1, a), (1, b), (2, c), (3, b)\}$
2. $\{(a, a), (b, b), (c, c)\}$
3. $\{(x, y): x, y \text{ are real numbers and } y - x = 1\}$
4. $\{(x, y): x, y \text{ are positive integers and } y - x = 1\}$

Solution

1. This is not a function because $(1, a)$ and $(1, b)$ are both in the relation.
2. This is a function, with domain, codomain, and range equal to $\{a, b, c\}$.
3. Since $y - x = 1$ if and only if $y = x + 1$, it is easy to see that this is a function with domain and codomain equal to the set of real numbers. Also, for each real number y, there is a real number x such that $y = x + 1$. Therefore, the range of the function is the set of real numbers.
4. This is a function. The domain and codomain are both the set of positive integers. However, if y is a positive integer, there is a positive integer x such that $y = x + 1$

if and only if $y \geq 2$. Hence the range is $\{2, 3, \ldots\}$. Notice that this function is not equal to the function in (3) even though they both have the form $y = x + 1$.

Injections and Surjections

We introduced the idea of a relation by constructing pairs of one student and one course for which the student is registered. If there is one student taking more than one course, this relation is not a function. On the other hand, the association of students to the *set* of courses for which the student is registered does define a function. The domain of the function is the set of all registered students, and the codomain is the power set of the set of courses. For this function, there may be several students taking exactly the same set of courses; that is, it is not possible to make unique assignments of students to a set of courses. For a given set of courses, there does not need to be a unique student. Moreover, there will be sets of courses (certainly any of cardinality greater than 10) for which no student is registered. These two observations form the basis for the next definitions.

Let $f \colon A \to B$ be a function. The function f is called an **injective function**, or an **injection**, if $\forall x, y \in A$, $f(x) = f(y)$ implies $x = y$. In terms of the graphic representation, this means that if two arrows arrive at the same point in B, they must come from the same point in A, and therefore they are the same. The function f is called a **surjective function**, or a **surjection**, if for each $y \in B$ there exists $x \in A$ such that $f(x) = y$. In terms of the graphic representation, this means there must be an arrow arriving at each point of B.

There are other terms applied to functions with the properties mentioned. An injective function is also called a **one-to-one** function, or $1-1$ function. A surjective function is also called **onto**. Figure 3.23(c) is a function, but is neither $1-1$ nor onto. A function which is both injective and surjective is called a **bijection**. A bijection from a set A to itself is called a **permutation** of the set A.

EXAMPLE 3.20

Determine if each of the following functions is an injection, surjection, or bijection.

1. the identity function i_A, on a nonempty set A
2. the functions in Figure 3.24

Solution

1. For each set A, the identity function i_A is an obvious bijection, and is a trivial permutation of A.
2. The functions in Figure 3.24 show the independence of injection and surjection. The function in (a) is a surjection, but not an injection. The function in (b) is an injection, but not a surjection.

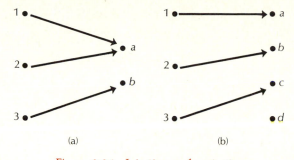

Figure 3.24. Injection and surjection

Functions and Cardinality

The functions in Figure 3.24 suggest that in order for the function $f: A \to B$ to be one-to-one, we need $|A| \leq |B|$, and that in order for f to be onto, we need $|A| \geq |B|$. There are several results relating set size and properties of functions. Theorem 3.3 contains five such results.

Theorem 3.3

Let A and B be finite sets. Let $f: A \to B$ be a function.

1. $|f(A)| \leq |A|$ and equality holds if and only if f is injective.
2. If f is an injection, then $|A| \leq |B|$.
3. If f is a surjection, then $|A| \geq |B|$.
4. If f is a bijection, then $|A| = |B|$.
5. The Pigeonhole Principle

Suppose that A and B are finite sets and $|A| = |B|$. Then f is an injection if and only if f is a surjection.

Proof

1. This result is intuitively obvious, but let's prove it in some detail. Let $|A| = k$, so that $A = \{a_1, \ldots, a_k\}$. For each i, $|\{f(a_i)\}| = 1$, and we obtain

$$|f(A)| = \left| \bigcup_{i=1}^{k} \{f(a_i)\} \right| \leq \sum_{i=1}^{k} |\{f(a_i)\}|$$
$$= k$$

Thus the inequality holds in general. In order for $|f(A)|$ to equal k, we must have $f(a_i) \neq f(a_j)$ for $i \neq j$. But this condition says precisely that f is an injection.

2. This statement follows from the first one and the fact that $f(A) \subseteq B$, so that when f is an injection

$$|A| = |f(A)| \le |B|$$

3. This statement also follows from the first one and the fact that when f is surjective, $f(A) = B$. Hence

$$|B| = |f(A)| \le |A|$$

4. This statement is an easy consequence of the second and third statements.

5. The Pigeonhole Principle is proved in two parts: if f is injective, then f is surjective; and if f is surjective, then f is injective. Suppose that f is injective. Then $|A| = |f(A)| = |B|$. But the only way a subset of a finite set B can have the same number of elements as the set B is if the subset equals B. In this case, that means that $f(A) = B$, so f is surjective. The converse is proved by applying the first statement of this theorem.

•   •   •

The Pigeonhole Principle gets its name from the following interpretation. Suppose there are 100 mail boxes (pigeonholes) and 100 letters. Think of the assignment of boxes to letters as the function f. In order to put letters in the mail boxes so that there is only one letter in each box, it is necessary to use all the boxes. This corresponds to the implication that if f is injective, it is a surjection. On the other hand, in order to make sure that each mail box gets at least one letter, it is clearly necessary to use all the letters, with only one letter per box. This corresponds to saying that if f is a surjection, then f is also an injection. The Pigeonhole Principle does not apply to infinite sets. In the fourth part of Example 3.19, the function $f(x) = x + 1$ maps positive integers to positive integers and is an injection. It is not a surjection, however, since the range does not equal the codomain. Here is another useful form of the Pigeonhole Principle that applies to infinite sets as well:

If $f: A \to B$ is a function and $|A| > |B|$, then
there exists $x, y \in A$ such that $x \ne y$ and $f(x) = f(y)$.

Invertible Functions

Any function f has an inverse relation, f^{-1}. The inverse relation does not need to be a function, as is the case in the fourth part of Example 3.19. If the inverse relation of a function is a function, we say that the function is **invertible**.

EXAMPLE 3.21

Determine if each of the functions in Figure 3.25 is invertible.

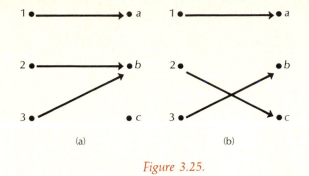

Figure 3.25.

Solution

a. The inverse relation does not contain a pair of the form (c, i), $i \in S_3$, since the function itself is not a surjection.

b. The function is an injection and a surjection, so the inverse relation is a function. This is formalized in Theorem 3.4.

EXAMPLE 3.22

Consider a function $e: B^m \to B^{m+1}$ used to encode binary data for transmission so that certain types of transmission errors may be detected. For $a \in B^m$ we have

$$e(a_1, a_2, \ldots, a_m) = (a_1, a_2, \ldots, a_m, a_{m+1}), \text{ where}$$

$$a_{m+1} = \begin{cases} 0 \text{ if the number of 1's in } a \text{ is even} \\ 1 \text{ if the number of 1's in } a \text{ is odd} \end{cases}$$

Show that there exists a function $d: B^{m+1} \to B^m$ such that $d \circ e = i_{B^m}$, but e is not a bijection.

Solution

e is an injection, but not a surjection. The function d will simply drop off the last term of each $(m+1)$-tuple. Thus

$$d(a_1, \ldots, a_m, a_{m+1}) = (a_1, \ldots, a_m).$$

d is a surjection, but not an injection.

Theorem 3.4

Let $f: A \to B$ be a function. The function f is invertible if and only if f is a bijection.

Proof

Suppose f is invertible. To show that f is a bijection, we need to show it is an injection and a surjection. To show it is an injection, we need to show that if $f(x) = f(z)$, then $x = z$. Let $y = f(x) = f(z)$. Then (x, y) and (z, y) are in f, so (y, x) and $(y, z) \in f^{-1}$. Since f^{-1} is a function, it follows that $x = z$. To show that f is a surjection, we need to show that for all $y \in B$, there exists $x \in A$ such that $y = f(x)$. But f^{-1} is a function, so for each $y \in B$, there exists $x \in A$ such that $(y, x) \in f^{-1}$, which means that $(x, y) \in f$.

The proof of the converse is left as an exercise.

• • •

Figure 3.26 suggests that the inverse function undoes the work of the function. This is in fact the case, and this is stated in Theorem 3.5. The proof is omitted.

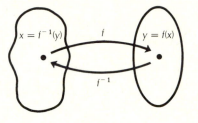

Figure 3.26.

Theorem 3.5

Let $f: A \to B$ be a function. Then f is invertible if and only if there exists a function $g: B \to A$ such that $g \circ f = i_A$ and $f \circ g = i_B$. Moreover, if g exists, $g = f^{-1}$.

Binary Operations

An important class of functions is the class of functions known as binary operations. A **binary operation on a set A** is a function $op: A \times A \to A$. Thus a binary operation takes two elements of A and maps them to a third element of A.

Rather than write $op(a, b)$ for the value of the operation, it is more common to write $a \, op \, b, a, b \in A$. The former notation is called **prefix notation** because the operator precedes the arguments. The latter notation is called **infix notation**. We have encountered binary operations before. We now list some of them. You should check their definitions to be sure that they are really functions.

Here are some examples of binary operations:

- The set operations union ($\cup$), intersection ($\cap$), and relative complement ($-$)
- The composition operation ($\circ$) on the set $F_A = \{f: f \text{ is a function from } A \to A\}$, A a set
- The logical connectives **and** and **or** on a set of propositions

SELF TEST 4

1. State whether each diagram of Figure 3.27 defines a function from $\{a, b, c\}$ to $\{x, y, z\}$.

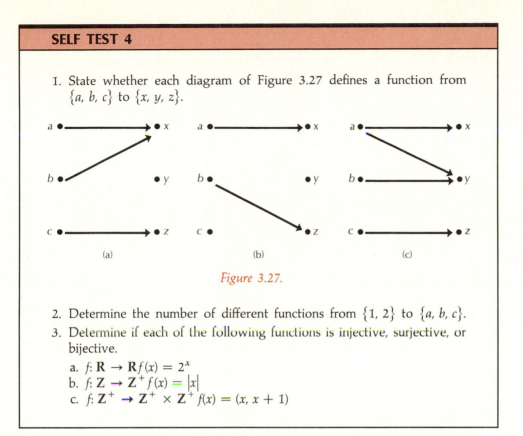

(a)   (b)   (c)

Figure 3.27.

2. Determine the number of different functions from $\{1, 2\}$ to $\{a, b, c\}$.

3. Determine if each of the following functions is injective, surjective, or bijective.
   a. $f: \mathbf{R} \to \mathbf{R}\ f(x) = 2^x$
   b. $f: \mathbf{Z} \to \mathbf{Z}^+\ f(x) = |x|$
   c. $f: \mathbf{Z}^+ \to \mathbf{Z}^+ \times \mathbf{Z}^+\ f(x) = (x, x + 1)$

Functions in Computer Languages

Modern programming languages provide facilities for isolating logically separate calculations and calculations performed in the same way with different values from the domain. These facilities are generally called *subprograms*. Functions are subprograms that return a single value. Thus they typically conform to our definition of *function* in that invoking one of them several times with the same value(s) from the domain yields the same result. Examples of functions available in many languages are the common mathematical routines of *sin*, *log*, and *abs* (absolute value). Most languages also allow users to write their own functions. Typically, the first line of the function definition specifies the name as well as the domain and codomain of the function. For example,

```
function Cube (x : real) : real;
begin
      Cube := x * x * x
end.
```

defines a function to compute the cube of a real number. The function *Cube* is written in the programming language Pascal. Given a value x, it will return x^3. The domain of this function is **R**, as is the codomain. In many languages, however, users may write functions that do not conform to the pure definition of *function* given here. Consider the simple function *TooBad*, written in Pascal, which makes use of an externally defined variable. Its domain is **Z** (integers) as is its codomain. Note that it uses a variable y, whose value is defined outside of the function somewhere else in the program.

```
function TooBad (x : integer) : integer;
begin
      TooBad := x * y;
end;
```

Consider the following program fragment:

```
y := 10;
x := 5;
z := TooBad(x);
   ⋮
y := 20;
w := TooBad(x);
```

Assuming there is no change in x between the invocations of *TooBad*, we might expect that z and w have the same value. This is not the case, however, because $z = 50$ and $w = 100$ after the fragment of the program terminates.

The source of the problem is the external variable y. Some languages eliminate this difficulty by not allowing external variables in function definitions, but in most cases this is regarded as too restrictive. The ability to access and change variables from outside the function definition is useful for writing functions such as random number generators. In fact, such functions are often written without specifying the domain in the function definition—only the codomain is specified. In this case, the function relies on variables defined elsewhere in the program. The function

```
function Random : real;
begin
      Seed := (Seed * 13077 + 6925) mod 32768;
      Random := Seed/32768.0
end.
```

uses the variable *Seed* and returns a random number. It also modifies *Seed* so that subsequent calls of the function return a different random number. This kind of change in the external environment is known as a *side-effect* of a function and is usually considered harmful because it can obscure the cause of a value change for a globally accessible variable. Some functions, however, such as the generation of random numbers, rely on the ability to modify external variables to perform their work, and it is not uncommon to find functions in some languages that allow changes to the external environment.

Exercises 3.4

1. Determine if each of the following relations is a function. For those that are functions, give the domain, codomain, and range of the function. Also determine whether each function is an injection, a surjection, or a bijection.
   a. $\{(x, y): x, y \in \mathbf{R} \text{ such that } x^2 + y^2 = 1\}$
   b. $\{(a, \alpha), (b, \delta), (c, \alpha), (d, \eta), (e, \eta)\}$
   c. The relations in Figure 3.28.

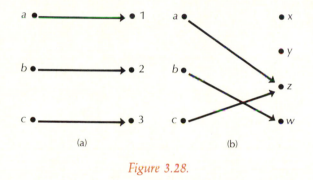

Figure 3.28.

2. a. Show that the composition of two injections is an injection.
   b. Show that the composition of two surjections is a surjection.
3. Let $f: A \to B$ and $g: B \to C$ be functions. Is it possible for the composition $g \circ f$ to be a bijection even if f and g are not? Give a reason for your answer.
4. Use the Pigeonhole Principle to conclude the following.
   a. Any 30-digit number must contain a repeated digit.
   b. There must be two people in Boston who have the same birthday.
5. Give an example of a function $f: \mathbf{Z} \to \mathbf{Z}$ that is surjective but not injective.
6. Prove the rest of Theorem 3.4.
7. Let $f: S_n \to S_n$ be given by

$$f(1) = n, \text{ and } f(i) = i - 1, \text{ for } 2 \le i \le n$$

Verify that f is a bijection and find its inverse. Draw the graphical representation of f in the case $n = 4$.
8. Let $f: \mathbf{Z} \to \mathbf{Z}$ be given by $f(i) = i - 1$, $i \in \mathbf{Z}$. Find f^{-1}.

Database Management Systems

What exactly is a database management system? Basically, it is nothing more than a computer-based record-keeping system for storing and maintaining information. Data that are stored in a computer are called a **database**. The collection of programs that allows people to use and modify these data is a **database management system (DBMS)**. The structure of a database management system is illustrated in Figure 3.29.

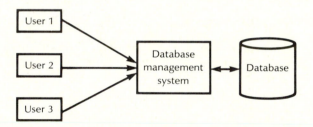

Figure 3.29. Structure of a database management system

A major role of the DBMS is to allow users to deal with data in abstract terms rather than in the way the computer actually stores the data. Thus users can specify what to do with data with little regard to detailed algorithms and data structures used by the system. One approach is to visualize the data as organized into tables, which, from an abstract viewpoint, are relations—the subject of this chapter. Table 3.1 is an example of a table representing customer orders.

Mathematically, we can think of the table as describing associations between objects from an arbitrary number of sets. Such relations are merely subsets of a product set $A_1 \times A_2 \times \ldots \times A_n$, and are called **$n$-ary** relations.

Here, A_1 is the set of valid order numbers, A_2 is the set of valid supplier names, A_3 is the set of items stocked, and A_4 is the set of positive integers. The elements of the relation are **4-tuples** such as (1004, Compcomp, Solder, 10). Tables of this kind represent one way to organize large varieties of data so that they can be manipulated effectively. Such tables and the underlying n-ary relations they represent are the basis of the approach to database organization called the **relational database model**.

TABLE 3.1
Customer Orders

Order Number	Supplier	Item	Quantity
1001	Acme, Inc	Printer	2
1003	Acme, Inc	Serial Interface	2
1004	Compcomp	Solder	10
1008	Hack, Ltd.	Disk, Floppy	999
1010	Hack, Ltd.	IC Socket	12

Operations on *n*-ary Relations

Since tables representing *n*-ary relations are used to store information, we define operations on them to make it easier to extract information stored there. We will illustrate three basic operations on *n*-ary relations: **selection**, **projection**, and **join**.

Selection.   The *selection* operation allows us to extract a row (or rows) of a table that satisfies a set of selection criteria. For instance, using the data in Table 3.1, we might want to find all orders for more than ten items. To do this, we would write something like the statement below.

select from Customer Orders all rows with Quantity > 10

This request would return Table 3.2.

TABLE 3.2
Large Customer Orders

Order Number	Supplier	Item	Quantity
1008	Hack, Ltd.	Disk, Floppy	999
1010	Hack, Ltd.	IC Socket	12

Table 3.1, Customer Orders, represents the *n*-ary relation on which the selection operation is carried out. Table 3.2, Large Customer Orders, is the result of this operation.

Projection.   Sometimes we need to extract columns of interest from a table representing an *n*-ary relation to create a new relation by specifying those columns. This operation is called *projection*. For example, in the Customer Order relation, we might want to extract the supplier and item supplied by that supplier only. Then we would write something like

project Supplier and Item from Customer Orders

This would return Table 3.3.

TABLE 3.3
Supplied Items

Supplier	Item
Acme, Inc	Printer
Acme, Inc	Serial Interface
Compcomp	Solder
Hack, Ltd.	Disk, Floppy
Hack, Ltd.	IC Socket

Join.   The third operation we consider is a way of combining two tables into a larger one. In general, one can join two tables only if they have one or more columns in common. For relations having no common columns, to join them would really be just to find the product of the relations. For tables with common columns, the *join* operation matches all rows in both tables that have common entries in the specified column(s) and creates a new, longer row in the joined relation. Of course, if the match column has no common entries in the tables to be joined, the resulting table will be empty. For instance, we could have the relation Suppliers shown in Table 3.4 and the Customer Orders relation shown in Table 3.1. Table 3.4 contains the supplier's name, city, and status (1 for preferred source, 2 for secondary source).

TABLE 3.4
Suppliers

Supplier	City	Status
Acme, Inc	Chicago	2
Compcomp	San Francisco	1
Hack, Ltd.	Boston	2
Hack, Ltd.	Miami	1
Infosys	New York	2

The join of the Customer Order and Supplier relations on the Supplier column is shown in Table 3.5.

TABLE 3.5
Results of the Join

Order Number	Supplier	Item	Quantity	City	Status
1001	Acme, Inc	Printer	2	Chicago	2
1003	Acme, Inc	Serial Interface	2	Chicago	2
1004	Compcomp	Solder	10	San Francisco	1
1008	Hack, Ltd.	Disk, Floppy	999	Boston	2
1008	Hack, Ltd.	Disk, Floppy	999	Miami	1
1010	Hack, Ltd.	IC Socket	12	Boston	2
1010	Hack, Ltd.	IC Socket	12	Miami	1

Exercises 3.5

1. A database is to contain information about students and the courses they have taken. At a particular point in time, the students and the courses they have taken are as follows:

Student	Courses
Andrew	cs101, cs108
Dianne	cs101, cs104
Erika	cs218, cs301, cs547
Kenneth	cs108, cs208, cs248, cs301

For each student, the database contains age, sex, address, and degree major. For each course, it contains a course number, year and semester course was taken, and the grade received. Sketch a relational structure for this data by describing appropriate tables to represent the information.

2. Given the *n*-ary relations you constructed for exercise 1, describe the operation(s) required to do each of the following tasks.

   a. Find the names and addresses of all male students.
   b. Construct a new table describing only the names and addresses of all students.
   c. Construct a new table describing only the names and addresses of all female students.
   d. Find the names and addresses of all female students who received an *A* in cs108.

SUMMARY

1. A **binary relation** between sets A and B is a subset R of $A \times B$.

2. Let R be a relation on a set A. We say that R is
   a. **reflexive** if for all $x \in A$, xRx.
   b. **symmetric** if for all $x, y \in A$, $xRy \Rightarrow yRx$.
   c. **transitive** if for all $x, y, z \in A$, xRy and $yRz \Rightarrow xRz$.

3. Relations that are reflexive, symmetric, and transitive are called **equivalence relations**.

4. Let S be a nonempty set. Let R be an equivalence relation on S. For each $x \in S$, we define the **equivalence class of x relative to R**, denoted $[x]$, by $[x] = \{y \in S: xRy\}$.

5. **Theorem 3.1**. Let S be a nonempty set. Let R be an equivalence relation on S. Then $\{[x]: x \in S\}$ is a partition of S. Conversely, given a partition π of S, we can define the relation R_π on S by $R_\pi = \{(x, y): [x] = [y]\}$. Then R_π is an equivalence relation on S.

6. A relation R on a set S is **irreflexive** if $x\bar{R}x$ for all $x \in S$.

7. A relation R is **antisymmetric** if for all $x, y \in S$, xRy and yRx implies that $x = y$.

8. A relation on a set A that is reflexive, antisymmetric, and transitive is called a **partial ordering** on the set. A set A with a partial ordering is called a partially ordered set, or **poset**.

9. A relation that is transitive is called an **ordering**.

10. Let A be a set, and let R be a relation on A. R is a **linear order** on A if and only if R has the properties of (1) antisymmetry, (2) transitivity, and (3) **trichotomy** (for any $x, y \in A$ either xRy, yRx, or $x = y$).

11. Let A be a set of sequences of the form $a = (a_1, a_2, \ldots)$, and let a linear ordering R_i be defined for each position i of the sequence. Then the sequence $x = (x_1, x_2, \ldots)$ precedes the sequence $y = (y_1, y_2, \ldots)$ in **lexicographic order** if and only if for the smallest integer k such that $x_k \neq y_k$, $x_k R_k y_k$.

12. Let R be a relation on a set S. Define the relation R^* on S as follows: for all $x, y \in S$ we say xR^*y if and only if there exists a sequence $x_1, x_2, \ldots, x_k \in S$, $k \geq 2$, such that $x = x_1$, $y = x_k$, and for all $i = 1, \ldots, k-1$, $x_i R x_{i-1}$. R^* is called the **transitive closure** of the relation R.

13. Let R be a relation between sets A and B, and let S be a relation between B and C. The **composition** of R and S is the relation between A and C, denoted $S \circ R$, given by $S \circ R = \{(x, z): x \in A, z \in C$ and there exists $y \in B$ such that xRy and $ySz\}$.

14. **Theorem 3.2**. If R is a relation between sets A and B, then $R \circ i_A = i_B \circ R = R$, where i_A and i_B are, respectively, the identity relations on A and B.

15. The **inverse relation** of R, denoted R^{-1}, is the relation between B and A given by $R^{-1} = \{(y, x): (x, y) \in R\}$.

16. A **function** from set A to set B is a relation F between A and B satisfying two additional conditions:

   a. For all $x \in A$, there exists $y \in B$ such that $(x, y) \in F$.

   b. For all $x \in A$, y, $z \in B$, $(x, y) \in F$ **and** $(x, z) \in F$ implies $y = z$.

17. The unique element $y = f(x)$ of B assigned to $x \in A$ by f is called the **image** of x under f. The **range** of f, denoted by $f[A]$, is the set of all images, i.e., $f[A] = \{f(x): x \in A\}$.

18. Two functions f and g are said to be **equal** if they have the same domain and codomain, and for all x in the domain $f(x) = g(x)$.

19. Let $f: A \to B$ be a function. f is called an **injective** function or an **injection** if $\forall x$, $y \in A$, $f(x) = f(y)$ implies $x = y$.

20. The function f is called a **surjective** function, or a **surjection**, if for each $y \in B$ there exists $x \in A$ such that $f(x) = y$.

21. A function which is both injective and surjective is called a **bijection**. A bijection $f: A \to A$—that is, from a set A to itself—is called a **permutation** of the set A.

22. **Theorem 3.3**. Let A and B be finite sets. Let $f: A \to B$ be a function.

   a. $|f(A)| \leq |A|$, and equality holds if and only if f is injective.

   b. If f is an injection, then $|A| \leq |B|$.

   c. If f is a surjection, then $|A| \geq |B|$.

   d. If f is a bijection, then $|A| = |B|$.

   e. **The Pigeonhole Principle**. Suppose $|A| = |B|$ and A is finite. Let $f: A \to B$. The function f is an injection if and only if f is a surjection.

23. If the inverse relation of a function is a function, we say that the function is **invertible**.

24. **Theorem 3.4**. Let $f: A \to B$ be a function. The function f is invertible if and only if f is a bijection.

25. **Theorem 3.5**. Let $f: A \to B$ be a function. Then f is invertible if and only if there exists a function $g: B \to A$ such that $g \circ f = i_A$ and $f \circ g = i_B$. Moreover, if g exists, $g = f^{-1}$.

26. A **binary operation** on a set A is a function $op: A \times A \to A$. Thus a binary operation takes two elements of A and maps them to a third element of A.

27. If $|A| > |B|$, then there exists x, $y \in A$ such that $x \neq y$ and $f(x) = f(y)$.

REVIEW PROBLEMS

1. $A = \{a, b, c\}$ and $B = \{1, 2, 3, 4\}$.

   a. Find $A \times B$.

   b. Find $B \times A$.

2. Let $A = S_4$, and let R be the following relation on A.

$$R = \{(1, 2), (1, 4), (2,1), (2, 2), (2, 4), (3, 3), (4, 2)\}$$

    a. Find all the elements of A that are related to 2; that is, find $\{x: (x, 2) \in R\}$.

    b. Find all those elements in A to which 4 is related; that is, find $\{x: (4, x) \in R\}$.

    c. Find $\bar{R}$.

    d. Find R^{-1}.

3. Let $A = \{a, b, c, d\}$, and let $B = \{x, y, z\}$. Consider relation R between A and B.

$$R = \{(a, y), (a, z), (c, y), (d, x), (d, z)\}$$

    a. Draw the graph of R.

    b. Show the logical matrix representation of the relation.

    c. Find $\bar{R}$, and determine its logical matrix.

    d. Find R^{-1}, and determine its logical matrix.

4. Suppose we know that only the following statements are true about the relation R: xRy, zRz, xRz, zRy, and yRz. Write the relation R as a set of ordered pairs.

5. Let $A = \{1, 3, 5, 7\}$, and let $B = \{a, b, c\}$. Represent the relations below in graphical form.

    a. $R_1 = \{(1, b), (1, c), (5, a), (7, b), (7, a)\}$

    b. $R_2 = \{(1, b), (3, b), (5, b), (7, b)\}$

    c. $R_3 = \{(5, a), (5, b), (5, c)\}$

6. Let Q be the relation on S_4 given by xQy if $x \neq y$. Represent Q and its complement, $\bar{Q}$, in each of the following ways.

    a. as sets of ordered pairs

    b. in graph form

    c. in matrix form

7. For each of the following relation matrices, list the set of ordered pairs that belong to the relation, and then draw the graph form.

    a. $[\text{F} \quad \text{T} \quad \text{F} \quad \text{T}]$
    b. $\begin{bmatrix} \text{T} & \text{F} \\ \text{F} & \text{T} \end{bmatrix}$
    c. $\begin{bmatrix} \text{T} & \text{F} & \text{T} \\ \text{T} & \text{F} & \text{T} \\ \text{T} & \text{F} & \text{T} \end{bmatrix}$

8. How many relations are there between a finite set A and itself?

9. Decide whether each of the following relations is reflexive, symmetric, or transitive.

    a. as-deep-as   e. parallel-to

    b. brother-of   f. subset-of

    c. married-to   g. divisible-by

    d. not-equal-to   h. perpendicular-to

10. Which of the relations of problem 9 are equivalence relations?

11. Let $x, y \in \mathbf{Z}$. We say that x is **congruent mod 5** to y if the same remainder is obtained from each after dividing by 5. When x is congruent mod 5 to y, we write $x =_5 y$. In short, $x =_5 y$ means that x mod 5 $= y$ mod 5.

a. Show that $=_5$ is an equivalence relation.

b. How many distinct equivalence classes are there?

12. Suppose f is a function with domain A and for $x, y \in A$, we agree that xRy means $f(x) = f(y)$.

a. Is R an equivalence relation?

b. Suppose f is defined as $f(x) = x \bmod 5$ for each $x \in Z$. What relation is R?

13. If A and B are nonempty, disjoint sets, is $(A \times A) \cup (B \times B)$ an equivalence relation on $A \cup B$? Explain.

14. Answer the question in problem 13 assuming A and B are not disjoint sets.

15. Let R be the following relation between S_3 and S_4.

$$R = \{(1, 1), (1, 3), (2, 3), (2, 4), (3, 1), (3, 4)\}$$

Let S be the following relation between S_4 and S_2.

$$S = \{(1, 1), (1, 2), (2, 1), (4, 2)\}$$

Find the composite relation $S \circ R$ between S_3 and S_2.

16. Let M_1 and M_2 be relation matrices.

$$M_1 = \begin{bmatrix} T & F & F \\ T & F & T \end{bmatrix} \qquad M_2 = \begin{bmatrix} T & F & T & T \\ T & T & F & F \\ F & T & T & T \end{bmatrix}$$

a. Compute $M_3 = M_1 \cdot M_2$.

b. If M_1 represents relation R and M_2 represents relation S, interpret M_3 as the composition of R and S.

17. Let R be the following relation between S_3 and S_4.

$$R = \{(1, 1), (2, 3), (2, 4), (3, 1), (3, 4)\}$$

Let S be the following relation between S_4 and S_2.

$$S = \{(1, 1), (1, 2), (2, 1), (3, 1), (4, 2)\}$$

a. Find R^{-1}.

b. Find S^{-1}.

c. Find $(S \circ R)^{-1}$.

18. Determine if each of the following relations is reflexive, symmetric, or transitive.

a. "Perpendicular to," for the set of lines in a plane

b. "Perpendicular to," for the set of lines in space

c. The relation $R = \{(1, 1), (2, 2), (3, 3)\}$, on the set S_3

d. The relation $R = \{(a, c), (c, d)\}$, on $\{a, b, c, d\}$

19. Let $S = S_6$. For $x, y \in S$, let xRy if $x - y$ is divisible by 4. Show that R is an equivalence relation. Find the equivalence classes relative to R.

20. Let $S = \{a, b, c, d, e, f\}$. Let π be the following partition of S.

$$\pi = \{\{a, d, f\}, \{b\}, \{c, e\}\}$$

Find the equivalence relation R_π defining the partition.

21. a. Show that the relation "greater than or equal to" on $\mathbf{Z}$ is antisymmetric.
    b. Show that the relation on the set $\{a, b, c\}$ given by the set of ordered pairs $\{(a, b), (b, c), (c, a)\}$ is irreflexive.

22. Let R be the relation $\{(a, b), (c, c), (b, a)\}$ on the set $S = \{a, b, c\}$. Find the transitive closure of R.

23. Which of the relations depicted in Figure 3.30 are functions?

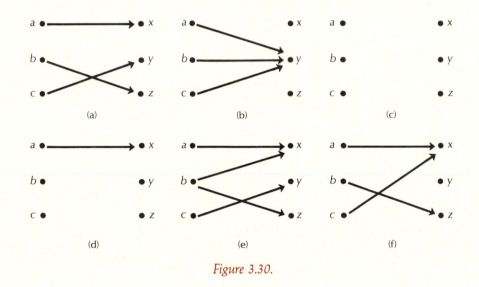

Figure 3.30.

24. Determine if each of the following relations is a function. For those that are functions, give the domain, codomain, and range of the function.
    a. $\{(1, a), (1, b), (2, c), (3, b)\}$
    b. $\{(a, a), (b, b), (c, c)\}$
    c. $\{(x, y): x, y \text{ are real numbers and } 3x + 2y = 1\}$
    d. $\{(x, y): x, y \text{ are positive integers and } 3x + 2y = 1\}$

25. Determine if each of the functions in Figure 3.31 is an injection, a surjection, or a bijection.

26. Given that $f: A \to B$ and $g: B \to C$ are surjections, prove that the composition $g \circ f: A \to C$ is a surjection.

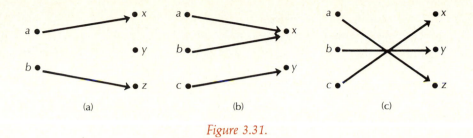

Figure 3.31.

27. Given that $f: A \to B$ and $g: B \to C$ are bijections, prove that the composition $g \circ f: A \to C$ is a bijection.

28. Let $A = \{1, 2, 3, 4\}$, and let $f: A \to A$ be $f = \{(1, 2), (2, 4), (3, 1), (4, 3)\}$. Show that f is a bijection and find f^{-1}.

29. Determine if each of the functions in Figure 3.32 is invertible.

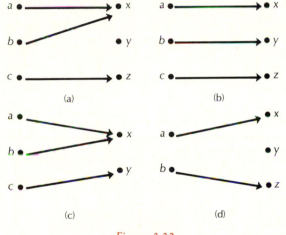

Figure 3.32.

30. Prove that if $f: A \to B$ and $g: B \to A$ satisfy $g \circ f = i_A$, then f is an injection and g is a surjection.

31. Let R and S be the n-ary relations shown below.

A	B	C	D
1	2	3	4
1	1	1	1
4	3	2	1
2	1	2	1
3	4	2	1
3	4	1	2

Relation R

A	B
1	2
1	5
4	5
2	1
3	1
3	5

Relation S

Compute each of the following.

a. The join of R and S on column B

b. The selection from R of all n-tuples whose value of A is greater than or equal to 2

c. The projection of A, C, and D from R

Programming Problems

32. Write a program that takes a description of a set and the ordered pairs of a binary relation on it as input and prints its matrix representation.

33. For each of the following properties, write a program that accepts the matrix representation of a binary relation and determines if the relation has the property.

a. reflexive

b. symmetric

c. irreflexive

34. Write a program to compute the product of two logical matrices.

*35. For each of the following types of order, write a program that accepts a description of a set and the set of ordered pairs of a relation on the set and determines if the relation is of that order.

a. a partial order

b. a linear order

SELF TEST ANSWERS

Self Test 1

1.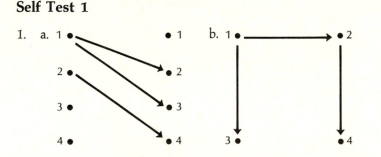

c. The matrix representation is the following 4×4 logical matrix.

$$
\begin{array}{c c c c c}
 & 1 & 2 & 3 & 4 \\
1 & \begin{bmatrix} F & T & T & F \\ F & F & F & T \\ F & F & F & F \\ F & F & F & F \end{bmatrix} \\
2 & & & & \\
3 & & & & \\
4 & & & &
\end{array}
$$

2.

$$\begin{array}{c c c c c c c} & 1 & 2 & 3 & 4 & 5 & 6 \\ 1 & \begin{bmatrix} T & T & T & T & T & T \\ 2 & T & T & T & T & T & T \\ 3 & T & T & T & T & T & T \\ 4 & F & F & F & F & T & T \\ 5 & T & T & T & T & T & T \\ 6 & F & F & F & F & F & F \end{bmatrix} \end{array}$$

3. $\{(1, 2), (2, 1), (2, 3), (3, 1), (3, 2), (3, 3)\}$

Self Test 2

1. a. transitive
   b. reflexive and symmetric, but not transitive
   c. reflexive and antisymmetric
2. R is reflexive, symmetric, and transitive.
3. $[a] = \{a, b\} = [b]$ and $[c] = \{c\}$
4.

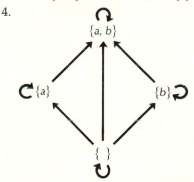

5. The transitive closure is $\{(b, a), (b, c), (c, a), (a, c), (c, c), (a, a)\}$. Note (c, c) is added, since cRa and aRc. Similarly, (a, a) is added, since aRc and cRa.

Self Test 3

1. a. $S \circ R = \{(1, 1), (1, 2), (2, 2), (3, 1), (3, 2)\}$
   b. The logical matrices for R and S are, respectively, the following 3×5 and 5×2 matrices.

$$R = \begin{bmatrix} T & F & T & F & F \\ F & F & T & T & F \\ T & F & F & T & T \end{bmatrix} \qquad S = \begin{bmatrix} T & T \\ T & F \\ F & F \\ F & T \\ T & T \end{bmatrix}$$

The composition is thus given by the following 3×2 matrix.

$$\begin{bmatrix} T & T \\ F & T \\ T & T \end{bmatrix}$$

2. The inverse of R is the relation obtained by interchanging the elements of each ordered pair in R. In this case, we get

$$\{(1, 1), (3, 1), (3, 2), (4, 2), (1, 3), (4, 3)\}$$

3. a. The product of a 3×2 and a 2×3 matrix is a 3×3 matrix.
   b. The product of a 2×3 and 3×2 matrix is a 2×2 matrix.

$$M_1 \cdot M_2 = \begin{bmatrix} T & T & T \\ T & F & F \\ T & T & T \end{bmatrix} \qquad M_2 \cdot M_1 = \begin{bmatrix} F & T \\ T & T \end{bmatrix}$$

Self Test 4

1. a. function
   b. not a function (There is no arrow leaving point c.)
   c. not a function (There are two arrows leaving point a.)
2. We have three choices of matching point 1 with an element of $\{a, b, c\}$ and three choices of matching point 2 with an element of $\{a, b, c\}$. Hence, there are nine possible functions from $\{1, 2\}$ to $\{a, b, c\}$.
3. a. f is injective since if $2^x = 2^y$, then $x = y$. The function f is not surjective, and hence is not a bijection. For example, there is no x such that $2^x = -3$.
   b. The function f is not injective, since, for example, $|-1| = 1 = |1|$. The function f is surjective since for all nonnegative integers y, $f(y) = y$.
   c. The function f is injective since $(x, x + 1) = (z, z + 1)$ only if $x = z$. The function f is not surjective since, for example, $(10, 15)$ is not in the image of f.

4 Combinatorics

Combinatorics is the branch of discrete mathematics that seeks to answer questions of the form "Is it possible to . . . ?" and "How many ways are there to . . . ?" Such questions are fundamental to probability and algorithm analysis. Such questions as "How many operations does this algorithm perform?" are fundamental to computer science. We have already asked about the number of elements belonging to given sets, and we have provided an answer in Theorem 2.7. We will use other results from Chapter 2 in our counting arguments. Recurrence relations and factorials also play an important role in combinatorics.

In the first section, we discuss four methods of selecting elements from a set, including combinations and permutations, and the counting rules associated with them. The second section contains an elaboration of the rules for counting patterns and partitions. In the third section, we look at methods of algorithm analysis.

Many counting problems can be solved by a direct application of the product rule. This rule is known as the Fundamental Principle of Counting.

The Fundamental Principle of Counting

Assume that some event can occur in n_1 ways, that a second event can occur in n_2 ways, and so on, up to event k, which can occur in n_k ways. Then the events can occur in the order given in exactly $n_1 \cdot n_2 \cdots \cdot n_k$ ways.

To see what this means, consider the five slots shown below.

$$\underline{\hspace{2em}} \quad \underline{\hspace{2em}} \quad \underline{\hspace{2em}} \quad \underline{\hspace{2em}} \quad \underline{\hspace{2em}}$$

Suppose that we are to record the number of ways event 1 can occur in the first slot, the number of ways event 2 can occur in the second slot, and so on. Then there are n_1 ways to fill the first slot, n_2 ways to fill the second slot, and so on. The Fundamental Principle of Counting says that the number of ways to fill all the slots is the product of the numbers of ways to fill each.

EXAMPLE 4.1

Find the number of license plates containing six characters that must be three letters followed by three digits.

Solution

Here there are six events. Event 1 is the selection of the first letter, event 2 is the selection of the second letter, and so on. Each of the first three events can happen in 26 ways, and each of the latter three events can happen in 10 ways. In the notation of the counting principle, $n_1 = n_2 = n_3 = 26$, and $n_4 = n_5 = n_6 = 10$. Then there must be $26 \cdot 26 \cdot 26 \cdot 10 \cdot 10 \cdot 10$ ways to construct the license plates.

EXAMPLE 4.2

A variable name in the programming language BASIC must be either a letter or a letter followed by a digit. How many different variable names are possible in BASIC?

Solution

First we consider variable names one character in length. Since such names must consist of a single letter, there is only one event, the selection of the letter. The event can happen in 26 ways. Hence there are 26 variable names of length 1.

Next we consider variable names two characters in length. Here there are two events. Event 1 is the selection of the letter, event 2 is the selection of the digit. The first event can happen in 26 ways, and the second event can happen in 10 ways. In the notation of the counting principle, $n_1 = 26$ and $n_2 = 10$. Then there must be $26 \cdot 10$, or 260, ways to construct variable names two characters in length. Hence, there are $26 + 260$, or 286, possible variable names in BASIC.

4.1   Selecting Elements from a Set

Our first concern is with methods of selecting elements from a given set. Given the task of selecting a fixed number of elements from a set S, we can allow the list to contain repetitions, or we can insist that the list not contain repetitions. For instance, given $S = \{x, y, z, w\}$, we could select three elements of S, with repetitions allowed. The selections would include xyz, zyx, zzx, xzz, and so on. We could also select elements without repetitions, such as xyz, xzy, wzx, and so on.

The second thing to decide when selecting elements is whether the order of elements within the list is important. For example, should we distinguish the list xzz from the list zzx? By combining the two issues of repetition and order, we see that there are four methods of selecting elements from a given set, as shown in Table 4.1. Explanations of the four selection methods and the lists that result from them follow the table.

TABLE 4.1

	Order matters	Order doesn't matter
Elements repeated	k-sample	k-selection
Elements not repeated	k-permutation	k-combination

Definitions

Each of the four methods for extracting lists of elements from a set occur in various situations. The formal definitions of the methods, along with examples of each, appear below.

k-Samples.   Select k things from a set S. Allow repetitions, but distinguish among lists in which the same elements appear in different order. This kind of list is called a *k-sample*. Here are the k-samples from $S = \{k, y, z, w\}$, for $k = 3$:

$$
\begin{array}{llllllll}
xxx & xxy & xxz & xxw & xyx & xyy & xyz & xyw \\
xzx & xzy & xzz & xzw & xwx & xwy & xwz & xww \\
yxx & yxy & yxz & yxw & yyx & yyy & yyz & yyw \\
yzx & yzy & yzz & yzw & ywx & ywy & ywz & yww \\
zxx & zxy & zxz & zxw & zyx & zyy & zyz & zyw \\
zzx & zzy & zzz & zzw & zwx & zwy & zwz & zww \\
wxx & wxy & wxz & wxw & wyx & wyy & wyz & wyw \\
wzx & wzy & wzz & wzw & wwx & wwy & wwz & www
\end{array}
$$

In the license plate problem of Example 4.1, there were two 3-samples linked together—the first from the set of letters, and the second from the set of digits.

k-Selections.   Select k things from a set S. Allow repetitions, but equate lists containing the same elements in any order. These lists are known as *k-selections* from a set S. For instance, here are the 3-selections from the set $\{x, y, z, w\}$:

$$
\begin{array}{llllllllll}
xxx & xxy & xxz & xxw & xyy & xyz & xyw & xzz & xzw & xww \\
yyy & yyz & yyw & yzz & yzw & yww & zzz & zzw & zww & www
\end{array}
$$

k-selections are useful for describing things such as outcomes of multiple tosses of a coin. (The order of the heads and tails is not important, but the number of heads and tails is.)

Samples and selections both allow repetitions. Thus k-samples and k-selections from a set S need not correspond to subsets of the set. In fact, it is possible that k be larger than $|S|$. The next two methods do correspond to subsets of S, so k must be no larger than n.

k-Permutations.   Select k things from a set S. Do not allow repetitions, and distinguish lists by the order in which elements are selected. These lists are called k-*permutations* from the set S.

Here are the 3-permutations for the set $S = \{x, y, z, w\}$:

xyz	*xzy*	*yxz*	*yzx*	*zxy*	*zyx*
xyw	*xwy*	*yxw*	*ywx*	*wxy*	*wyx*
xzw	*xwz*	*zxw*	*zwx*	*wxz*	*wzx*
yzw	*ywz*	*zyw*	*zwy*	*wyz*	*wzy*

k-permutations describe such things as the set of "words" that can be formed with k distinct letters chosen from a set of n letters.

k-Combinations.   Select k things from a set S. Do not allow repetitions, and equate lists containing the same elements in any order. These lists are called k-*combinations* from S, or combinations of the elements of S taken k at a time.

It is easy to see that the k-combinations of a set S correspond exactly to the subsets of S with k elements. Here are the 3-combinations for the set $S = \{x, y, w, z\}$:

xyz	*xyw*	*xzw*	*yzw*

k-combinations describe such things as the set of hands that can be dealt in various card games.

Counting Formulas

We have named the four ways of selecting elements from a set and constructing lists of the elements. Now we want to count how many ways the different types of lists can be constructed. In the formulas that follow, S is a fixed set, n is the number of elements of S, and k is the size of the list we are extracting from S. We start with the easiest case first.

k-Samples.   To count the number of k-samples from a set S containing n elements we apply the Fundamental Principle of Counting to the idea of filling slots.

$$\underline{\hspace{1.5em}} \quad \underline{\hspace{1.5em}} \quad \underline{\hspace{1.5em}} \quad \underline{\hspace{1.5em}} \cdots \underline{\hspace{1.5em}}$$
$$1 \qquad\quad 2 \qquad\quad 3 \qquad\quad 4 \qquad\quad k$$

Any element of S can be in the first slot. Since k-samples allow repetitions, we can put any element of S in the second slot, any element of S in the third slot, and so on, up to the kth slot. Hence, there are n ways to fill each slot.

$$\frac{n}{1} \qquad \frac{n}{2} \qquad \frac{n}{3} \qquad \frac{n}{4} \cdots \frac{n}{k}$$

Since there are k slots, there are n^k ways to fill all the slots. Thus there are n^k k-samples from S.

EXAMPLE 4.3

A computer represents integers with n binary digits, using one bit (binary digit) to indicate the sign and the remaining $n - 1$ bits to represent the magnitude of the integer. This is called *sign-magnitude* representation of integers. How many distinct integers can be represented in this notation?

Solution

There are n slots to fill, and each slot can be filled two different ways. Hence, there are 2^n distinct bit patterns. However, 0 is represented as $^+0$ and as $^-0$, so there are $2^n - 1$ distinct integers that can be represented using n bits in the sign-magnitude representation.

---

k-Permutations.   We use the notation $P(n, k)$ to indicate the number of k-permutations from a set with n elements. $P(n, k)$ is read "the permutations of n things taken k at a time." Again, it is convenient to consider the counting problem as one of filling slots.

$$\frac{}{1} \qquad \frac{}{2} \qquad \frac{}{3} \qquad \frac{}{4} \cdots \frac{}{k}$$

In the first slot, we can put any of the n elements of S. Because repetitions are not allowed, there are only $n - 1$ choices for the second slot. For the third slot, we have only $n - 2$ choices, and so on. By the time we get to the kth slot, we have only $(n - k + 1)$ elements to choose from.

$$\frac{n}{1} \qquad \frac{n-1}{2} \qquad \frac{n-2}{3} \qquad \frac{n-3}{4} \cdots \frac{n-k+1}{k}$$

By the Fundamental Principle of Counting, we get the formula

$$P(n, k) = n \cdot (n - 1) \cdot (n - 2) \cdot \cdots \cdot (n - k + 1) \qquad (4.1)$$

Rewriting the formula using factorials, we get

$$P(n, k) = \frac{n!}{(n - k)!}$$

Observe that the requirement $k \leq n$ guarantees that $n - k \geq 0$.

EXAMPLE 4.4

Find the number of "words" with four distinct letters that can be made from the letters c, a, b, i, n, e, t.

Solution

In this case, we are extracting lists from the set $S = \{c, a, b, i, n, e, t\}$, so $n = 7$. The lists are to consist of four symbols chosen without replacement from S. Hence, the lists are 4-permutations from S. The number of the words selected is

$$
\begin{aligned}
P(7, 4) &= \frac{7!}{3!} \\
&= \frac{7 \cdot 6 \cdot 5 \cdot 4 \cdot 3 \cdot 2 \cdot 1}{3 \cdot 2 \cdot 1} \\
&= 7 \cdot 6 \cdot 5 \cdot 4 \\
&= 840
\end{aligned}
$$

EXAMPLE 4.5

How many different ways are there to encode the 26 uppercase letters of the alphabet as binary sequences of length 7?

Solution

There are $2^7 = 128$ binary sequences of length 7. Representations for the 26 letters $A, B, C, \ldots, Z$ can be chosen in $P(128, 26)$ ways.

EXAMPLE 4.6

Your computer science professor has bet the class that no one can find a word using each of the ten letters $a, b, c, e, i, n, r, s, t$, and u exactly once. Having wasted several sheets of paper trying to unscramble the letters, you decide to use some of your 30 minutes of cpu time to do a systematic search. Your method is simple. First you implement an algorithm that will generate all possible permutations of the ten letters.

Then you have each permutation checked against a large dictionary on your school's system. Assume that generating each permutation and looking it up in the dictionary takes 1 millisecond (1/1,000 of a second). How long will it take your program to check all permutations?

Solution

There are $P(n, n) = n!$ permutations of n letters, so your program will have an approximate running time of

$$\frac{10!}{1000} \text{ seconds} = \frac{10!}{1000 \cdot 60} \text{ minutes}$$

$$= \frac{9 \cdot 8 \cdot 7 \cdot 4 \cdot 3}{100} \text{ minutes}$$

$$= 60.48 \text{ minutes}$$

Observe that even if the operation on each permutation occurs twice as quickly, 30 minutes of cpu time is still not quite enough. A situation like this one, in which a large number results from a simple operation on a small set, is known as a "combinatorial explosion."

---

k-Combinations.   The symbol for the number of k-combinations from a set with n elements is $C(n, k)$. $C(n, k)$ is read "the combinations of n things taken k at a time." To find $C(n, k)$, we first observe that for each k-combination, there are many k-permutations. For instance, given the 3-combination xyz, there are six permutations of the symbols x, y, z. For each k-combination, there are $k! = k \cdot (k - 1) \cdot \cdots \cdot 1$ corresponding k-permutations. Hence, we easily obtain the formula

$$P(n, k) = C(n, k) \cdot k!$$

Solving for $C(n, k)$ yields

$$C(n, k) = \frac{n!}{(n - k)!k!} \tag{4.2}$$

Because they appear in expansions of binomials, the quantities $C(n, k)$ are called **binomial coefficients**. You might recall the following result from algebra:

$$(a + b)^n = a^n + C(n, 1)a^{n-1}b^1 + \cdots + C(n, n - 1)a^1b^{n-1} + b^n$$

$$= \sum_{k=0}^{n} C(n, k)a^{(n-k)}b^k \tag{4.3}$$

EXAMPLE 4.7

A menu in a Chinese restaurant allows you to order exactly two of eight main dishes as part of the dinner special. How many different combinations of main dishes could you order?

Solution

There are $C(8, 2)$ combinations of the eight main dishes taken two at a time. Thus, you could choose one of

$$C(8, 2) = \frac{8!}{6!2!} = \frac{8 \cdot 7}{2} = 28$$

different combinations.

There are two important recurrence relations involving combinations. They are stated in Theorem 4.1.

Theorem 4.1

1. For all integers $n \geq 1$ and $0 \leq k \leq n$,

$$C(n + 1, k + 1) = C(n, k) + C(n, k + 1) \tag{4.4}$$

2. For all integers $n \geq 1$ and $k \geq 1$,

$$C(n + k, k) = 1 + C(n, 1) + C(n + 1, 2) + \cdots + C(n + k - 1, k) \tag{4.5}$$

Proof

1. First consider Equation (4.4). Recall that $C(n, k)$ gives the number of k-element subsets from a set with n elements. We use this interpretation to justify the equation. The left side of Equation (4.4) represents the number of subsets with $k + 1$ elements from a set S with $n + 1$ elements. Let S be the set

$$S = \{a_1, a_2, \ldots, a_n, a_{n+1}\}$$

Let S' be the n-element subset of S given by

$$S' = \{a_1, a_2, \ldots, a_n\}$$

Extracting a subset of S with $k + 1$ elements can be done in two ways. First, we can extract a subset of S' that has k elements and then add a_{n+1} to the subset. This can be done in $C(n, k)$ ways. Second, we can extract a subset of S' that has $k + 1$ elements. This can be done in $C(n, k + 1)$ ways. Hence, the total number of ways to extract a subset with $k + 1$ elements from S is

$$C(n, k) + C(n, k + 1)$$

For an alternate method of proving Equation (4.4), see exercise 10 on page 142.

2. To establish Equation (4.5), we use an induction argument, with n fixed and $k \geq 1$. Let $Q(k)$ be the statement of Equation (4.5). Then $Q(1)$ is just the equation

$$C(n + 1, 1) = 1 + C(n, 1)$$

Both sides of the equation yield $n + 1$, so $Q(1)$ is *true*.

For the induction step, we must assume that $Q(k)$ is *true* and derive $Q(k + 1)$; that is, we assume the equality

$$C(n + k, k) = 1 + C(n, 1) + C(n + 1, 2) + \cdots + C(n + k - 1, k)$$

and we must derive the equality

$$C(n + k + 1, k + 1) = 1 + C(n, 1) + \cdots + C(n + k - 1, k) + C(n + k, k + 1)$$

Applying the inductive assumption to the right side of the last equation reduces it to

$$C(n + k, k) + C(n + k, k + 1)$$

By Equation (4.4), this sum is just $C(n + k + 1, k + 1)$, from which it follows that $Q(k) \Rightarrow Q(k + 1)$, for $k \geq 1$. The induction argument is complete.

•   •   •

The results of Theorem 4.1 can be illustrated graphically. Consider Equation (4.4).

$$C(n + 1, k + 1) = C(n, k) + C(n, k + 1)$$

We can arrange the number of combinations in a table as in Table 4.2.

Starting with $C(n, 0) = 1$, we can fill the table by using Equation (4.4) and the arrangement of $C(n, k)$ as shown in Table 4.2. The resulting triangular table, known as *Pascal's Triangle*, is shown in Table 4.3.

TABLE 4.2

n	k					
	0	1	2	3	4	5
0	C(0, 0)					
1	C(1, 0)	C(1, 1)				
2	C(2, 0)	C(2, 1)	C(2, 2)			
3	C(3, 0)	C(3, 1)	C(3, 2)	C(3, 3)		
4	C(4, 0)	C(4, 1)	C(4, 2)	C(4, 3)	C(4, 4)	
5	C(5, 0)	C(5, 1)	C(5, 3)	C(5, 3)	C(5, 4)	C(5, 5)

TABLE 4.3

n	k					
	0	1	2	3	4	5
0	1					
1	1	1				
2	1	2	1			
3	1	3	3	1		
4	1	4	6	4	1	
5	1	5	10	10	5	1

Each entry in the table is the sum of the entry directly above it and the entry to the left of the one directly above it. If there is no entry directly above a position in the triangle, the value for that entry is 0.

SELF TEST 1

1. Compute $P(20, 17)$ and $C(13, 9)$.
2. How many different 7-card hands can be drawn from a standard 52-card deck?
3. How many different 3-note sequences can be formed from an 8-note scale?

k-Selections.   We compute the number of k-selections from a set S last because the formula depends on our results about k-combinations. Recall that in k-selections, as in k-combinations, order does not matter. But in contrast to k-combinations, repetitions *can* occur in k-selections.

Let $S'(n, k)$ denote the number of k-selections from a set with n elements. We claim that the following recurrence relation holds for $S'(n, k)$ for all $n \geq 1$ and $k \geq 1$:

$$S'(1, k) = 1 \text{ for all } k \geq 1$$

and

$$S'(n + 1, k) = 1 + S'(n, 1) + \cdots + S'(n, k) \tag{4.6}$$

The first part of the claim follows directly from the definition of a k-selection. Since we have only one element to choose, there is only one way to select it. For the second part, observe that the left side of the recurrence relation in Equation (4.6) represents the number of k-selections from a set with $n + 1$ elements. The right side sums all selections from a set with n elements. The justification for Equation (4.6) follows the pattern of the justification for Equation (4.4). Let

$$A = \{a_1, a_2, a_3, \ldots, a_n, a_{n+1}\}$$

be a set with $n + 1$ elements. Let

$$A' = \{a_1, a_2, \ldots, a_n\}$$

A k-selection from A can be obtained by taking an i-selection from A' and filling the remaining $(k - i)$ positions with a_{n+1} where $i = 0, 1, \ldots, k$. (You should try this with $k = 3$, $n = 3$, and $i = 2$.) But this says that the number of k-selections from A is the sum of all the numbers of i-selections from A', $i = 0, 1, \ldots, k$, which is exactly what Equation (4.6) claims.

The result could be left in this form. To do so means that evaluating, say, $S'(5, 3)$ requires first evaluating the terms $S'(4, 1)$, $S'(4, 2)$, and $S'(4, 3)$. Each of these requires evaluating even more terms. We can do better with the closed form expression for $S'(n, k)$ given in Theorem 4.2.

Theorem 4.2

For all $n \geq 1$ and $k \geq 1$, the number of k-selections from a set with n elements is given by

$$S'(n, k) = C(n + k - 1, k)$$

Proof

We argue by induction on n. Let $Q(n)$ be the statement of Theorem 4.2 for all $k \geq 1$. $Q(1)$ is just $S'(1, k) = C(k, k)$, for all $k \geq 1$. Both sides of the equation have value 1 for all $k \geq 1$, so $Q(1)$ is *true*.

For the induction step, assume that $S'(n, k) = C(n + k - 1, k)$, for all $k \geq 1$. Then we must show that

$$S'(n + 1, k) = C(n + k, k), \text{ for all } k \geq 1$$

But by formula (4.6),

$$S'(n + 1, k) = 1 + S'(n, 1) + \cdots + S'(n, k)$$
$$= 1 + C(n, 1) + C(n + 1, 2) + \cdots + C(n + k - 1, k)$$

Now we can apply part (2) of Theorem 4.1, which yields the equality

$$S'(n + 1, k) = C(n + k, k), \text{ for } k \geq 1$$

EXAMPLE 4.8

Compute $S'(4, 3)$.

Solution

Observe that to use Equation (4.6) we have to compute

$$S'(4, 3) = 1 + S'(3, 1) + S'(3, 2) + S'(3, 3)$$

and then apply the formula again to each of the three S' terms. This tedious approach is unnecessary if we apply Theorem 4.2 directly. According to Theorem 4.2,

$$S'(4, 3) = C(4 + 3 - 1, 3) = \frac{6!}{(3!)(3!)}$$
$$= \frac{6 \cdot 5 \cdot 4 \cdot 3 \cdot 2 \cdot 1}{(3 \cdot 2 \cdot 1)(3 \cdot 2 \cdot 1)}$$
$$= 20$$

This agrees with the number of 3-selections listed on page 131.

EXAMPLE 4.9

Four dice are to be tossed. How many different outcomes are possible if the order does not matter?

Solution

If the first die shows 4 and the rest of the dice show 2, it is the same outcome as if the first three dice show 2 and the last one shows 4. Hence, order does not matter. But since all dice could show the same value, repetitions are allowed. Thus the outcomes are the 4-selections from the set $\{1, 2, 3, 4, 6\}$, and there are

$$
\begin{aligned}
C(9, 4) &= \frac{9!}{4!5!} \\
&= \frac{9 \cdot 8 \cdot 7 \cdot 6}{4 \cdot 3 \cdot 2 \cdot 1} \\
&= 9 \cdot 7 \cdot 2 \\
&= 126
\end{aligned}
$$

possible outcomes.

SELF TEST 2

1. Compute $C(7, 5)$, $P(7, 5)$, and $S'(7, 5)$.
2. A hockey team has ten players who want to play in the front line. How many distinct 3-player lines can be formed?
3. A certain software package has three main modules. To configure the package, users must choose among options given for each module. If the first module has four options, the second five options, and the third three options, how many different sets of options are supported?

Exercises 4.1

1. Let $S = \{1, 2, 3, 4\}$. List all the 2-samples, 2-selections, 2-permutations, and 2-combinations from S. Verify that you have the correct number of each by applying the appropriate formula.
2. How many distinct 5-card poker hands are there?
3. Are there more 4-samples or more 4-selections from a set with six elements?
4. A salesperson is to visit eight different cities exactly once. In how many different ways can this be done?
5. a. Let $S = \{1, 2, 3, 4, 5, 6\}$. For each $i = 0, 1, 2, \ldots, 6$, how many subsets of S are there with i elements? Verify that the sum of all seven answers is 2^6.

b. Verify Equation (4.3) in the case that $n = 3$ by expanding both sides of the equation.

c. Verify

$$C(n, 0) + C(n, 1) + C(n, 2) + \cdots + C(n, n) = 2^n$$

for $n = 0, 1, 2, \ldots$.

6. How many elements of B^5 have an even number of 0's?

7. In the dictionary problem of Example 4.6, assume that the permutations are to be printed ten to a line, with 60 lines per page. How many pages will be printed?

8. Three exams are to be scheduled during a four-day period, with at most one exam per day. In how many ways can this be accomplished?

9. Generate the seventh row of Pascal's Triangle.

*10. Verify the recurrence relation

$$C(n, k) + C(n, k + 1) = C(n + 1, k + 1), \text{ for } n \geq 1, k \geq 0$$

by applying Equation (4.2) directly to the left side.

4.2   Patterns and Partitions

There are many cases in which we have to extract a list from a collection of objects that are not all distinct. For instance, we may have a collection of letter tiles in a word game. The tiles themselves are distinct, but the letters on them may be duplicates. In effect, the tiles are partitioned into groups with the same face value. In addition to counting the number of distinct patterns possible when making lists from sets with duplicate elements, we will count the number of ways a set can be partitioned. Solving these problems requires us to use variations of the formulas discussed in the first section of this chapter.

Patterns

EXAMPLE 4.10

Suppose you have four tiles containing the letters T, O, O, T. There are 24 permutations of the four tiles. How many spell the name of Dorothy's little dog, TOTO?

Solution

Just once, we'll do a problem such as this by brute force. To keep our sanity, we will pretend that the four tiles have been colored so that the red and blue tiles contain the letter T and that the green and yellow tiles contain the letter O. (This will help us keep track of the permutations of the distinct tiles.) Here are the 24 permutations of the four tiles:

green, yellow, red, blue	green, red, yellow, blue	green, red, blue, yellow
green, yellow, blue, red	green, blue, yellow, red	green, blue, red, yellow
yellow, green, red, blue	yellow, red, green, blue	yellow, red, blue, green
yellow, green, blue, red	yellow, blue, green, red	yellow, blue, red, green
red, green, yellow, blue	red, yellow, green, blue	red, yellow, blue, green
red, green, blue, yellow	red, blue, green, yellow	red, blue, yellow, green
blue, green, yellow, red	blue, yellow, green, red	blue, yellow, red, green
blue, green, red, yellow	blue, red, green, yellow	blue, red, yellow, green

Here are the 24 patterns of O's and T's if we look at the letters on the tiles instead of the colors:

OOTT	OTOT	OTTO
OOTT	OTOT	OTTO
OOTT	OTOT	OTTO
OOTT	OTOT	OTTO
TOOT	TOOT	TOTO
TOTO	TTOO	TTOO
TOOT	TOOT	TOTO
TOTO	TTOO	TTOO

There are four occurrences of *TOTO* among the 24 permutations. In fact, there are four occurrences of each of the six distinct patterns. This is because there are 2! ways to arrange each pair of indistinguishable letters once we have chosen positions for them. Since there are two pairs of two letters, there are (2!)(2!) ways to get the same pattern.

Theorem 4.3 generalizes the result of Example 4.10. The key difference between choosing a permutation from n distinct objects and choosing a pattern from n objects partitioned according to type is that when we choose from the partitioned objects we can factor out the permutations of elements of the same type.

Theorem 4.3

Suppose there are n objects, of which n_1 objects are of type 1, n_2 objects are of type 2, and so on, up to n_r objects of type r. Then there are

$$\frac{n!}{(n_1!)(n_2!) \ldots (n_r!)}$$

distinct patterns that can be formed with the n objects. Moreover, each pattern appears exactly

$$(n_1!)(n_2!) \ldots (n_r!)$$

times among the $n!$ permutations.

EXAMPLE 4.11

Use the result of Theorem 4.3 to solve the problem of Example 4.10.

Solution

There are four objects—two T's and two O's. By the formula of Theorem 4.3, the four objects can be arranged in

$$\frac{4!}{(2!)(2!)} = \frac{4 \cdot 3}{2} = 6$$

distinct patterns, each of which appears $(2!)(2!)$, or 4, times among the 24 permutations. These results agree with the counts laboriously obtained in the solution to Example 4.10.

EXAMPLE 4.12

A puzzle has three squares, two triangles, and four circles. How many patterns can be formed by laying these nine shapes out in a row?

Solution

By Theorem 4.3, the pieces can be arranged into

$$
\begin{aligned}
\frac{9!}{(3!)(4!)(2!)} &= \frac{9 \cdot 8 \cdot 7 \cdot 6 \cdot 5 \cdot 4 \cdot 3 \cdot 2 \cdot 1}{(3 \cdot 2 \cdot 1)(4 \cdot 3 \cdot 2 \cdot 1)(2 \cdot 1)} \\
&= \frac{9 \cdot 8 \cdot 7 \cdot 5}{2} \\
&= 1260
\end{aligned}
$$

distinct patterns.

Partitions

We also seek a formula for counting the number of partitions that can be formed from a given set. Let's start by considering an example.

EXAMPLE 4.13

1. There are seven programming projects to be assigned to three programmers. If three projects are to be assigned to the senior programmer, and two projects to each of the other two programmers, how many ways can the projects be assigned?
2. How many ways can seven programming projects be grouped into sets of three, two, and two projects?

Solution

At first glance, parts (1) and (2) may seem to be identical problems. However, careful thought reveals a significant difference between them. In (2) we are interested only in forming sets of the projects, whereas in (1) the sets have to be assigned (ordered) in some way, according to which programmer gets each set of projects.

1. To clarify our thinking, consider the projects as slots, numbered 1 to 7, to which programmers A (the senior programmer), B, and C are to be assigned. Here are two possible assignments:

A	A	A	B	B	C	C
1	2	3	4	5	6	7

A	B	C	A	B	C	A
1	2	3	4	5	6	7

Observe how this problem is similar to previous problems with patterns. There are three identical objects (programmer A) and two each of two more identical objects (programmers B and C) that are to be assigned to the seven slots. Thus we can apply Theorem 4.3 and find that there are

$$\frac{7!}{(3!)(2!)(2!)} = 210$$

distinct ways to make the assignment.

2. This part of the problem is more straightforward because order in the partition does not matter. The problem is simply to find the number of ways seven elements can be partitioned into three blocks, where one block contains three elements and each of the other two blocks contain two elements. To help us grasp the situation, we list two distinct partitions of the elements 1 through 7 that meet the required conditions.

$$\{\{1, 2, 3\}, \{4, 5\}, \{6, 7\}\}$$
$$\{\{2, 4, 5\}, \{1, 6\}, \{3, 7\}\}$$

One way to analyze the problem is as follows: From the seven original elements, we have to extract a subset with three elements, which we can do in $C(7, 3)$ ways. From the remaining four elements, we have to extract all distinct pairs of subsets with two elements each. There are $C(4, 2)$ subsets of four elements taken two at a time, so there are $C(4, 2)/2$ distinct pairs of these subsets. Hence the number of partitions we seek is

$$C(7, 3) \cdot C(4, 2)/2 = \frac{7!}{(3!)(4!)} \cdot \frac{4!}{(2!)(2!)(2)}$$

$$= \frac{7!}{(3!)(2!)(2!)(2)}$$

$$= 105$$

In part (2) of Example 4.13, we had to count the number of partitions of a set, where the partitions have a *fixed* number of elements in each block. Another question often asked about partitions is "How many ways can a set be partitioned into k blocks?" Let $S(n, k)$ denote the number of partitions of n elements into k blocks, without restriction on the sizes of the blocks. There is a simple recurrence relation for $S(n, k)$, which we derive in Theorem 4.4. Unfortunately, there is not a simple closed form expression for $S(n, k)$.

Theorem 4.4

The number of ways to partition a set with n elements into k blocks is given by $S(n, k)$, where $S(n, k)$ satisfies the following recurrence relation:

1. $S(n, 1) = 1$ and
   $S(n, n) = 1$, for all $n \geq 1$
2. $S(n + 1, k + 1) = (k + 1)S(n, k + 1) + S(n, k)$,
   for $n \geq 1$ and $1 \leq k \leq n$

Proof

The equations in (1) are obvious from the definition of partition. Suppose that a set A has $n + 1$ elements, and we extract one element, x, from A. Let $A' = A - \{x\}$. Then A' has n elements. There are two ways to use partitions of A' to form partitions of A. First, if A' is partitioned into $k + 1$ blocks, then a partition of A with $k + 1$ blocks can be formed by putting x into one of the blocks from A'. This can be done in $(k + 1)S(n, k + 1)$ ways.

Secondly, A' can be partitioned into k blocks in $S(n, k)$ ways. For each one of these, we can form a partition of A into $k + 1$ blocks by adding the set $\{x\}$ to the partition of A'. Hence, there are $(k + 1)S(n, k + 1) + S(n, k)$ ways to partition A into $k + 1$ blocks. Equation (2) of the theorem follows.

EXAMPLE 4.14

In how many ways can a set with five elements be partitioned into three blocks?

Solution

We use Theorem 4.4 and write

$$S(5, 3) = 3S(4, 3) + S(4, 2)$$

Now we must apply (2) again to obtain

$$S(4, 3) = 3S(3, 3) + S(3, 2)$$
$$S(4, 2) = 2S(3, 2) + S(3, 1)$$
$$S(3, 2) = 2S(2, 2) + S(2, 1)$$

Now by (1),

$$S(2, 1) = S(3, 1) = 1$$
$$S(3, 3) = 1$$
$$S(2, 2) = 1$$

Putting these values in and moving back up the equations, we get

$$S(3, 2) = 2 + 1 = 3$$
$$S(4, 2) = 2 \cdot 3 + 1 = 7$$
$$S(4, 3) = 3 + 3 = 6$$

Finally,

$$S(5, 3) = 3 \cdot 6 + 7 = 25$$

There is no known closed form solution to the recurrence relation of Theorem 4.4. Because the recurrence relation for $S(n, k)$ has a form similar to the recurrence relation for the binomial coefficients, it is possible to construct a triangle for the computation of successive terms of $S(n, k)$. This triangle is called *Stirling's Triangle*, and the numbers $S(n, k)$ are called *Stirling's numbers*.

---

1. Find the number of patterns that can be formed by permuting the letters of the word *reentrant*.
2. How many ways can a set with three elements be partitioned?
3. Compute $S(6, 4)$.

Exercises 4.2

1. Find the number of distinct patterns that can be formed by permuting the letters of the following words.
   a. *unicorn*   d. *entities*
   b. *madam*   e. *focus*
   c. *insipid*
2. How many ways can a committee be selected from four males and five females if there are to be an equal number of males and females?
3. How many ways can ten students be assigned to three teams, if one team is to have four members and each of the other two teams are to have three members?
4. How many partitions of a set with five elements are there?
5. How many ways can six books be given to three students, so that
   a. each student gets two books.
   b. one student gets one book, one student gets two books, and one student gets three books.
6. Generate the first four rows of Stirling's triangle.

4.3   Algorithm Analysis (Optional)

The primary efficiency criterion for analyzing the efficiency of an algorithm is the *running time* of the algorithm as a function of the number of values it processes. We cannot compute the seconds or minutes for the running time of an algorithm coded in a particular language on a particular machine. The actual running time depends on many things, including the machine used to execute a program. We will only be able to *estimate* the running time by counting the number of elementary operations performed (the number of multiplications or the number of comparisons, for example), given a particular number of input values.

Generally, given an algorithm that performs a task, we will be interested in estimating the running time as a function of the problem size. For example, given a list and an algorithm to search the list, we will want to ask: "Using the algorithm provided, how quickly can we search the list to see if a given item is in it?" Here we say the problem size is n, the number of items in the list. The answer to the question will be given as $T(n)$, a function of n. Since $T(n)$ is a measure of the time required to execute an algorithm on a problem of size n, it is called the **time complexity function** of the algorithm. As n increases, we expect the search time to increase. If n is sufficiently small, then even an inefficient algorithm will not have a long running time. The choice of algorithm for small problems is usually not too critical, but the choice of algorithm for a large problem can be. For this reason, the critical question is: "How fast does $T(n)$ increase as n increases?" This is called the **asymptotic behavior** of the time complexity function. In contrast with the tidy, exact results in the earlier sections of this chapter, we deal with estimates and inequalities in this section.

For many algorithms, the running time is a function of the particular input as well as the input size. For this reason, we will restrict our analysis to the **worst case** behavior of an algorithm; that is, the case requiring the greatest number of elementary operations for a given problem size.

EXAMPLE 4.15

Estimate $T(n)$, the time complexity function for the following algorithm.

```
begin
    i ← 1
    p ← 1
    while i < n do
            begin
                p ← p · i
                i ← i + 1
            end
end.
```

Solution

We count the elementary operations. Prior to entering the loop, it takes two assignment statements to initialize the variables i and p. The loop is executed n times, and each time it executes three statements: the test for the condition of repetition and the two assignment statements in the body of the loop. Thus, the time complexity of the algorithm is given by

$$T(n) = 3n + 2$$

Order Classes

In Example 4.15 we developed a precise expression for the time complexity of the algorithm. What usually interests us, however, is the *order of magnitude* of the complexity function. We now examine the fundamental concept of the "order" of a function. Let $f, g: \mathbf{N} \to \mathbf{R}$ be functions. When we assert that "f asymptotically dominates g," we mean that there exist constants $C \geq 0$ and $n_0 \geq 0$ such that $g(n) \leq Cf(n)$ for all $n \geq n_0$. The set of all functions that are asymptotically dominated by the function f is denoted by $O(f(n))$. If $g(n) \in O(f(n))$, then g is said to be $O(f)$, which is read "g is order at most f," or simply "g is big-oh of f."

If f asymptotically dominates g and $f(n) \neq 0$, then $g(n)/f(n) \leq C$ for all but a finite number of values of n, none of which are greater than n_0. In other words, if $f(n)$ and $g(n)$ are time complexity functions for two algorithms, then for problems of size n_0 or larger, execution of the algorithm with time complexity $g(n)$ will never be more than C times the execution of the algorithm with time complexity $f(n)$.

EXAMPLE 4.16

Show that $T(n) = 3n + 2$ is $O(n)$.

Solution

To show that $T(n) \in O(n)$, we must produce the constants C and n_0 such that $T(n) \leq Cn$ when $n \geq n_0$. Let $C = 4$. The inequality $3n + 2 \leq 4n$ is true when $n \geq 2$. Thus $T(n) = 3n + 2$ is of the order of n.

EXAMPLE 4.17

Show that $n^2 \in O(n^3)$, but $n^3 \notin O(n^2)$.

Solution

Since $n^2 \leq n^3$ for all $n \in \mathbf{N}$, $n^2 \in O(n^3)$. All we need to do is select $C = 1$ and $n_0 = 0$. To show that $n^3 \notin O(n^2)$, we must show that for all $C \geq 0$ there exists n such that

$n^3 > Cn^2$. Given C, all we need to do is to choose a value of n such that $n > C$. Then $n^3 = n \cdot n^2 > Cn^2$, as desired.

From Example 4.17 we see that the bounds must hold for *all $n \geq n_0$*. If, in Example 4.17, we take $C = 100{,}000$ and $n_0 = 1$, then for $n_0 \leq n \leq C$, $n^3 \leq Cn^2$. However, this is not enough to conclude that $n^3 \in O(n^2)$ since only a finite number of values of n are covered. This is the essence of an asymptotic upper bound: *eventually* (that is, from some point on) the upper bound dominates.

Theorem 4.5

Let $f, g: \mathbf{N} \to \mathbf{R}$ be functions.

1. $g(n) \in O(f(n))$ and $f(n) \in O(g(n))$ if and only if $O(f(n)) = O(g(n))$.
2. If $f(n) \in O(g(n))$ and $g(n) \in O(h(n))$, then $f(n) \in O(h(n))$.

Proof

We will prove part (2) and leave the proof of part (1) as an exercise. Since $g(n)$ asymptotically dominates $f(n)$ and $h(n)$ asymptotically dominates $g(n)$, there exist C_1, C_2, n_1, $n_2 \geq 0$, such that $f(n) \leq C_1 g(n)$ for $n \geq n_1$ and $g(n) \leq C_2 h(n)$ for $n \geq n_2$. Choose $n_0 = \max\{n_1, n_2\}$ and $C = C_1 C_2$. Then we have $f(n) \leq C_1 g(n) \leq Ch(n)$ for $n \geq n_0$. It follows that $f(n) \in O(h(n))$.

Common Order Classes

There are a few order classes that occur frequently enough to warrant detailed discussion. These are, in increasing magnitude, **constant**, **logarithmic**, **linear**, **polynomial**, and **exponential** order. We will discuss each of these, but we will spend more time on logarithmic order because logarithms are likely to be less familiar to you. We say an algorithm is *of the order $f(n)$* if the time complexity function $T(n) \in O(f(n))$.

An algorithm in $O(1)$ is said to have *constant complexity*. The running time of such an algorithm is independent of the problem size. As an example of such an algorithm, consider one that prints one element of an array. The running time of the algorithm is independent of the size of the array.

Let a and b be positive real numbers, and assume $a \neq 1$. We define the **logarithm base** a of b to be the real number c such that $a^c = b$. We write $c = \log_a(b)$. The *existence* of the number c is guaranteed by some theorems of continuous functions, which are beyond the scope of this text. Another way to define the logarithm base a is to say $\log_a(b)$ is the power to which a must be raised to give b. Let's look at some simple examples of logarithms.

EXAMPLE 4.18

Find $\log_a(b)$ for each of the following pairs of values for a and b.

1. $\log_2(8)$
2. $\log_2(1/8)$
3. $\log_a(1)$, for any number $a \neq 1$
4. $\log_3(243)$
5. $\log_{10}(1,000,000)$

Solution

1. Since $2^3 = 8$, $\log_2(8) = 3$.
2. Since $2^{-3} = 1/8$, $\log_2(1/8) = -3$.
3. Since $a^0 = 1$, $\log_a(1) = 0$ for $a \neq 1$.
4. Since $3^5 = 243$, $\log_3(243) = 5$.
5. Since $10^6 = 1,000,000$, $\log_{10}(1,000,000) = 6$.

In computer science, the most commonly used base for logarithms is base 2. Unless explicitly stated, all logarithms in this book are to base 2. If $T(n) \in O(\log(n))$, we say that $T(n)$ has *logarithmic complexity*.

If $T(n) \in O(n)$, then we say $T(n)$ has *linear complexity*. The running time of such an algorithm is dominated by a multiple of n.

We say a function $T(n)$ is of *polynomial complexity* if $T(n) \in O(n^p)$ for some integer $p \in \mathbf{N}$. Any function that is of linear complexity is of polynomial complexity. The matrix multiplication algorithm described in Chapter 3 has polynomial running time (with $p = 3$). A function $T(n)$ is said to be of *exponential complexity* if $T(n) \in O(a^n)$, for some $a > 1$.

EXAMPLE 4.19

The algorithm given below is called the *fast exponentiation* algorithm. Show that it is of $O(\log(n))$. The notation $[k/2]$ represents the greatest integer in $k/2$.

Fast Exponentiation Algorithm

{Compute a^n, where a is a real number and n is a positive}
{integer. The algorithm proceeds by successively dividing}
{the value of n by 2, until zero is reached. The result}
{is returned in the variable z.}

begin
    $z \leftarrow 1.0$
    $t \leftarrow a$
    $k \leftarrow n$
    while $k > 0$ **do**
        begin
            if k is odd **then**
                $z \leftarrow z \cdot t$
            $k \leftarrow [k/2]$
            if $k \neq 0$ **then**
                $t \leftarrow t^2$
        end
end.

Solution

We are interested in counting the number of multiplications involved in finding a^n. There are two assignment statements involving multiplication within the **while** loop. One is conditional on k being odd, so the worst case will occur if k is odd every time we enter the loop. We need to count the number of times the loop is entered. The value of k is initially n. After the first assignment, k becomes $n/2$, then $n/4$, then $n/8$, and so on. The algorithm stops after m iterations, where m is the smallest integer such that $n/2^m < 1$, or $2^{m-1} \leq n < 2^m$. The algorithm thus has worst case behavior $2\log(n) + 2 \in O(\log(n))$.

Putting It All Together

So far we have looked at different order classes, discussed algorithm analysis, and studied a few simple examples. In this section, we complete the analysis by comparing the running times for different classes. Suppose we have five algorithms that perform the same task, and after analysis, we find that they have running times of $1 + \log(n)$, n, n^2, n^4, and 2^n. Suppose that the elementary operation we count in the analysis takes 1 millisecond to execute. Table 4.4 shows the estimated running times for problem sizes $n = 1$, 10, 100, and 1000. Notice the units of time used change from milliseconds (ms) to centuries (c.). It is no wonder that problems for which no polynomial time algorithm exists are called *intractable*.

TABLE 4.4
Approximate Running Times

	$1 + \log(n)$	n	n^2	n^4	2^n
1	1 ms	1 ms	1 ms	1 ms	2 ms
10	4 ms	10 ms	100 ms	10 sec	1 sec
100	7 ms	100 ms	10 sec	1.2 day	4×10^{18} c.
1000	10 ms	1 sec	0.28 hr	4 c.	4×10^{88} c.

The relative growth rates are illustrated in Figure 4.1.

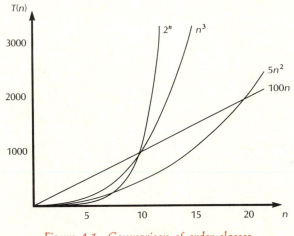

Figure 4.1. Comparison of order classes

SELF TEST 4

1. Compute $\log(16)$ and $\log_5(25)$.
2. Show that $T(n) = n \log(n) \in O(n^2)$.
3. Show that for $2 < n < 16$, $2^n \leq n^4$. What does this say about choosing an algorithm with exponential running time over an algorithm with running time of n^4 for "small" values of n?
4. Show that $T(n) = 1/n + 20 \in O(1)$.

Exercises 4.3

1. Verify that $(1/2)n^2 + 10n \in O(n^2)$.

2. Find each of the following logarithms.
   a. $\log_5(625)$
   b. $\log_{1/3}(27)$
   c. $\log_2(4096)$

3. Prove part (1) of Theorem 4.5.

4. It should be apparent to you by now that the logarithm base a is the inverse function of the exponential base a. Formally, verify that

$$a^{\log_a(b)} = b, \text{ for all } b > 0$$

and

$$\log_a(a^b) = b, \text{ for all } b$$

5. By following the steps of the *fast exponentiation* algorithm given on page 153, find the actual number of multiplications performed when $n = 7$ and when $n = 8$. Compare these to the worst case estimate.

6. For a publicity stunt, the Pennies from Heaven Curio Shop will give customers little bags of pennies for change on the shop's centennial. Ten bags of pennies are to be available to make change for any purchase less than ten dollars, assuming that the customer pays with a ten-dollar bill. How many pennies should each of the ten bags contain?

7. The *fast exponentiation* algorithm given on page 153 can easily be modified to find powers of a logical matrix. All that is necessary within the main loop is to replace the real number multiplications by matrix multiplications. What should be the initial value of z? If the matrix involved is $k \times k$, and we compute its nth power using fast exponentiation, how many **and** operations will be required?

*8. From the known additive property of exponents and the inverse property given in exercise 4, verify that $\log_a(xy) = \log_a(x) + \log_a(y)$, for all $x, y > 0$.

*9. Show that for any $a > 1$, $O(\log_a(n)) = O(\log(n))$.

*10. Prove the correctness of the *fast exponentiation* algorithm given on page 153.

How Fast Can We Sort?

Sorting pieces of information into a prescribed order is a major application of a computer system. The information is usually described as a set of *records*, each of which has a *key* used to determine its position in the sorted list. The keys are assumed to belong to a set, and the sorting criteria is a linear ordering. One instance in which data is sorted is a data dictionary consisting of terms relevant to a particular project and their associated definitions. We might want to sort the terms in the dictionary alphabetically to make searching easier and to make sure there are no duplicates. Another instance is a set of employee records, containing employees' names, addresses, department numbers, and payroll identification numbers. We might want to obtain a list of employees sorted by department, by payroll identification number, or by name. We are not likely to want to sort by street address, but we might want to sort by zip code for a mailing label program.

Thus there seem to be three kinds of uses for sorting:

- Collecting records with common keys together, as in sorting a data dictionary to find duplicates or as in sorting employee records by department
- Comparing two lists of records sorted in the same order, as in comparing an old version of a data dictionary with a new version to discover which records have been deleted and which have been inserted
- Searching a list of records, as in looking up definitions in the data dictionary

Sorting problems have been a spur to researchers and designers, who have developed numerous methods to handle sorting problems for different circumstances. Knuth [1] devotes an entire volume to presenting and analyzing algorithms for sorting and searching. Sorting problems are generally divided into two classes: *internal sorting*, in which all records to be sorted can be maintained in a given machine's memory, and *external sorting*, in which records to be sorted reside on storage media such as disks or tape external to the main memory.

The time complexity of internal sorting algorithms is usually measured by the number of comparisons between keys, expressed as a function of the number of records to be sorted. There are sort methods, known as *radix sorts*, that do not require direct key comparisons. Exercise 3 on page 157 should familiarize you with the difference between a radix sort algorithm and comparison-based sorting methods.

The techniques used to derive the theoretical time complexity for each sorting algorithm include results about permutations (which represent all possible unsorted inputs), recurrence relations (which can express the time complexity for sorting sets of a certain size in terms of the time complexity for sets of smaller sizes), and data structures (which tell us about the requirements for organization and storage of the records to be sorted).

So, how fast can we sort? There are two answers to the question: The first is that if we are sorting internally and want to minimize the number of key comparisons, the best we can do in terms of our abstract time complexity is $n\log n$, where n is the number of records to be sorted. The other answer is "It depends." The number of sorting algorithms is as large as it is because the criteria of internal sorting and number of comparisons is not always an adequate basis for choosing a sorting algorithm. Some algorithms perform well for the average case, but perform poorly for the worst case. Others perform well for the worst case, but require significant overhead in terms of internal storage requirements and managing data movements that in a particular implementation, can make them perform less well than simpler algorithms, particularly for small sets of records. It is up to a software engineer to supplement understanding of the theoretical results with knowledge of the particular context in order to weigh the trade-offs between the various choices.

Exercises 4.4

Construct a stack of twenty cards labeled with the 2-digit numbers from 10 to 29. Shuffle the cards to put them in random order. Sort the cards using each of the following methods (reshuffle them after each sort).

1. Search through the cards until you find the greatest index (29), and put this card face-up on the table. Then search, starting at the front, for the greatest remaining index (28), and place this card on top of the first one. Continue this process until all cards are in the stack. Are the cards now in order? How many passes through the card deck did you need? What's the worst order in which the cards could be initially? What's the best order in which the cards could be initially? Why? What's the time complexity as a function of the number of cards?

2. Deal the cards into two stacks: the first stack contains only cards whose index has a first digit of 1; the second stack contains only cards whose index has a first digit of 2. Then use the second digit of the indexes to sort each of the shorter stacks into increasing numerical order, using the same method as in exercise 1. Place the first stack on top of the second. Are the cards now in order? How many times did you have to compare the index of one card with the index of another card?

3. Look at the second digit of the indexes and deal the cards into ten stacks, arranged from left to right by increasing the second digit (0 through 9). Pick up the cards: 0's first, then 1's on top, and so on. Now turn the cards over and deal them into two stacks, based on the first digit. Put the stack with first digit 1 on top of the stack with first digit 2. The cards should be in order. How many times did you have to compare the index on one card with the index on another card?

REFERENCES

[1] D. Knuth, *The Art of Computer Programming*, Vol. 3 (Reading, Mass.: Addison-Wesley, 1973).

SUMMARY

1. **The Fundamental Principle of Counting**. Assume that some event can occur in n_1 ways, that a second event can occur in n_2 ways, and so on, up to event k, which can occur in n_k ways. Then the events can occur in the order given in exactly $n_1 \cdot n_2 \cdots \cdot n_k$ ways.

2. Select k things from a set S, allowing repetitions, but distinguishing among lists in which the same elements appear in different order. This kind of list is called a **k-sample**.

3. Select k things from a set S, allowing repetitions, but equating lists containing the same elements in any order. These lists are known as **k-selections** from a set S.

4. Select k things from a set S, without repetition, and distinguish lists by the order in which elements are selected. These lists are called **k-permutations** from the set S.

5. Select k things from a set S, without repetition, and equate lists containing the same elements in any order. These lists are called the **k-combinations** from S, or combinations of the elements of S taken k at a time.

6. There are n^k k-samples from a set S.

7. The number of k-permutations from a set of size n is

$$P(n, k) = \frac{n!}{(n-k)!}$$

8. The number of k-combinations from a set of size n is

$$C(n, k) = \frac{n!}{(n-k)!k!}$$

9. **Theorem 4.1**.

   a. For all integers $n \geq 1$ and $0 \leq k \leq n$,

   $$C(n+1, k+1) = C(n, k) + C(n, k+1)$$

   b. For all integers $n \geq 1$ and $k \geq 1$,

   $$C(n+k, k) = 1 + C(n, 1) + C(n+1, 2) + \cdots + C(n+k-1, k)$$

10. **Theorem 4.2**. For all $n \geq 1$ and $k \geq 1$, the number of k-selections from a set with n elements is given by

$$S'(n, k) = C(n + k - 1, k) \qquad (4.7)$$

11. **Theorem 4.3**. Suppose there are n objects, of which n_1 objects are of type 1, n_2 objects are of type 2, and so on, up to n_r objects of type r. Then there are

$$\frac{n!}{(n_1!)(n_2!) \ldots (n_r!)}$$

distinct patterns that can be formed by permuting the n objects. Moreover, each pattern appears exactly

$$(n_1!)(n_2!) \ldots (n_r!)$$

times among the $n!$ permutations.

12. **Theorem 4.4**. The number of ways to partition a set with n elements into k blocks is given by $S(n, k)$, where $S(n, k)$ satisfies the recurrence relation:
    a. $S(n, 1) = 1$ and
       $S(n, n) = 1$, for all $n \geq 1$
    b. $S(n + 1, k + 1) = (k + 1)S(n, k + 1) + S(n, k)$,
       for $n \geq 1$ and $1 \leq k \leq n$

13. Let $f, g: \mathbf{N} \rightarrow \mathbf{R}$ be functions. When we assert that "f **asymptotically dominates** g," we mean that there exist constants $C \geq 0$ and $n_0 \geq 0$ such that $g(n) \leq Cf(n)$ for all $n \geq n_0$. The set of all functions that are asymptotically dominated by the function f is denoted by $O(f(n))$. If $g(n) \in O(f(n))$, then g is said to be $O(f)$, which is read "g is order at most f," or simply "g is big-oh of f".

14. **Theorem 4.5**. Let $f, g: \mathbf{N} \rightarrow \mathbf{R}$ be functions.
    a. $g(n) \in O(f(n))$ and $f(n) \in O(g(n))$ if and only if $O(f(n)) = O(g(n))$.
    b. If $f(n) \in O(g(n))$ and $g(n) \in O(h(n))$, then $f(n) \in O(h(n))$.

REVIEW PROBLEMS

1. Compute each of the following expressions.
   a. $\dfrac{16!}{14!}$

   b. $\dfrac{6!}{9!}$

   c. $\dfrac{13!}{11!}$

2. Write each of the following expressions in terms of factorials.

    a. $13 \cdot 12 \cdot 11$

    b. $\dfrac{1}{13 \cdot 12 \cdot 11}$

3. Let $S = \{\alpha, \beta, \gamma, \delta\}$. List all the 2-samples, 2-selections, 2-permutations, and 2-combinations from S.

4. Use the Fundamental Principle of Counting to answer the following questions.

    a. A club is to elect three officers: a president, a vice president, and a secretary/treasurer. If the club has 33 members, how many ways can the officers be elected? (Assume no one is elected to more than one office.)

    b. There are five roads between A and B and four roads between B and C. How many ways can one drive from A to C?

    c. Given the roads in (b), how many ways can one drive from A to B to C, back to B, and back to A.

5. Consider the following algorithm:

```
begin
    if x < 0 then
        if y < 0 then
            z ← x + y
        else
            if x + y = 0 then
                z ← 0
            else
                z ← x − y
    else
        if y < 0 then
            z ← x + y²
        else
            z ← y²
end.
```

In order to test the algorithm, we would like to exercise every possible assignment statement. How many data sets will be needed?

6. Numbers are to be formed from the digits 1, 2, 3, and 4. Repetitions are not permitted.

    a. How many 3-digit numbers can be formed?

    b. How many even 3-digit numbers can be formed?

    c. How many odd 3-digit numbers can be formed?

    d. How many 3-digit numbers greater than 200 can be formed?

7. Answer the questions in problem 6 assuming repetitions of digits are allowed.

8. A palindrome is a word that reads the same forwards as backwards. How many

4-letter single-word palindromes can be formed from the letters in the English alphabet?

9. In a particular programming language, a variable name must start with a letter, but subsequent characters may be letters or digits.

   a. How many distinct identifiers of length 2 are possible?

   b. How many distinct identifiers of length less than or equal to 3 are possible?

10. In some versions of Pascal, only eight characters are significant for variable names. A variable name in Pascal must begin with a letter. The remaining characters may be letters or digits. How many variable names are possible in Pascal?

11. A list consisting of ten items is to be sorted. How many different arrangements of the original list are possible?

12. Show that

$$n \cdot (n-1) \cdot (n-2) \cdot \cdots \cdot (n-k+1) = \frac{n!}{(n-k)!}$$

13. How many ways can a party of eight persons arrange themselves in a row?

14. How many ways can a party of k persons arrange themselves in a row? Leave the answer in factorial form.

15. Find n if $P(n, 2) = 6$.

16. Compute each of the following.

   a. $C(16, 3)$

   b. $C(12, 10)$

   c. $C(15, 5)$

17. How many ways can a committee consisting of two men and three women be chosen from five men and five women?

18. A student is to do three out of five problems on a test.

   a. How many choices does the student have?

   b. The student must do the first problem. How many choices does the student have for the remaining two problems?

   c. The student must do one of the first two problems and two of the remaining problems. How many ways can this be done?

19. The eighth row of Pascal's Triangle is given below.

$$1 \quad 7 \quad 21 \quad 35 \quad 35 \quad 21 \quad 7 \quad 1$$

Compute the ninth and tenth rows.

20. Show that

$$C(n, 0) - C(n, 1) + C(n, 2) - C(n, 3) + \cdots \pm C(n, n) = 0$$

21. A person is dealt a 5-card poker hand from an ordinary deck of playing cards. How many ways can the following poker hands be dealt?

    a. diamond flush (five diamonds)

    b. three aces

    c. full house (three of one kind and two of another)

22. Compute $S'(5, 4)$.

23. Consider the set $S = \{a, b, c, d, e, f, g, h, i\}$. Determine if each of the following classes of subsets forms a partition of S. If it is not a partition, state why.

    a. $\{\{a, c, e\}, \{b, f\}, \{d, h, i\}\}$

    b. $\{\{a, c, e\}, \{b, d, f, h\}, \{e, g, i\}\}$

    c. $\{\{a, c, e\}, \{b, d, f, h\}, \{g, i\}\}$

24. Suppose you have five tiles containing the letters A, A, D, R, R. How many permutations will spell *RADAR*?

25. How many ways can a committee be selected from five males and five females if there are to be an equal number of males and females?

26. In how many ways can ten toys be divided among three children if the youngest is to receive four toys and the other two are to have three toys each?

27. How many partitions of a set with four elements are there?

28. How many ways can a club with fifteen members be partitioned into three committees containing five members each?

29. How many ways can eight books be given to four students, so that

    a. each student gets two books?

    b. one student gets two books, two students get one book, and one student gets four books?

30. A block set has three squares, four triangles, and five circles among its pieces. How many patterns can be formed by laying these twelve shapes out in a row?

31. Show that for all $n \in \mathbf{N}$, $\log(n^2) = 2 \log(n)$.

32. How many times is the assignment statement in the following algorithm fragment executed?

```
begin
     for i = 1 to n do
          for j = −m to m do
               sum ← sum + 1
end.
```

What is the order of the algorithm? If the initial value of *sum* is 0, what is its final value?

*33. Which of the following statements about order classes are true?

    a. $n^n \in O(n!)$

    b. $n! \in O(2^n)$

    c. $n^2 \in O(\log(n^2))$

▣ Programming Problems

*34. Write a program to print all the permutations of a given list of n elements.

35. Write a program to print all the k-combinations of a given list of n elements.

36. Write a program to compute $C(n, k)$ given n and k.

37. Write a program to print the first n rows of Pascal's Triangle.

38. Write a program to evaluate $S(n, k)$.

SELF TEST ANSWERS

Self Test 1

1. $P(20, 17) = \dfrac{20!}{3!}$

   $C(13, 9) = \dfrac{13!}{9!4!} = \dfrac{13 \cdot 12 \cdot 11 \cdot 10}{4 \cdot 3 \cdot 2} = 13 \cdot 11 \cdot 5 = 715$

2. We are interested in selecting sets of cards in which the order the cards are dealt does not matter.

$$C(52, 7) = \frac{52!}{45!7!}$$
$$= \frac{52 \cdot 51 \cdot 50 \cdot 49 \cdot 48 \cdot 47 \cdot 46}{7 \cdot 6 \cdot 5 \cdot 4 \cdot 3 \cdot 2 \cdot 1}$$
$$= 52 \cdot 17 \cdot 10 \cdot 7 \cdot 47 \cdot 45$$

3. Since the order of the notes in the sequence is significant, and notes can be repeated, we need the number of 3-samples from a set with eight elements. This is given by $8^3 = 512$.

Self Test 2

1. $C(7, 5) = \dfrac{7!}{5!2!} = \dfrac{7 \cdot 6}{2!} = 21$

   $P(7, 5) = \dfrac{7!}{2!} = 7 \cdot 6 \cdot 5 \cdot 4 \cdot 3 = 2420$

   $S'(7, 5) = C(7 + 5 - 1, 5) = C(11, 5) = 462$

2. The front line consists of a set of players. The order in which the players are selected does not matter. Hence, there are $C(10, 3)$, or 120, different 3-player lines.

3. We use the Fundamental Counting Principle directly. There are $4 \cdot 5 \cdot 3$, or 60, different sets of options available.

Self Test 3

1. The word *reentrant* has five distinct letters. The letter a occurs once, and e, n, r, and t each occur twice. Using Theorem 4.3, we see that there are

$$\frac{9!}{1!2!2!2!2!} = 9 \cdot 7 \cdot 6 \cdot 5 \cdot 4 \cdot 3 = 22{,}680$$

distinct patterns that can be formed from the letters in *reentrant*.

2. A set with three elements can be partitioned in the following ways:

   • one block with three elements
   • two blocks, one block with two elements and the other with one element
   • three blocks, each with one element

   Hence, there are $S(3, 1) + S(3, 2) + S(3, 3)$ partitions. Using our previously computed values, we have $1 + 3 + 1$, or 5, distinct ways to partition a set with three elements.

3. Using the second part of Theorem 4.4, we see that

$$S(6, 4) = 4S(5, 4) + S(5, 3)$$

Now

$$S(5, 4) = 4S(4, 4) + S(4, 3)$$

By the first part of Theorem 4.4, we know that $S(4, 4) = 1$, and in Example 4.14 we found that $S(5, 3) = 25$ and $S(4, 3) = 6$. Hence,

$$S(5, 4) = 4 + 6 = 10$$

and

$$S(6, 4) = 40 + 25 = 65$$

Self Test 4

1. Since $2^4 = 16$, $\log_2(16) = 4$.
   Since $5^2 = 25$, $\log_5(25) = 2$.

2. $\log n < n$ for all $n \geq 1$, so $n \log n < n^2$ for $n \geq 1$. Hence, $n \log n \in O(n^2)$.

3.

n	n^4	2^n
3	81	8
4	256	16
5	625	32
6	1296	64
7	2401	128
8	4096	256
9	6561	512
10	10,000	1024
11	14,641	2048
12	20,736	4096
13	28,561	8192
14	38,416	16,384
15	50,625	32,768
16	65,536	65,536

For small values of n, the actual running time for the more complex algorithm can be much less.

4. Use the fact that $1 + 20n \leq 21n$ for $n \geq 1$.

5 Undirected Graphs

The theory of graphs had rather humble origins in a paper by the eighteenth-century mathematician Leonhard Euler. Euler was trying to solve a problem that is similar to many mathematical puzzles still popular today. The problem is known as the **Königsberg Bridge problem**:

There are seven bridges connecting two islands and the banks of a river, as shown in Figure 5.1. Find a path that crosses each bridge exactly once.

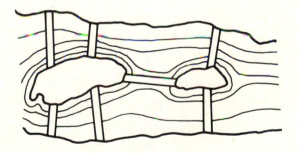

Figure 5.1. The Königsberg Bridges

Try several paths yourself. If you had found a path that was a solution, then Euler probably would have found one, too, and graph theory might have had a very different history. The difficult thing, after futile attempts to find a path, is to prove conclusively that *no* such path exists in this case.

A method of solving this problem is to model it as a **multigraph**, as shown in Figure 5.2. A multigraph consists of two things: a set of elements called **vertices**, or **nodes**, and a set of elements called **edges** connecting the vertices. A multigraph allows for *multiple edges* between the vertices.

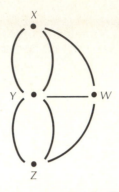

Figure 5.2. Multigraph representation of the Königsberg Bridge problem

The dots labeled *W*, *X*, *Y*, and *Z* in Figure 5.2 represent the vertices, and the lines between the dots represent the edges. Euler devised a rule, applicable to all multigraphs, that determines whether or not a path, as described in the bridge problem, can be found. A path that uses all edges in a multigraph exactly once is called an **Eulerian path**. Here is Euler's rule stated in terms of multigraphs:

> Assume we have a multigraph such that for any pair of vertices there is a path between them. The multigraph has an Eulerian path if and only if there are only 0 or 2 vertices that have an odd number of edges leaving them.

The multigraph for the Königsberg Bridge problem has four vertices with an odd number of edges leaving them, so there is no Eulerian path.

Euler's rule is a deceptively simple result. From it the theory of graphs has grown. Graphs are among the most useful models within computer science, appearing in data flow models, state machines, scheduling algorithms, circuit layouts, communication networks, flowcharts, and sorting and searching algorithms. Other applications outside of computer science include transportation networks, chemistry, process control systems, and the social sciences. At its core, graph theory is an extension of the theory of relations on a set. However, you will recall we used arrows to connect vertices in a graph of a relation. Such graphs are called **directed graphs**. The multigraph in Figure 5.2 does not have arrows. Such graphs are called **undirected graphs**. In this chapter, we will focus on undirected graphs, and Chapter 6 will focus on directed graphs. In both chapters, several of the major results will be expressed as algorithms.

We start this chapter by discussing undirected graphs and the key concepts of paths, cycles, and connectivity. After a brief discussion of Eulerian paths, we introduce Hamiltonian circuits and the closely related traveling salesperson's problem. In the second part of this chapter, we give a detailed account of a special class of undirected graphs called trees, and we show how they arise naturally as a tool in sorting and searching.

5.1   Simple Graphs

It is customary in graph theory to restrict the discussion to those graphs that have at most one edge between a pair of vertices. Some of the concepts we present for graphs have extensions to multigraphs. A **simple graph** (or **undirected graph**) is a pair $G = (V, E)$ where:

1. V is a finite set whose elements are called **vertices**.
2. E is an irreflexive, symmetric relation on V.

We shall refer to a simple graph $G = (V, E)$ as a *graph*. The ordered pairs in E are called the **edges** of the graph. The irreflexivity of E implies that there are no edges from a vertex to itself. The symmetry of E implies that $(u, v) \in E$ if and only if $(v, u) \in E$. More specifically, if $e = (u, v) \in E$, we say that the edge e is between u and v (respectively v and u), and that u is **adjacent to** v. Moreover, we say that e is **incident to** u (respectively, v). Because of this symmetry, we can denote e as the unordered pair $\{u, v\}$ rather than as either of the two distinct ordered pairs (u, v) or (v, u). So far when we have drawn the graphical representation of a relation, we have used arrows for edges between vertices. In the case of simple graphs, we will omit the arrows. Figure 5.3 shows two simple graphs.

Figure 5.3. Two simple graphs

The symmetry and irreflexivity of E allow us to simplify the matrix representation of a simple graph. Let $G = (V, E)$ be a graph, say with $V = \{v_1, v_2, \ldots, v_n\}$. Let M be the $n \times n$ (logical) relation matrix for E. We will refer to M as the **adjacency matrix** for the graph. By symmetry, the upper-right portion of M is symmetric to the lower-left portion. The irreflexivity of the relation implies that the diagonal entries are all F's. Hence, it is sufficient to use only the lower triangular portion of the relation matrix.

EXAMPLE 5.1

1. Expand the lower triangular matrix given below to the full 4×4 form.
2. Sketch the graph defined by the following lower triangular matrix.

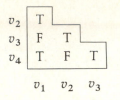

Solution

1. The best way to read the lower triangular form is from the bottom up. In this case, the first column says that v_1 is adjacent to v_2 and v_4. In the second column, we see that v_2 is also adjacent to v_3. And from column 3, we see that v_3 is adjacent to v_4. In general, the ith column tells us about connections between v_i and v_j, where $j > i$. We use this information and symmetry to complete the full 4×4 matrix.

$$\begin{bmatrix} F & T & F & T \\ T & F & T & F \\ F & T & F & T \\ T & F & T & F \end{bmatrix}$$

2. The graph is shown in Figure 5.4.

Figure 5.4.

Let $G = (V, E)$ be a graph. Let u, v be vertices. The **degree** of v, denoted $\delta(v)$, is the number of edges incident to v. Since each edge must be incident to two vertices, Theorem 5.1 is apparent.

Theorem 5.1

Let $G = (V, E)$ be a graph.

$$\sum_{v \in V} \delta(v) = 2|E|$$

• • •

A **subgraph** of a graph $G = (V, E)$ is a graph $G' = (V', E')$ such that $V' \subseteq V$ and $E' \subseteq E$. The subgraph G' is a **spanning** subgraph if $V = V'$. Figure 5.5 shows two subgraphs, G_1 and G_2, of graph G.

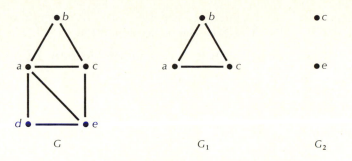

Figure 5.5. Graph G and two subgraphs, G_1 and G_2

A graph $G = (V, E)$ is said to be **complete** if for all vertices $u, v \in V \{u, v\} \in E$. A complete graph with n vertices is denoted by K_n. Every graph is a spanning subgraph of a complete graph.

Paths, Cycles, and Connectivity

A **path** of length k in a graph is a sequence of vertices $v_0, v_1, v_2, \ldots, v_k$ such that for $i = 1, 2, \ldots, k$, $\{v_{i-1}, v_i\} \in E$. The sequence c, a, d, e is a path of length 3 in the graph G in Figure 5.5. We can regard the path as being *from c to e* because of the order in which the vertices were listed. Of course, by reversing the order of the listing, we get a path from e to c.

A path in an undirected graph is called a **cycle** (or **circuit**) if the first and last vertices in the path are the same vertex and no edges in the path are repeated. Thus a cycle in a graph must have at least three edges. A graph that does not contain any cycles is called **acyclic**. The path a, b, c, e, a is a cycle in graph G of Figure 5.5. While a cycle may not contain a repeated edge, it may contain a repeated vertex. For instance, the path a, b, c, a, e, d, a is a cycle in graph G of Figure 5.5. The subgraph G_2 in Figure 5.5 is acyclic.

A graph is said to be **connected** if there is a path between every pair of vertices in the graph. The graphs G and G_1 in Figure 5.5 are connected, but G_2 is not. The phrase *is connected to by a path* defines a relation, C, on the set of vertices of a graph. The underlying relation, E, is symmetric, so C is symmetric and transitive. Assuming every vertex is connected to itself *by a path of length 0*, we conclude that C is an equivalence relation. Hence, the set of vertices is partitioned into equivalence classes. Giving each equivalence class all the edges connecting its vertices, we construct connected subgraphs of G, which we call **connectivity components**, or *components*, of the graph. The number of components of a graph G is called the **connectivity number** of G, which is denoted by $C(G)$. Evidently, G is connected if and only if $C(G) = 1$.

The next algorithm computes, for a given undirected graph $G = (V, E)$, the connectivity number $C(G)$. We start at any vertex and proceed to find all vertices that are neighbors of the original vertex, and then all their neighbors, and so on, until there are no more neighbors left. This means we have found an entire component.

Then we jump to the next component and carry out the same process. Since the graph is finite, eventually all vertices will be accounted for.

Algorithm for Computing the Connectivity Number of a Graph

{Let $G = (V, E)$ be a graph. This algorithm computes the value}
{of $c = C(G)$, the number of components of G.   }

begin
   $V' \leftarrow V$
   $c \leftarrow 0$
   while $V' \neq \emptyset$ **do**
      begin
         Choose $y \in V'$.
         Find all vertices connected to y.
         Remove these from V'.
         $c \leftarrow c + 1$
      end.
end.

EXAMPLE 5.2

Trace the connectivity algorithm on the graph in Figure 5.6.

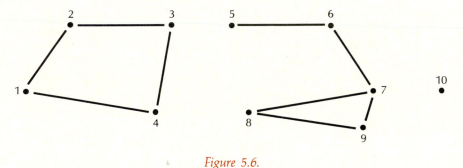

Figure 5.6.

Solution

Initially, $V' = \{1, \ldots, 10\}$. We arbitrarily choose $y = 5$. The vertices connected to y are 6, 7, 8, and 9. The component we have found is shown in Figure 5.7(a). Removing these from V' reduces V' to $\{1, 2, 3, 4, 10\}$. We increment c, and return to the top of the loop. Since $V' \neq \emptyset$, we continue, this time selecting, say, $y = 1$. The vertices connected to y this time are 2, 3, and 4, as shown in Figure 5.7(b). Removing these from V' leaves $V' = \{10\}$. We increment c again, and return to the top of the loop. This time there is only one choice for y, namely 10. Since there is nothing connected

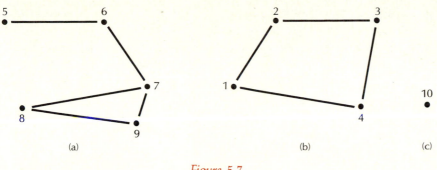

(a)   (b)   (c)

Figure 5.7.

to 10, the component found has only one vertex, as shown in Figure 5.7(c). We re-move 10 from V', leaving $V' = \emptyset$, and increment c a third time. Since $V' = \emptyset$, the algorithm terminates with $c = 3$.

SELF TEST 1

Consider the graph in Figure 5.8.

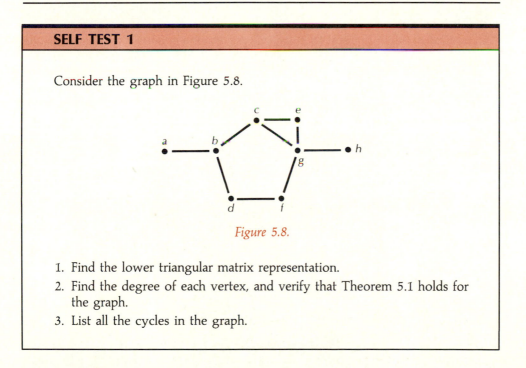

Figure 5.8.

1. Find the lower triangular matrix representation.
2. Find the degree of each vertex, and verify that Theorem 5.1 holds for the graph.
3. List all the cycles in the graph.

Eulerian Paths

We now return to the problem of finding a path in a graph that uses each edge exactly once. These are the Eulerian paths discussed in relation to the Königsberg Bridge problem. We restate Euler's rule in Theorem 5.2.

Theorem 5.2

A connected graph with at least two vertices has an Eulerian path if and only if there are 0 or 2 vertices of odd degree. The path is a cycle if and only if each vertex has even degree.

Proof

First, suppose the graph $G = (V, E)$ has an Eulerian path. Since each edge is used once, the circuit must pass through each vertex at least once. For each "visit" to a vertex that is not the first or the last, there will be an edge used to get to the vertex, and another used to leave. Hence, edges incident to an intermediate vertex come in pairs, and every vertex except the first and last has even degree. If the path is a cycle, the first vertex also has even degree.

   For the other direction, we will consider only the case in which all vertices have even degree and show how an Eulerian circuit can be constructed. Assume that every vertex has even degree. Pick any vertex, v, as a starting point. Since there are an even number of edges incident to v, departing v on one edge will leave at least one edge on which to return to v. Similarly, when the path reaches any vertex on a given edge, there will be an edge on which to leave. So, starting at v, select a path through the graph, and remove the edges used from further consideration. If the process terminates at v and all edges have been used, the Eulerian circuit is complete. If not, the remaining subgraph consists of vertices of even degree, and must have a vertex v_1 in common with the circuit already found. The process can be repeated again and again until all edges have been used. The resulting interconnected cycles together make up an Eulerian circuit. Figure 5.9 illustrates the construction.

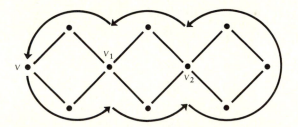

Figure 5.9. An Eulerian circuit made up of three pieces

   If there are two vertices of odd degree, we must pick one of them as the starting point. The rest of the construction is the same, except the path ends at the other vertex of odd degree, since its extra edge will allow us to arrive there without leaving again.

EXAMPLE 5.3

1. Show that the graph in Figure 5.10(a) has an Eulerian path but that it does not have an Eulerian circuit. Construct the Eulerian path.

2. Show that the graph in Figure 5.10(b) does not have an Eulerian path.

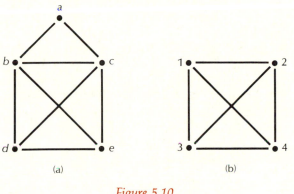

(a)   (b)

Figure 5.10.

Solution

1. Since there are exactly two vertices (*d* and *e*) of odd degree, we know that there is an Eulerian path, but not an Eulerian circuit. To construct the path, we must begin at either *d* or *e* and then end at the other. Starting at *d*, we obtain the path *d, e, c, a, b, c, d, b, e*. Figure 5.11 shows the sequence in which the edges of the graph are traversed. Since each edge is labeled exactly once, the path is Eulerian.

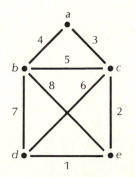

Figure 5.11. Eulerian path

2. All four vertices in the graph of Figure 5.10(b) have degree 3, so it is impossible to find an Eulerian path for this graph.

Hamiltonian Circuits

A problem similar to finding a Eulerian circuit is the problem of finding a Hamiltonian circuit. A **Hamiltonian circuit** is a cycle in a graph that passes through each vertex exactly once. Figure 5.12(a) shows a graph with a Hamiltonian circuit, and Figure 5.12(b) shows a graph without a Hamiltonian circuit. Finding the right path amounts to tracing the graph (without lifting your pen off the paper) so that you reach each vertex just once. There is no simple rule for determining the existence of Hamiltonian circuits as there is for Eulerian circuits. Certainly a graph with a Hamiltonian circuit will be connected, but not every connected graph will have a Hamiltonian circuit.

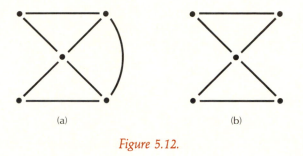

(a)   (b)

Figure 5.12.

If a graph represents a set of cities to be visited, then a path that visits each city exactly once is a Hamiltonian circuit. Assuming that every city is connected by an edge to every other city, there will always be a Hamiltonian circuit.

The problem becomes more difficult if the edges are assigned positive weights (which could represent distances or travel costs) and the objective is to find a Hamiltonian circuit of the least total weight. This latter problem is called the **traveling salesperson problem**. There are algorithms that find suboptimal solutions to this problem. This means that the algorithm finds a Hamiltonian circuit that tends to be better than most, but does not guarantee finding the best circuit. For large sets of vertices, the problem is sufficiently complex so that no known algorithm computes the best solution in a reasonable time. One algorithm that finds a suboptimal solution to the traveling salesperson problem is the *nearest neighbor method*. We assume that the graph is complete.

Nearest Neighbor Method for Traveling Salesperson Problem

```
{This algorithm finds a suboptimal solution to the traveling     }
{salesperson problem; that is, the algorithm finds a Hamiltonian}
{circuit in a complete graph with positive weights, but does    }
{not guarantee finding the one with least total weight.         }
{The path is represented as a list of vertices. The weight is   }
{assigned to the variable w.                                    }
```

begin
  Choose any $v1 \in V$.
  $v' \leftarrow v1$
  $w \leftarrow 0$
  Add v' to the list of vertices in the path.
  while unmarked vertices remain **do**
   begin
    Mark v'.
    Choose any unmarked vertex, u, that is closest to v'.
    Add u' to the list of vertices in the path.
    $w \leftarrow w +$ the weight of the edge $\{v', u\}$
    $v' \leftarrow u$
   end
  Add $v1$ to the list of vertices in the path.
  $w \leftarrow w +$ the weight of the edge $\{v', v1\}$
end.

EXAMPLE 5.4

Apply the *nearest neighbor* algorithm to the weighted graph of Figure 5.13.

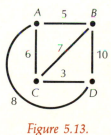

Figure 5.13.

Solution

We start, arbitrarily, at B. Initially, $w = 0$ and the path list is B. The unmarked vertex closest to B is A, so we add A to the path, as shown in Figure 5.14(a), and we mark

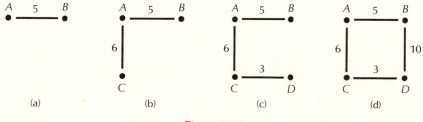

  (a)   (b)   (c)   (d)

Figure 5.14.

B. Then we add 5 to *w*. The unmarked vertex closest to *A* is *C*, so we add the vertex *C* to the path, as shown in Figure 5.14(b). We increase *w* by 6. The unmarked vertex closest to *C* is *D*, so we add *D* to the path, as in Figure 5.14(c). We increase *w* by 3. Since no unmarked vertices remain, the cycle is completed by adding the vertex *B* to the path. Then we increase *w* by 10. The resulting circuit is shown in Figure 5.14(d). The weight of the circuit is 24. Figure 5.15 shows all three Hamiltonian circuits for

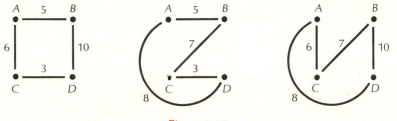

Figure 5.15.

the graph, and reveals that the best one has total weight 23. Observe that starting the nearest neighbor algorithm at *A* or *C* will yield the best path, but starting at *B* or *D* will yield the next best path.

Isomorphisms

We have a great deal of leeway in drawing graphs. For instance, we readily accept the fact that the two graphs in Figure 5.16 are the same. If pressed for a precise reason for their sameness, we can refer to their lower triangular matrix representations, which are identical for the two graphs. The only difference between the graphs is the way the line representing one of the edges is drawn. The structure of a graph is captured in the set of vertices and the set of edges. In general mathematical terms, when two objects have essentially the same structure, we say that they are **isomorphic**. Let $G_1 = (V_1, E_1)$ and $G_2 = (V_2, E_2)$ be graphs. We define an **isomorphism** from G_1 to G_2 to be a bijection $f: V_1 \to V_2$ such that for all $u, v \in V_1$, $\{u, v\} \in E_1$ if and only if $\{f(u), f(v)\} \in E_2$. In other words, the isomorphism maps V_1 onto V_2, and maps the set of vertices and the set of edges in a bijective way.

Figure 5.16. *Two representations of the same graph*

EXAMPLE 5.5

Show that the two graphs in Figure 5.17 are isomorphic.

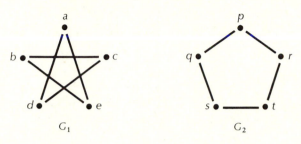

Figure 5.17.

Solution

We have to find a bijection f from the set $\{a, b, c, d, e\}$ to the set $\{p, q, r, s, t\}$ that preserves edges in the sense we have described. We will describe the function as a table of the form

v	a	b	c	d	e
$f(v)$					

where the entries in the second row will give the image of each of the elements in the first row. We start by arbitrarily assigning $f(a) = p$. Now a is adjacent to d and e, so p must be adjacent to $f(d)$ and $f(e)$. This requires that $\{f(d), f(e)\} = \{q, r\}$ since q and r are the vertices adjacent to p. Arbitrarily, we select $f(d) = r$ and $f(e) = q$. Then our table, so far, looks like

v	a	b	c	d	e
$f(v)$	p			r	q

Our remaining task is to map the set $\{b, c\}$ onto the set $\{s, t\}$. At this stage, our choices are limited. Observe that b is adjacent to c and e. Since $f(e) = q$, we must have $f(b)$ adjacent to q. Vertices p and s are adjacent to q, but we already have $f(a) = p$, so we are forced to take $f(b) = s$ to be sure that f is a bijection. Then we have to let

$f(c)$ equal t. The complete table is

v	a	b	c	d	e
$f(v)$	p	s	t	r	q

While we used some of the adjacency information to construct f, we have not checked all the conditions necessary for isomorphism. First, f is certainly a bijection. Next we need to check that the mapping $\{u, v\} \rightarrow \{(f(u), f(v)\}$ is a bijection between the sets of edges. We can check this by inspection in this case, since there are only five edges in each graph. The following table shows the correspondence between edges.

$\{u, v\}$	$\{f(u), f(v)\}$
$\{a, d\}$	$\{p, r\}$
$\{a, e\}$	$\{p, q\}$
$\{b, c\}$	$\{s, t\}$
$\{b, e\}$	$\{s, q\}$
$\{c, d\}$	$\{t, r\}$

It is also possible to give a geometric interpretation to the isomorphism. What the isomorphism f prescribes is a way to pick up G_1, rearrange its vertices, and place it exactly on top of G_2. In this example, we place a on top of p, move the edge $\{b, c\}$ onto the edge $\{s, t\}$, and then uncross the edges $\{b, e\}$ and $\{c, d\}$ by laying them on top of the edges $\{s, q\}$ and $\{t, r\}$, respectively.

When we believe that two graphs are isomorphic, we can justify that belief by constructing an isomorphism, as we did in Example 5.5. When we suspect that two graphs are not isomorphic, we can substantiate that suspicion only by showing that no isomorphism exists. In some cases this is easy. For instance, if the graphs do not have the same number of vertices, then there can be no bijection between the sets of vertices, and hence no isomorphism. Similarly, if the graphs do not have the same number of edges, the graphs cannot be isomorphic. There are other properties that must be common between isomorphic graphs that we can check (connectivity, for example). Otherwise, verifying that two graphs are nonisomorphic can be difficult. Suppose $G_1 = (V_1, E_1)$ and $G_2 = (V_2, E_2)$. If $|V_1| = |V_2| = n$, then there are $n!$ possible bijections between V_1 and V_2. To show that G_1 is not isomorphic to G_2, we must show that every one of the $n!$ bijections fails to map the edges of G_1 onto the edges of G_2. In Example 5.6 we show how the complexity of the task can be reduced.

EXAMPLE 5.6

Show that the graphs G_1 and G_2 in Figure 5.18 are not isomorphic.

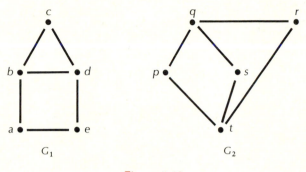

G_1 G_2

Figure 5.18.

Solution

We cannot easily show this by inspecting the graphs because each of the graphs is connected and has five vertices and six edges. However, before we start checking the 120 bijections of $\{a, b, c, d, e\}$ onto $\{p, q, r, s, t\}$, let's look at the way the graphs are put together. G_1 has a cycle b, c, d, b of length 3. An isomorphism from G_1 to G_2 must map this cycle, vertex by vertex, onto a cycle in G_2 of the same length. But G_2 has no cycle of length 3, so there can be no isomorphism of G_1 onto G_2.

SELF TEST 2

1. Is there an Eulerian circuit in the graph of Figure 5.19? Is there a Hamiltonian circuit?

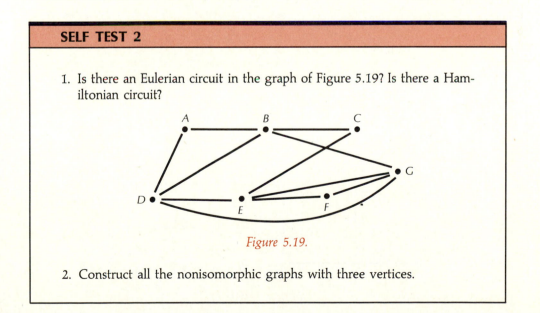

Figure 5.19.

2. Construct all the nonisomorphic graphs with three vertices.

Exercises 5.1

1. Verify the following properties of a complete graph, K_n.
   a. K_n is connected.
   b. Each vertex of K_n has degree $n - 1$.
   c. K_n has $n(n - 1)/2$ edges.
   d. K_n has a Hamiltonian circuit.

2. The algorithm for counting the components of a graph included the following statements:

   Choose $y \in V'$.
   Find all vertices connected to y.
   Remove these from V'.

   Write a detailed version of the algorithm that expands the statement "Find all vertices connected to y."

3. Verify, by direct count, that Theorem 5.1 holds for the graphs in Figure 5.5.

4. Find the lower triangular matrix for each of the graphs in Figures 5.5 and 5.6.

5. List all the cycles in the graph G in Figure 5.5.

6. Find Eulerian and Hamiltonian circuits for the graph in Figure 5.20.

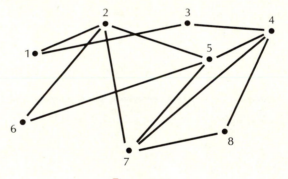

Figure 5.20.

7. For which values of n does the complete graph K_n have an Eulerian path? When is the path a circuit?

Figure 5.21.

8. Are the graphs in Figure 5.21 isomorphic or nonisomorphic? If they are isomorphic, exhibit the isomorphism. If not, give a reason for your answer.

5.2   Trees

A graph $G = (V, E)$ is called a **tree** if G is connected and acyclic. Figure 5.22 shows three graphs that are trees. Trees are an extremely important class of graphs. From a theoretical point of view, trees have a very clean set of properties, which are summarized in Theorem 5.3. In terms of applications, tree structures appear naturally as a representation vehicle for expression evaluation in sorting and searching problems, and in numerous situations where hierarchical representations are needed.

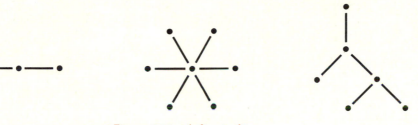

Figure 5.22. A forest of trees

Theorem 5.3

The following statements are equivalent for a graph $G = (V, E)$ with n vertices and m edges.

1. G is a tree.
2. There exists exactly one path between any pair of vertices in G.
3. G is connected and $m = n - 1$.
4. G is connected and removing any edge disconnects G.
5. G is acyclic and $m = n - 1$.
6. G is acyclic and adding any edge creates a cycle.

Proof

We will prove only three of the required implications, and we will suggest methods for proving some of the others in the exercises on page 195. The important thing to get from the theorem is an intuitive feel for the various equivalent statements. Drawing several graphs will be helpful.

(1) ⇒ (2).   This implication follows from two separate statements about connected graphs and acyclic graphs. The first is that G is connected if and only if there exists at least one path between any pair of vertices in G. The second is that G contains a cycle if

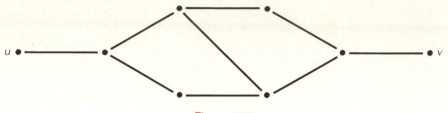

Figure 5.23.

and only if there exist vertices u and v with two paths between them. Figure 5.23 illustrates the latter statement. Observe that u and v need not be part of the cycle. Hence, if G is connected and acyclic, then there will be exactly one path between any pair of vertices.

(2) $\Rightarrow$ (3).   If there is exactly one path between any pair of vertices in G, then G is connected. To show that $m = n - 1$, we will use an induction argument on n, the number of vertices in G. To do so, we let $P(n)$ be the following assertion, for $n \geq 1$:

> $P(n)$: If G has k vertices, $1 \leq k \leq n$, and if there is exactly one path between any pair of vertices, then the number of edges in G is $k - 1$.

For the base step, observe that if $n = 1$, G has 0 edges, so $P(1)$ holds. For the induction step, assume that $P(n)$ is true for some $n \geq 1$. Let G be a graph with $n + 1$ vertices and suppose there exists exactly one path between any pair of vertices. Let $e = \{u, v\}$ be an edge in G. Since there is only one path from u to v, removing e from the graph G disconnects G, creating a graph G' with two components and $m - 1$ edges. Let G_1 and G_2 be these components. Suppose G_1 has n_1 vertices and m_1 edges, and suppose G_2 has n_2 vertices and m_2 edges. The number of vertices in G' is $n = n_1 + n_2$, and the number of edges in G' is $m_1 + m_2$. Both G_1 and G_2 are subgraphs of G, so any pair of vertices in each is connected by exactly one path. Applying the inductive hypothesis $P(n)$, we conclude that $m_1 = n_1 - 1$ and $m_2 = n_2 - 1$. Now G' has one less edge than G. Then

$$m = \text{(the number of edges in } G') + 1$$
$$= m_1 + m_2 + 1 = n_1 - 1 + (n_2 - 1) + 1$$
$$= n_1 + n_2 - 1 = n - 1$$

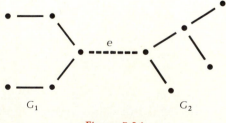

Figure 5.24.

which completes the proof. The removal of the edge e and the decomposition of the resulting graph into two components is illustrated in Figure 5.24.

(6) $\Rightarrow$ (1).   Assume that (6) is true. We need to show that G is connected. Suppose u and v are vertices. If we add the edge $\{u, v\}$ to the graph, our hypothesis implies that we have created a cycle. There must have been a path from u to v already. Hence, G is connected.

$\bullet$ $\qquad$ $\bullet$ $\qquad$ $\bullet$

We can now use Theorem 5.3 to show that every connected graph contains a *spanning tree*—a spanning subgraph that is a tree. There are two solutions, both constructive. The first procedure starts with the connected graph and successively removes edges so that at each stage the resulting subgraph is connected. Eventually, deleting any of the remaining edges will disconnect the subgraph. By statement (4) of Theorem 5.3, the subgraph remaining at this point is a spanning tree. Observe that the spanning tree will have $n - 1$ edges, where n is the number of vertices in the original graph.

The second procedure works in the opposite direction, starting without edges and adding them one at a time so that at each stage no cycle has been created. Eventually, adding any of the remaining edges will create a cycle. Hence the graph constructed will be a tree.

EXAMPLE 5.7

Use the first procedure described above to find a spanning tree for the graph in Figure 5.25.

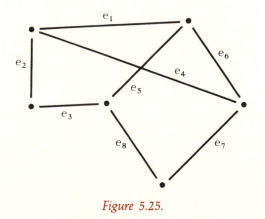

Figure 5.25.

Solution

Since there are eight edges and six vertices, we need to eliminate three edges. Now any edge can be removed without disconnecting the graph. Remove e_3 first. Then any edge but e_2 can be removed. Remove e_4. Next, any edge but e_1 and e_2 can be removed. Remove e_7. We have reached a point where removing any of the remaining edges will disconnect the graph, so the resulting subgraph, shown in Figure 5.26, is the desired spanning tree. Note that it has five edges, as required by statement (3) of Theorem 5.3.

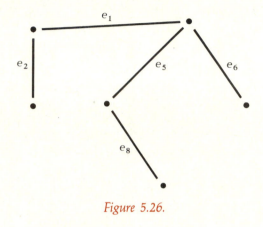

Figure 5.26.

Minimal Spanning Trees

In Chapter 1, where we introduced abstract models, we used a graph representation of a highway system. Each edge represented a highway between cities, and the numbers on the edges represented the distances. The problem there was to find a minimal set of edges connecting all cities together; we were looking for a spanning tree which had the least total weight among all the spanning trees. We call this a **minimal spanning tree (MST)**. This problem bears some similarity to the traveling salesperson problem, but turns out to have a simple algorithmic solution. The algorithm uses the second method for constructing a spanning tree discussed on page 185. At each stage, we *optimize locally* by adding the edge of least weight whose addition does not create a cycle. To prove that this sequence of local optimums yields the global optimum is difficult and involves more than we want to get into in this text. While such a proof may seem unnecessary to you, remember that in the *nearest neighbor* algorithm we also picked the "best" edge available at each step, but we could not guarantee that the final result would be globally optimal.

Algorithm for Finding the Minimal Spanning Tree

```
{Let G = (V, E) be a graph. The algorithm will obtain a        }
{MST for G by adding edges of least weight whose addition}
{does not create a cycle. At the end of the calculation,       }
{the MST is described by the set of edges, T.                  }
```

begin
    $e \leftarrow$ an edge in E with the smallest weight
    $T \leftarrow \{e\}$
    $E' \leftarrow E - \{e\}$
    while $E' \neq \emptyset$
        begin
            $e' \leftarrow$ edge $\in E'$ with smallest weight
            $T \leftarrow T \cup \{e'\}$
            $E' \leftarrow$ set of edges in $E' - T$ whose addition to T
                does not create a cycle
        end
end.

EXAMPLE 5.8

Use the above algorithm to find the MST for the graph in Figure 5.27.

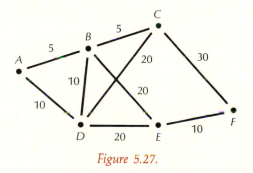

Figure 5.27.

Solution

The first step in the algorithm is to select an edge with the least weight. In this case, there are two edges with weight 5, $\{A, B\}$ and $\{B, C\}$. We are free to choose either of these two. The process of building the tree is shown in the diagrams of Figure 5.28. In each picture, the edges belonging to T are drawn as solid lines, and the edges in E' are drawn as dashed lines. After adding the fifth edge, $E' = \emptyset$, which causes termination of the loop.

Rooted Trees

Trees support decision structures very well. We begin the discussion of this application with an example.

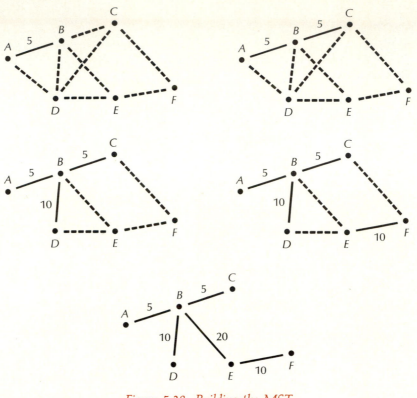

Figure 5.28. Building the MST

EXAMPLE 5.9

There is a simple guessing game that goes like this: Someone thinks of a number between 1 and 31. You promise to guess the number correctly in a small number of guesses. You ask a question like, "Is the number x?" Then the other person responds with "Yes," "Less than x," or "Greater than x." Show that you can guarantee guessing the number in no more than five guesses.

Solution

The trick is always to guess the midpoint of the range of numbers remaining. Then even a wrong guess reduces the range by half, until finally only one number will remain. Figure 5.29 shows how the guessing proceeds, beginning at 16. Each vertex in the tree is a decision point. Either it holds the correct value, or the correct value is in one of its two subtrees. The subtree on the left holds the remaining lesser values, and the subtree on the right holds the remaining greater values. Since there are only four levels in the tree, no more than five guesses will be needed.

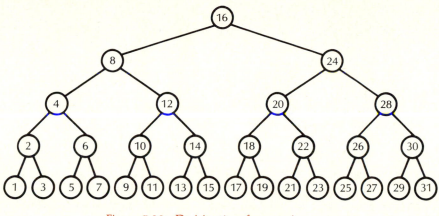

Figure 5.29. Decision tree for guessing game

The tree constructed in Example 5.9 is an example of a binary rooted tree. A **rooted tree** is a tree with vertex set V such that one vertex $v \in V$ is designated the **root** of the tree. For any other vertex u in the tree, there is a unique path from v to u. The length of the path is called the **level of the vertex** u. The vertices at level 1, the vertices adjacent to the root, are called the **descendants** of the root. Recursively, the descendants of the root can be designated as the roots of the subtrees obtained by deleting the root from the original tree.

The length of the longest path from the root is called the **level of the rooted tree**. If each vertex in the rooted tree has at most two subtrees, the tree is called a **binary tree**. The tree used in Example 5.9 is a binary tree of level 4.

Any tree can be regarded as a rooted tree by redrawing it to accentuate the designated root. Rooted trees are usually drawn with the roots at the top and the subtrees opening down towards the **leaves**, which are the vertices farthest from the root.

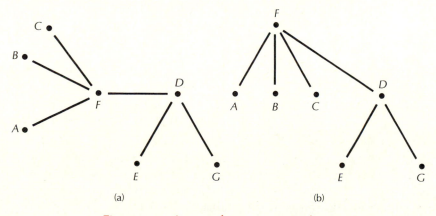

(a)   (b)

Figure 5.30. A tree redrawn as a rooted tree

Leaves must have degree 1. The graph in Figure 5.30(a) is redrawn in Figure 5.30(b) as a rooted tree with *F* as the root, and with *A, B, C, E,* and *G* as the leaves.

Sorting and Searching

Binary rooted trees, as in Example 5.9, are useful structures for handling decisions. In the rest of this section, we show how trees can be used as a tool for handling the kind of decisions required for sorting and searching. The general problem is to organize a collection of objects in order by some key value. For instance, a set of employee records may be organized alphabetically by the name of the employee, last name first. Consider three operations (among many) that must be performed on such a set: *searching* for a particular employee's record, *printing* the list in *sorted* order, and *inserting* a new employee's record. A simple algorithm is given for each operation. All are based on the form of the definition of a rooted tree; that is, when a vertex is encountered, it is simply the root of another subtree, so that the same procedure may be applied to the left and right subtrees as needed. Hence, each of the algorithms is a **recursive** algorithm in that it invokes itself.

For each of the tree algorithms, the same data elements are used. The main data objects are the vertices of the tree. For purposes of these algorithms, we handle the information about vertices and their connections a bit differently than in previous algorithms. Each vertex is assumed to have at least three components:

- a *key* that identifies the vertex
- a reference to the left subtree
- a reference to the right subtree

If the left or right subtree is nonexistent, the reference is given the value **null**.

The procedure *search* is quite simple. We give a verbal description and leave it to you to express *search* as a formal recursive algorithm. Starting with the root of the tree, if the search key equals the vertex key, the search halts. If the search key is less than the vertex key, the left subtree is searched, if it is not empty. Otherwise, the right subtree is searched, if it is not empty.

The invocation of the *insert* procedure has the form *insert(item, treename)*, where *item* is the item to be inserted, and *treename* is a reference to a subtree. Initially, *treename* is the root of the tree. Insertions in a tree will always take place by creating a new leaf in the tree and adding it as either a left or right subtree of a vertex in the tree. The search for the appropriate place for the insertion begins with the root of the tree. If at any point the key of the item to be inserted equals the key of the current vertex, the search halts, and no insertion takes place.

Insert(item, treename)

begin
    if the insertion key < the vertex key **then**
        if the vertex has no left subtree **then**
            Add the new item on the left of the vertex.

      else

         insert(*item*, left subtree of vertex)

    else

      if the insertion key > the vertex key **then**

        if the vertex has no right subtree **then**

          Add the new item on the right of the vertex.

        else

          insert(*item*, right subtree of the vertex)

      else

        print "The item is already in the tree."

end.

EXAMPLE 5.10

1. Trace the *search* algorithm on the tree in Figure 5.31, where the search key is first *A*, then *L*.

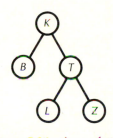

Figure 5.31. A search tree

2. Trace the *insert* algorithm on the tree in Figure 5.31, where the items to be inserted, successively, have keys *A*, *R*, and *N*.

Solution

1. The search starts at the root, *K*. Since *A* < *K*, the search moves down the left subtree of *K* to *B*. Since *A* < *B*, the search would move down to the left subtree of *B*. However, this is **null**, so the search terminates, with the item not found.

    Now consider searching for the key *L*. Since *L* > *K*, the search moves to the right subtree of *K*, which has root *T*. *L* < *T*, so the next move is down to the left subtree of *T*, which has root *L*. Since this is the same as the search key, the search terminates, with the item found.

2. First consider inserting *A*. Since the root is *K*, and *A* < *K*, we apply *insert* to the subtree with root *B*. Since *A* < *B*, and *B* has no left subtree, the item *A* is added as the left subtree of *B*, as shown in Figure 5.32(a).

    Next consider inserting *R*. *R* > *K*, so *R* is to be inserted in the right subtree with root *T*. *R* < *T*, so we move to the left subtree of *T*, which has root *L*. *R* > *L*, and *L*

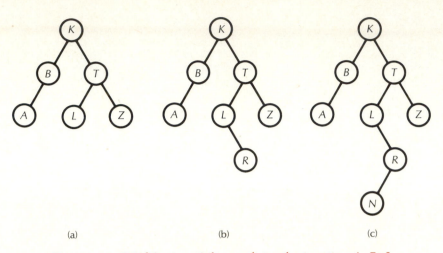

Figure 5.32. Modification of the search tree by inserting A, R, L

has no right subtree, so we add R as the right subtree of L. The resulting tree is depicted in Figure 5.32(b).

Finally, consider inserting N. Comparing N to the appropriate vertices, the procedure moves from K to T to L to R, and adds N as the left subtree of R. The result is shown in Figure 5.32 (c).

The original idea behind using a tree for storing information was to reduce the number of steps necessary in searching for a given item. The tree constructed for the number guessing game in Example 5.9 was optimal in that regard. It is a perfectly balanced tree, since every vertex, except the leaves, has exactly two subtrees. Thus the 31 numbers could be stored in four levels. The tree in Figure 5.32(c), constructed by applying the *insert* algorithm, did not come out as nicely balanced. In effect, it wastes some of the effort used to construct it by giving a deeper tree than needed. In the worst case, the algorithm *insert* could build a tree with $n - 1$ levels to hold n items. In that case, searching for an item in the tree would be no better than a straightforward linear search algorithm.

If there are n items to be stored in a binary tree, what is the optimal way to organize the tree to minimize the number of steps required to search for any of the items? Since the worst case requires searching from the root to the deepest level of the tree, we want to minimize the level of the tree by balancing it as it is constructed. A k-level binary tree can hold at most

$$\sum_{i=0}^{k} 2^i = 2^{k+1} - 1$$

items. Thus the ideal balanced tree for n items would have k levels, where $2^k \leq$

$n < 2^{k+1}$. Hence $k = [\log(n)] + 1$. There are tree-building algorithms more sophisticated than the *insert* algorithm that attain this optimum level.

The third algorithm, which prints all the vertices in the tree, is an example of a tree **traversal** algorithm. All vertices are to be visited in order. For this reason, the algorithm is called **in-order traversal**. The *in-order traversal* algorithm is a simple recursive procedure. We leave it as an exercise to express the algorithm using formal control structures. The algorithm works as follows: For each vertex in the tree, beginning with the root of the tree, print all elements in the left subtree. Then print the vertex. Then print all the vertices in the right subtree of the vertex.

EXAMPLE 5.11

Apply the *in-order traversal* algorithm to list the elements of the tree in Figure 5.32(c).

Solution

The difficult part of tracing the *in-order traversal* algorithm is keeping track of where we are in the tree. Before a given vertex can be printed, all the vertices in its left subtree must be printed. When a vertex is finally printed, the root of its right subtree is designated as the next vertex to be considered. Table 5.1 shows a column for vertices whose printing has been deferred, a column for the root of the next subtree, and a column for the vertex being printed. Observe that the printed items appear in alphabetical order, as desired.

TABLE 5.1

Deferred vertices	Root of the next subtree	Item to be printed
	K	
K	B	
K, B	A	
K, B		A
K		B
	T	K
T	L	
T	R	L
T, R	N	
T, R		N
T		R
	Z	T
		Z

SELF TEST 3

1. Prove the implication 2) ⇒ 1) of Theorem 5.3.
2. The numerical entries in the lower triangular matrix

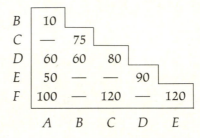

represent road distances between certain towns. A dash means there is no direct road between the towns. Find a MST for the road set and interpret its meaning.

3. Using the *insert* algorithm, build a tree to sort the words *comet, cupid, donner, blitzen, prancer, dancer,* and *rudolph*. Assume the words are presented in the order given.

Exercises 5.2

1. Verify Theorem 5.3 by inspection for the graph in Figure 5.33.

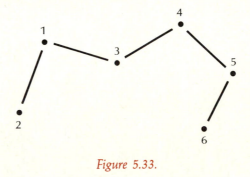

Figure 5.33.

2. Determine whether each of the following graphs is a tree.

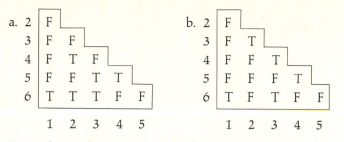

3. Prove the implication 3) ⇒ 4) of Theorem 5.3.

4. Prove the implication 5) ⇒ 6) of Theorem 5.3.

5. Draw all the nonisomorphic trees with five vertices.

6. A tree has three vertices of degree 3 and four vertices of degree 2. How many vertices of degree 1 must it have?

7. Suppose that the guessing game of Example 5.9 is to cover numbers in the range from 1 to 100. How many guesses will it take to guarantee finding the correct number? Show the first four levels of the decision tree you need to carry out the strategy in this case.

8. Assume that the tree shown in Figure 5.34 is used to keep a list of words in alphabetical order.

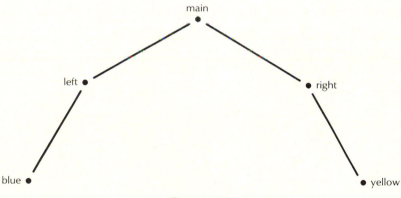

Figure 5.34.

   a. Show how successive insertions of the words *gold*, *mauve*, *aqua*, and *red* affect the tree.
   b. After the insertions of part (a), what is the maximum number of comparisons required to determine that a given item is not in the tree?

9. Write the algorithms *search* and *in-order traversal* as recursive procedures using formal algorithm control structures.

10. Give an example to show that the algorithm *insert* can produce a tree with four levels to hold five elements.

Syntax of Languages

English grammar provides a set of rules by which we can classify words in a sentence according to their function in the sentence. In the sentence, *The dog chases the cat, The dog* is the subject, and *chases the cat* is the predicate. The subject, in turn, consists of the article *the* followed by the noun *dog*. The predicate can also be decomposed into the verb *chases* and the object *the cat*, which is an article and a noun.

We can represent this structure by a tree known as a **parse tree**, as shown in Figure 5.35. There are two kinds of nodes in the parse tree. One kind represents the words of the original English sentence. These nodes appear as the leaves of the tree and are called **terminals**. The terminals, taken in left to right order, form the original English sentence. The other kind represents grammatical categories. These nodes are called **nonterminals**, or **syntactic categories**.

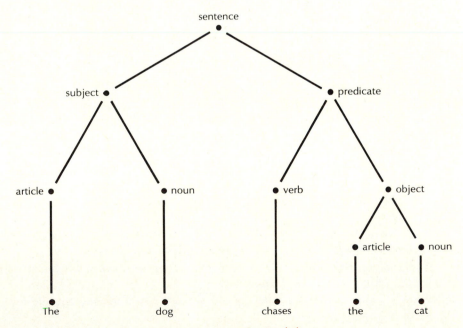

Figure 5.35. Parse tree for English sentence

Grammars

The grammar of a language consists of a set of rules that specify precisely what descendent nodes each nonterminal may have. Since the rules can be used to produce sentences, they are known as **production rules**. The grammar corresponding to the parse tree above is given below.

```
Sentence ::= Subject Predicate
Subject   ::= Article Noun
Predicate ::= Verb Object
Object    ::= Article Noun
Article   ::= the
Noun      ::= dog
Noun      ::= cat
Verb      ::= chases
```

The ::= symbol indicates that, on a parse tree, the item on the left has the items on the right as descendents. One symbol on the left of a ::= will appear as the root of a parse tree. This symbol is referred to as the **start** symbol.

The rules of the grammar can be used to generate all the possible sentences of the language. In our example, since the only possible terminal symbols are *the*, *dog*, *cat*, and *chases*, there are only four possible parse trees. They correspond to the following sentences:

The dog chases the cat.
The dog chases the dog.
The cat chases the dog.
The cat chases the cat.

These sentences are the entire language defined by this grammar.

Another way to use the rules is to analyze a sentence to see if it is a syntactically correct sentence in the language. This is done by using the rules to attempt to generate a parse tree whose terminal symbols are the sentence. If such a tree can be constructed, the sentence is part of the language. The analysis of a sentence in this way is called **parsing**.

Derivations

The repeated application of the production rules of the grammar results in sequences of terminal and nonterminal symbols that can be derived from the start symbol. Any sequence that contains only terminal symbols is a sentence of the language defined by the grammar. Figure 5.36 shows the partial derivation of the sentence *The dog chases the cat* as a sequence of increasingly larger parse trees. The leaves of the last parse tree would contain only terminal symbols.

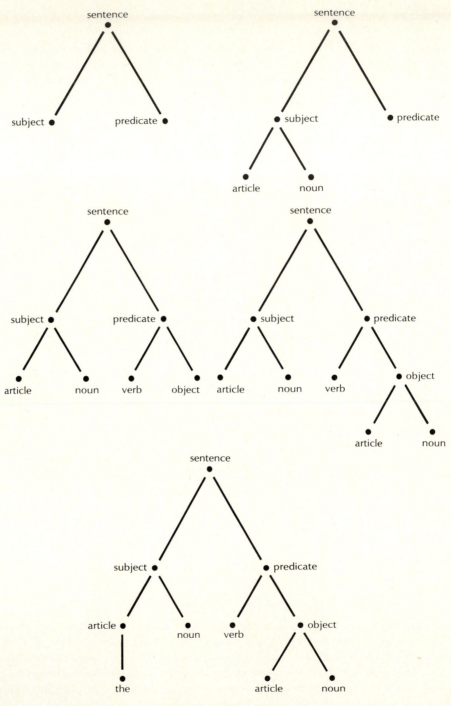

Figure 5.36. Derivation trees

Backus–Naur Form

Computer scientists use grammars to give formal definitions to programming languages. The formal definitions specify the "legal" statements that are part of a program written in a given language. But computer scientists need a language to define programming languages. A language used to describe other languages is called a **metalanguage**. Probably the most common metalanguage used by computer scientists to define programming languages is known as **Backus–Naur Form**, abbreviated **BNF**.

For example, the BNF specification of a grammar for simple arithmetic expressions may appear as:

$$\langle \text{Expression} \rangle ::= \langle \text{Expression} \rangle \langle \text{Operator} \rangle \langle \text{Expression} \rangle$$
$$|(\langle \text{Expression} \rangle)$$
$$|-\langle \text{Expression} \rangle$$
$$|0|1|2|3|4|5|6|7|8|9$$
$$\langle \text{Operator} \rangle ::= +|-|*|/|\uparrow$$

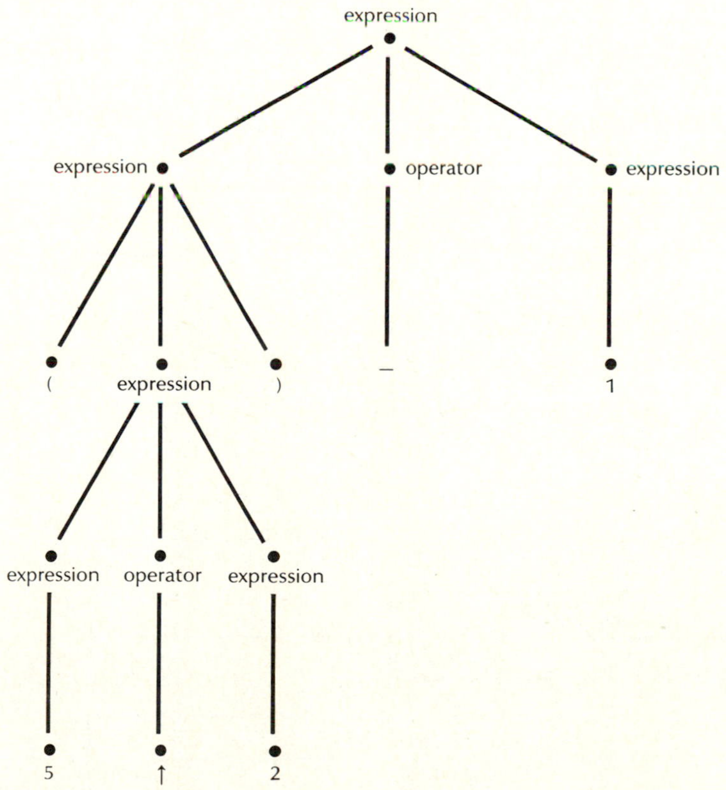

Figure 5.37. Parse tree for $(5 \uparrow 2) - 1$

The nonterminal symbols are ⟨Expression⟩ and ⟨Operator⟩. ⟨Expression⟩ is the start symbol. The terminal symbols are 0, 1, 2, 3, 4, 5, 6, 7, 8, 9, +, −, *, /, ↑, (, and). Note that the definition of this language is recursive because ⟨Expression⟩ is defined in terms of itself.

The symbols of BNF are:

::= "is defined as"
| "or"
⟨ ⟩syntactic category name

The parse tree in Figure 5.37 shows that the expression

$$(5 \uparrow 2) - 1$$

is a valid expression in this language. The expression

$$5 + \uparrow 2$$

is not valid in this language.

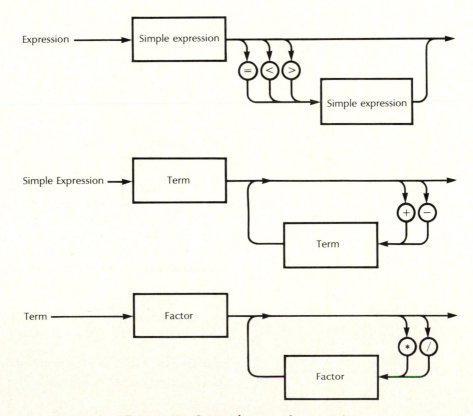

Figure 5.38. Syntax diagrams for expressions

Syntax Diagrams

Another method of defining a programming language is to use **syntax diagrams**. This method was used to define the programming language Pascal. The rules take the form of digraphs. The possible paths represent the possible sequence of symbols. In the diagram, rounded symbols represent terminal symbols, and symbols enclosed in rectangular boxes represent syntactic categories. The set of syntax diagrams in Figure 5.38 define an expression.

Exercises 5.3

1. Here is a simple grammar for expressions:

⟨Expression⟩ ::= ⟨Term⟩
                   |⟨Expression⟩ + ⟨Term⟩
⟨Term⟩   ::= ⟨Operand⟩
                   |⟨Term⟩ * ⟨Operand⟩
⟨Operand⟩ ::= A|B|C|

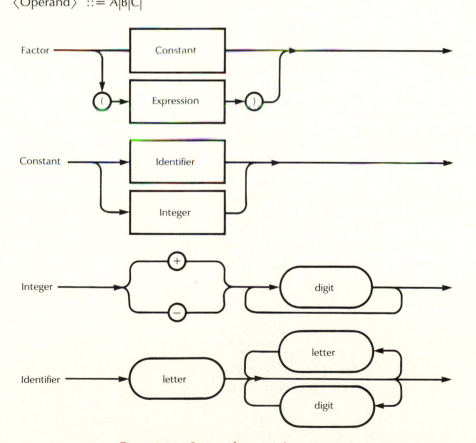

Figure 5.38. Syntax diagrams for expressions

a. List the syntactic categories and the terminal symbols. The starting symbol is ⟨Expression⟩.
b. Construct the parse tree for A + B * C.
c. Extend the grammar to include subtraction and division operators and the use of parentheses.

2. Describe the grammar in exercise 1 using syntax diagrams.

*3. Consider the following two grammars. What is the difference between these grammars? Illustrate your answer by showing two parse trees for expressions defined by the grammars.

Grammar 1

⟨Expression⟩ ::= ⟨Term⟩
                 |⟨Expression⟩ + ⟨Term⟩
                 |⟨Expression⟩ − ⟨Term⟩
⟨Term⟩   ::= ⟨Factor⟩
                 |⟨Term⟩ * ⟨Factor⟩
                 |⟨Term⟩ / ⟨Factor⟩
⟨Factor⟩   ::= 1|2|3|4|5
                 |−⟨Factor⟩
                 |⟨Power⟩
⟨Power⟩   ::= 1|2|3|4|5 ↑ ⟨Factor⟩

Grammar 2

⟨Expression⟩ ::= ⟨Term⟩
                 |⟨Term⟩ + ⟨Expression⟩
                 |⟨Term⟩ − ⟨Expression⟩
⟨Term⟩   ::= ⟨Factor⟩
                 |⟨Factor⟩ * ⟨Term⟩
                 |⟨Factor⟩ / ⟨Term⟩
⟨Factor⟩   ::= ⟨Power⟩ ↑ 1|2|3|4|5
                 |⟨Power⟩
⟨Power⟩   ::= 1|2|3|4|5
                 |−⟨Power⟩

SUMMARY

1. A **multigraph** consists of two things: a set of elements called vertices, or nodes, and a set of elements called edges connecting the vertices. A multigraph allows for multiple edges between the vertices.

2. A path that uses all edges in a multigraph exactly once is called an **Eulerian path**.

3. A multigraph has an Eulerian path if and only if there are only 0 or 2 vertices that have an odd number of edges leaving them.

4. A **simple graph** (or undirected graph) is a pair $G = (V, E)$ where
   a. V is a finite set whose elements are called **vertices**.
   b. E is an irreflexive, symmetric relation on V, whose elements are called **edges**.

5. **Theorem 5.1**. Let $G = (V, E)$ be a graph.

$$\sum_{v \in V} \delta(v) = 2|E|$$

6. A **path** of **length** k in a graph is a sequence of vertices $v_0, v_1, v_2, \ldots, v_k$ such that for $i = 1, 2, \ldots, k$, $\{v_{i-1}, v_i\} \in E$.

7. A path in an undirected graph is called a **cycle** (or circuit) if the first and last vertices in the path agree, and no edges in the path are repeated.

8. A graph G is said to be **connected** if there is a path between every pair of vertices in the graph.

9. **Theorem 5.2**. A connected graph with at least two vertices has an Eulerian path if and only if there are 0 or 2 vertices of odd degree. The path is a cycle if and only if each vertex has even degree.

10. A **Hamiltonian circuit** is a cycle in a graph that passes through each vertex exactly once.

11. Let $G_1 = (V_1, E_1)$ and $G_2 = (V_2, E_2)$ be graphs. An isomorphism from G_1 to G_2 is a bijection $f: V_1 \to V_2$ such that for all $u, v \in V_1$, $\{u, v\} \in E_1$ if and only if $\{f(u), f(v)\} \in E_2$. In other words, the isomorphism maps V_1 onto V_2, and maps, in a bijective way, the set of vertices and the set of edges.

12. A graph $G = (V, E)$ is called a **tree** if G is connected and acyclic.

13. **Theorem 5.3**. The following statements are equivalent for a graph $G = (V, E)$ with n vertices and m edges.
   a. G is a tree.
   b. There exists exactly one path between any pair of vertices in G.
   c. G is connected and $m = n - 1$.
   d. G is connected and removing any edge disconnects G.
   e. G is acyclic and $m = n - 1$.
   f. G is acyclic and adding any edge creates a cycle.

14. Every connected graph has a spanning tree.

15. If each vertex in a rooted tree has at most two subtrees, the tree is called a **binary** tree.

REVIEW PROBLEMS

1. Show that in any graph there is an even number of vertices with odd degree.
2. a. Draw a connected graph with eight vertices that has an Eulerian circuit but does not have a Hamiltonian circuit.
   b. Draw a connected graph with eight vertices that has a Hamiltonian circuit but does not have an Eulerian circuit.
3. For the graph in Figure 5.39:
   a. Verify, by inspection, that Theorem 5.1 holds.
   b. Find the lower triangular matrix representation.
   c. List all the cycles in the graph.

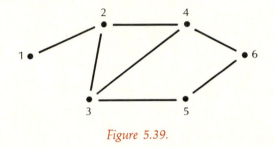

Figure 5.39.

4. Find all nonisomorphic graphs with four vertices. How many are trees?
5. Let $G = (V, E)$ be a graph. Suppose there exists an integer k such that $\delta(v) \geq k$ for all $v \in V$. Show that G contains a path of length k.
6. Find all the minimum spanning trees for the graph in Figure 5.40.

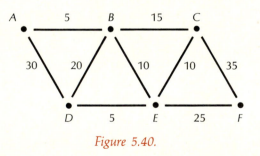

Figure 5.40.

7. Find all the spanning trees for the graph in Figure 5.41. How many are nonisomorphic?
8. Draw the complete graphs K_2, K_3, and K_6.
9. Use the connectivity algorithm to compute $C(G)$ for the graph given by the following matrix.

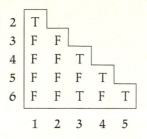

2	T				
3	F	F			
4	F	F	T		
5	F	F	F	T	
6	F	F	T	F	T
	1	2	3	4	5

10. A *clique* in a graph $G = (V, E)$ is a set of vertices $\{v_1, \ldots, v_k\}$ that are mutually adjacent; that is, for $i, j = 1, \ldots, k$, $i \neq j$, $\{v_i, v_j\} \in E$.

   a. Find the largest clique in the graph of Figure 5.41.
   b. If the vertices in the graph represent people, and the edge relation is "is a friend of," how do you interpret a clique?
   c. Suppose that G and G' are isomorphic graphs. Show that when G has a clique with k elements, G' also has a clique with k elements.

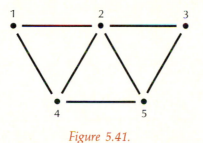

Figure 5.41.

11. Show that the two graphs in Figure 5.42 are not isomorphic.

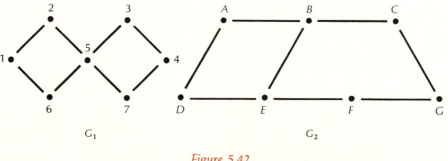

Figure 5.42.

12. A man arrives at a river with a goat, a dog, and a bag of cabbages. The boat he can use will hold himself plus at most one of the other three. However, he may not leave the dog with the goat, nor can he leave the goat with the cabbages. Using a graph model, where the vertices are possible arrangements of the

man and the objects on the sides of the river, and where the edges represent river crossings, show how the man can eventually get himself and all three objects safely across the river.

13. Show that two isomorphic graphs have the same number of components.

14. Find the level of each of the vertices of the tree in Figure 5.43, given that the root is the vertex D.

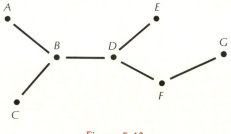

Figure 5.43.

15. A *trinary rooted tree* is a rooted tree in which the root of each subtree has at most three descendants. How many vertices can there be in a trinary rooted tree of level 2 (respectively, n, where $n \geq 1$)?

16. Show that any tree with two or more vertices must have at least two leaves.

*17. Show that if $G = (V, E)$ is an acyclic graph, then $|E| - |V| + C(G) = 0$.

18. Show that if $G = (V, E)$ is connected, then $|E| \geq |V| - 1$.

19. A graph $G = (V, E)$ is said to be *bipartite* if there is a partition $\{V_1, V_2\}$ of V such that if $\{u, v\} \in E$, then either $u \in V_1$ and $v \in V_2$, or $u \in V_2$ and $v \in V_1$.
    a. Is the graph of Figure 5.39 bipartite?
    b. Show that every tree is a bipartite graph.

20. Show that if every vertex in a graph has degree 2 or more, then the graph contains a cycle.

*21. Let G be a graph with n vertices. Show that if G has more than $(n - 1)(n - 2)/2$ edges, then G is connected.

*22. A *preorder* traversal of a binary rooted tree "visits" each vertex, then its left subtree, then its right subtree. A *postorder* traversal of a binary rooted tree "visits" the left and right subtrees of a vertex, then "visits" the vertex. Write recursive procedures for preorder and postorder traversals.

◼ Programming Problems

23. Write a program to compute the connectivity number of a graph.

24. Write a program that uses the *nearest neighbor* algorithm to solve the traveling salesperson problem.

25. Write a program to compute the minimal spanning tree for a graph.

SELF TEST ANSWERS

Self Test 1

1.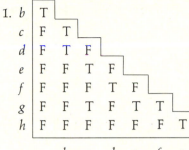

b	T						
c	F	T					
d	F	T	F				
e	F	F	T	F			
f	F	F	F	T	F		
g	F	F	T	F	T	T	
h	F	F	F	F	F	F	T
	a	*b*	*c*	*d*	*e*	*f*	*g*

2. The degrees of the vertices are given in the following table.

vertex	*a*	*b*	*c*	*d*	*e*	*f*	*g*	*h*
degree	1	3	3	2	2	2	4	1

The sum of the degrees is 18. The number of edges is 9, and $2 \cdot 9 = 18$.

3. In terms of the sets of edges used in the cycle, there are only three cycles in the graph. From these, all others can be derived easily.

   b, d, f, g, c, b

   b, d, f, g, e, c, b

   g, e, c, g

Self Test 2

1. An Eulerian circuit is *A, B, C, E, F, G, E, D, B, G, D, A*. The path *A, B, C, E, F, G, D, A* is a Hamiltonian circuit.

2. There are four nonisomorphic graphs with three vertices. These are shown in the figure by number of edges. (There can be 0, 1, 2, or 3 edges in a graph with three vertices.)

0 edges   1 edge

2 edges   3 edges

Self Test 3

1. Assume that (2) is true; that is, there exists exactly one path between any pair of vertices in G. We need to show that G is connected and acyclic. G must be connected, since there is a path between any pair of vertices. If G contains a cycle, then the vertices in the cycle have at least two paths between them. Hence, (2) also implies that G is acyclic.

2. Apply the MST algorithm: First select edge $\{A, B\}$. Then select edge $\{A, E\}$. Then there is a choice of $\{A, D\}$ and $\{B, D\}$. Arbitrarily pick $\{A, D\}$. Now you can't choose $\{B, D\}$, because this would create a cycle, so the edge $\{B, C\}$ must be next. At this stage, $\{C, D\}$ and $\{D, E\}$ must be passed over, because adding them creates a cycle. The edge $\{A, F\}$ is added next, and this completes the spanning tree.

3. Adding the words in the order given yields the following tree.

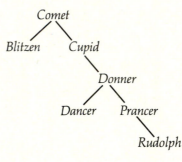

6 Directed Graphs

In this chapter, we cover the fundamental concepts of *directed graphs*, or *digraphs*. Digraphs are useful models in situations where there is an *ordering* between objects, say, in time or space. Digraphs appear in data flow diagrams, finite automata, task scheduling algorithms, and network analysis. For digraphs, unlike undirected graphs, we are concerned with the direction of edges between vertices. The underlying structure of a digraph is a relation on a set. We use both pictorial and logical matrix representations of relations in defining digraphs.

In the first section, we cover basic definitions of degree, paths, and cycles, and we present an algorithm for topological sorting. In the second section, we concentrate on two kinds of path problems in digraphs: (1) existence of paths, which we solve first by using powers of the logical matrix and then by using Warshall's algorithm, and (2) shortest path problems in weighted digraphs, for which we present Dijkstra's algorithm as an example solution.

6.1   Digraphs

A **directed graph**, or **digraph**, is a pair (V, E), where V is a finite set of elements called **vertices**, and E is a relation on V. The elements of E are called the **edges** of the digraph. For any pair of vertices $u, v \in V$, the set of edges E will contain at most one edge (u, v) from u to v, and at most one edge (v, u) from v to u. If $(u, v) \in E$, we say that u *precedes* v, or is an **antecedent** of v. The pictorial representations of digraphs are like those used to represent relations on a set. It is possible to represent a given digraph with different pictures. For example, Figure 6.1 shows two different pictures representing the same relation on the set $\{a, b, c, d\}$. We can also define the digraph in two other ways: as a set of ordered pairs or as a logical matrix of T's and F's (where T stands for *true* and F stands for *false*). This logical matrix is called the **adjacency matrix** for the digraph. The digraph in Figure 6.1 has the set definition

$$\{(a, b), (b, d), (d, c), (c, b)\}$$

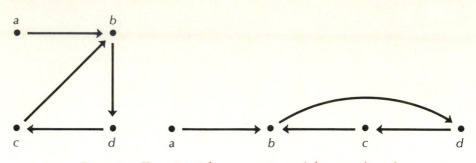

Figure 6.1. Two pictorial representations of the same digraph

and adjacency matrix

$$
\begin{array}{c c c c c}
 & a & b & c & d \\
a & \begin{bmatrix} F & T & F & F \\ b & F & F & F & T \\ c & F & T & F & F \\ d & F & F & T & F \end{bmatrix}
\end{array}
$$

The graph in Figure 6.1 has four vertices and four edges. Each vertex has one edge leaving it, which is reflected in the adjacency matrix by the fact that every row contains exactly one T. Similarly, there are no edges arriving at vertex a, two edges arriving at vertex b, and one edge arriving at each of the vertices c and d. Observe that because the matrix does not have symmetry, we cannot reduce it to a lower triangular form, as we did for undirected graphs.

Degrees, Paths, and Cycles

Given a graph $G = (V, E)$ and a vertex $v \in V$, the **indegree** of v is the number of edges arriving at v. That is, the indegree of v is given by

$$\text{indegree}(v) = \left|\{u \in V: (u, v) \in E\}\right|$$

Similarly, the **outdegree** of a vertex $v \in V$ is the number of edges leaving v. Thus the outdegree of v is given by

$$\text{outdegree}(v) = \left|\{u \in V: (v, u) \in E\}\right|$$

A vertex with indegree 0 is called a **source**, and a vertex with outdegree 0 is called a **sink**. In the digraph of Figure 6.1, vertex a is a source and there are no sinks.

We record some simple facts about indegrees and outdegrees in Theorem 6.1.

Theorem 6.1

Let $G = (V, E)$ be a digraph. Let M be the $n \times n$ adjacency matrix for E.

1. $\sum \{\text{indegree}(v): v \in V\} = \sum \{\text{outdegree}(v): v \in V\}$.

2. For all $v_i \in V$, indegree(v_i) = the number of T's in the ith column of M.
3. For all $v_i \in V$, outdegree(v_i) = number of T's in the ith row of M.

EXAMPLE 6.1

Find the indegree and outdegree of each of the vertices in Figure 6.2.

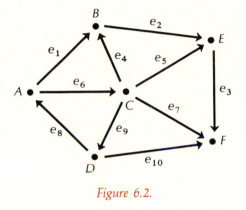

Figure 6.2.

Solution

Table 6.1 shows the indegree and outdegree for each of the vertices of the digraph in Figure 6.2.

TABLE 6.1

Vertex	Indegree	Outdegree
A	1	2
B	2	1
C	1	4
D	1	2
E	2	1
F	3	0

The indegree and outdegree columns sum to the same value (10), as required by Theorem 6.1. We see that F is a sink and that there is no source in this digraph.

/

A **path** from vertex u to vertex v is a sequence of vertices $v_0, v_1, \ldots, v_k$ such that $v_0 = u$, $v_k = v$, and for $i = 1, 2, \ldots, k$, the edge $e_i = (v_{i-1}, v_i) \in E$. Alternatively, the path can be defined as the sequence of edges $e_1, e_2, \ldots, e_k$. k is called the **length** of the path.

EXAMPLE 6.2

Consider the graph in Figure 6.2. Find two paths from A to F. Describe each as a sequence of vertices and as a corresponding sequence of edges.

Solution

Two paths from A to F are A, B, E, F and A, C, D, A, C, D, F. These paths can also be defined as these sequences of edges: e_1, e_2, e_3; and $e_6, e_9, e_8, e_6, e_9, e_{10}$.

A **cycle (circuit)** is a path from a vertex to itself. When a digraph does not contain a cycle, it is said to be **acyclic**. The graph in Figure 6.3 is an acyclic digraph. The left-to-right ordering apparent in the digraph is typical of acyclic digraphs, which are useful models for sets of tasks that need to have a particular sequence. In this case, it is important that there is not a cycle, since any task in a cycle would have to precede itself.

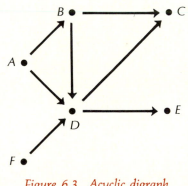

Figure 6.3. Acyclic digraph

EXAMPLE 6.3

Table 6.2 shows a (partial) list of tasks for building a house. For each task, T, the second column shows those tasks that must be completed before T can be started.

Construct a digraph model to show the relationships among the tasks. What would it mean if the digraph turned out to contain a cycle?

TABLE 6.2
Tasks for Building a House

Task	Prerequisite tasks
1. Make foundation	None
2. Add subflooring	Make foundation (1)
3. Frame walls/roof	Add subflooring (2)
4. Add siding	Frame walls/roof (3)
5. Install windows	Frame walls/roof (3)
6. Partition rooms	Frame walls/roof (3)
7. Shingle roof	Frame walls/roof (3)
8. Install wiring	Partition rooms (6)
9. Install plumbing	Partition rooms (6)
10. Install interior doors	Partition rooms (6)
11. Paint interior walls/ceiling	Install wiring (8), plumbing (9), windows (5), and interior doors (10)
12. Lay carpeting	Paint interior walls/ceiling (11)

Solution

Each task (as numbered) is a vertex, and each pair of tasks (T_1, T_2), where T_1 is a prerequisite of T_2, is an edge. The resulting digraph is shown in Figure 6.4. Task 1,

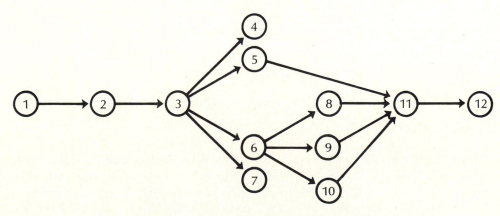

Figure 6.4. Digraph for house-building tasks

make foundation, is a source. The tasks add siding (4), shingle roof (7), and lay car-peting (12) are sinks. The digraph reveals many indirect prerequisites. For instance, we must add subflooring before we can partition rooms. On the other hand, many tasks, such as add siding and install wiring, can go on concurrently. When a task, such as paint interior walls/ceiling (11), has indegree greater than one, our interpre-tation is that all the antecedent tasks must be performed first. Thus, multiple edges arriving at a vertex correspond to an *and*. The digraph must be acyclic, since no task can be an indirect prerequisite for itself.

It is easier with the digraph than with the list to determine the order in which the tasks must be accomplished as well as which tasks can be carried out concurrently. When completion times for each task are available, the digraph can be an aid in com-puting the minimum amount of time needed to finish all the tasks. In exercise 9 on page 222, we ask you to carry out the computation for the house-building tasks.

SELF TEST 1

Consider the digraph with adjacency matrix given below.

$$\begin{array}{c|cccccc} & a & b & c & d & e & f \\ \hline a & T & F & F & T & F & F \\ b & T & F & T & F & F & T \\ c & F & T & F & F & T & F \\ d & F & F & T & F & F & T \\ e & F & T & F & T & F & T \\ f & F & F & F & F & F & F \end{array}$$

1. Draw the pictorial representation.
2. Find a path from a to b, if one exists.
3. Find a cycle of length 5, if one exists.
4. Find the indegree and outdegree of each vertex.

Consistent Labeling

The set of house-building tasks in Example 6.3 is small enough so that it is not difficult to find by inspection that there is not a cycle. In more complex situations, with more vertices, it is usually difficult to detect cyclic relationships. Determining that a digraph is acyclic is equivalent to finding a *consistent labeling* for the digraph. Roughly speaking, a consistent labeling is a way of numbering the vertices so that each vertex has a lower number than any vertex it precedes. However, the relation does not have to be a total ordering. We will give an algorithm for finding a consistent labeling, which is also known as **topological sorting**, on page 218. Let $G = (V, E)$ be a digraph. Let $\{v_1, v_2, \ldots, v_n\}$ be a labeling of the vertices. The labeling is said to be a **consistent labeling** if

$$(v_i, v_j) \in E \Rightarrow i < j$$

EXAMPLE 6.4

Find three different consistent labelings for the digraph in Figure 6.4.

Solution

Actually, the labeling as given is a consistent labeling, so we need to find only two more. This can be accomplished by switching the labels of, say, add siding (4) and shingle roof (7). We could not switch the labels for install plumbing (9) and install windows (5), because this would not be consistent with the fact that partition rooms must be completed before install plumbing. Using this reasoning, we get the two different consistent labelings shown in Table 6.3.

TABLE 6.3
Consistent Labelings for House-Building Tasks

Task	First	Second	Third
Make foundation	1	1	1
Add subflooring	2	2	2
Frame walls/roof	3	3	3
Add siding	4	7	11
Install windows	5	5	4
Partition rooms	6	6	5
Shingle roof	7	4	12
Install wiring	8	10	6
Install plumbing	9	9	7
Install interior doors	10	8	8
Paint interior walls/ceiling	11	11	9
Lay carpeting	12	12	10

Suppose the digraph G has a consistent labeling $v_1, v_2, \ldots, v_n$. If $v_{i_1}, v_{i_2}, \ldots, v_{i_k}$ is a cycle, then we must have

$$i_1 = i_k$$

and

$$i_1 < i_2 < \ldots < i_k$$

Since this is an obvious contradiction, it follows that a digraph with a consistent labeling must be acyclic.

An important property of an acyclic digraph is that it must have at least one source vertex. If a digraph does not have a source vertex, then for each vertex v, there exists a vertex u such that $(u, v) \in E$. Another way to say this is that every vertex $v \in V$ has an antecedent. This means that we can "back up" forever in the digraph. But the number of vertices is finite, so eventually some vertex must be repeated. Hence, if there is no source, we can "back into" a cycle. Exercise 5 on page 222 asks you to write a complete proof for this property. We are now ready to state Theorem 6.2, the main theorem concerning acyclic digraphs.

Theorem 6.2

Let $G = (V, E)$ be a digraph. G has a consistent labeling if and only if G is acyclic.

Proof

We have already proved that if G has a consistent labeling, then G is acyclic. For the other half of the proof, we use the *Topological Sort* algorithm, which constructs a consistent labeling for an acyclic digraph. The algorithm relies on the property that acyclic digraphs must have a source. The method of the algorithm is to label the set of sources in the digraph, and then remove them from consideration. Then the next level of sources is labeled, and so forth. One of two conditions will result: either all vertices will be labeled successfully or the remaining unlabeled vertices will not be sources, implying that the digraph contains a cycle (which is contrary to the assumption).

Topological Sort

```
{The objective is to find a consistent labeling }
{for a digraph (V, E). If there is no such        }
{labeling, the algorithm reports that there is a}
{cycle in the digraph. We denote the set of      }
{antecedents of a vertex v by A(v).              }
```

```
begin
    for v ∈ V do
        compute A(v)
    label ← 0
    while unlabeled vertices remain whose antecedent
            set is empty
        begin
            label ← label + 1
            v ← vertex with A(v) = ∅
            assign label to v
            for each unlabeled vertex u ∈ V do
                A(u) ← A(u) − {v}
        end
    if not all vertices are labeled then
        report there is a cycle in the digraph.
end. {algorithm}
```

EXAMPLE 6.5

Trace the operation of the *Topological Sort* algorithm on the digraph shown in Figure 6.5.

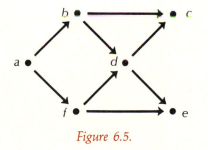

Figure 6.5.

Solution

The first step is to compute the antecedent sets $A(v)$ for $v \in V$:

$$A(a) = \emptyset \qquad A(b) = \{a\} \qquad A(c) = \{b, d\}$$
$$A(d) = \{b, f\} \qquad A(e) = \{d, f\} \qquad A(f) = \{a\}$$

The next step is to show the labeling that occurs within the main **while** loop. The progress is shown as the sequence of digraphs in Figures 6.6 (a) through (e). For each picture, we show the order in which the vertices have been labeled (for example, $a(1)$ indicates that vertex a was labeled first). Once a vertex has been labeled, we draw the edges leaving the vertex with dashed lines to represent its deletion from the antecedent sets of other vertices. At the outset, the only source is vertex a, so we assign it label 1. The result of deleting a from the antecedent sets (in this case, only for b and f), is shown in Figure 6.6(a). We let S be the set of all unlabeled vertices with empty antecedent sets. In Figure 6.6(a), we see that $S = \{b, f\}$. We can now choose either of these vertices to label next. Arbitrarily, we label f 2 and delete f from the antecedent sets in Figure 6.6(b). There $S = \{b\}$, so we have no choice but to label b

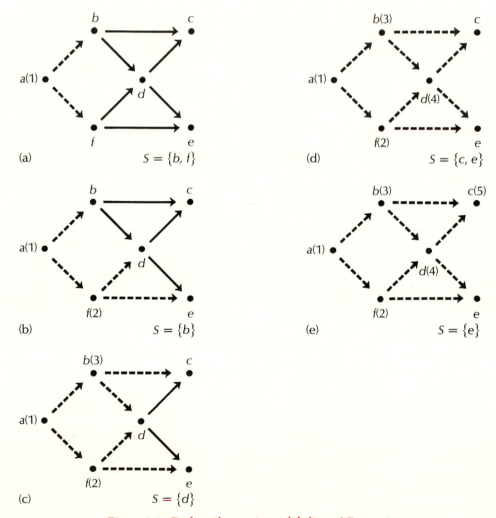

Figure 6.6. Finding the consistent labeling of Figure 6.5

3. As shown in Figure 6.6(c), S is now $\{d\}$, so we label d 4 and obtain Figure 6.6(d), with $S = \{c, e\}$. Then we can label either c or e. Choosing c, we obtain Figure 6.6(e), with $S = \{e\}$. We conclude the algorithm by labeling e 6.

SELF TEST 2

1. Apply the *Topological Sort* algorithm to the digraph in Figure 6.7.

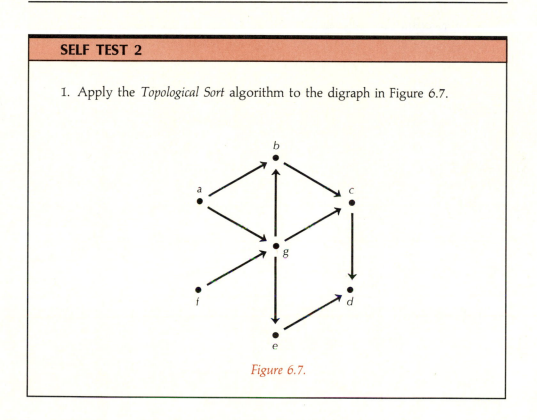

Figure 6.7.

Exercises 6.1

1. Let G be the digraph with vertices $\{1, 2, 3, 4, 5, 6\}$ and edges defined by the matrix below.

$$\begin{bmatrix} T & F & F & T & F & F \\ T & F & T & F & F & T \\ F & T & T & F & T & F \\ T & F & F & T & T & T \\ F & F & F & T & F & T \\ F & T & T & T & T & F \end{bmatrix}$$

a. Find paths from 1 to 2 and from 3 to 6.

b. Which vertices appear in a cycle?

2. Which of the lists of vertices given below are paths in the graph of exercise 1?

   a. 1, 2, 3, 4, 5   c. 4, 1, 4, 6, 2, 3, 5

   b. 1, 1, 4, 1   d. 5, 4, 6, 3

3. Find another consistent labeling for the graph in Figure 6.4. Is it possible to change the order of the first three vertices?

4. a. Apply the *Topological Sort* algorithm to the graph given by the matrix below.

$$
\begin{array}{c c c c c c c}
 & a & b & c & d & e & f \\
a & \begin{bmatrix} F & F & F & T & F & F \\ T & F & F & F & F & T \\ F & T & F & F & T & F \\ F & F & F & F & F & T \\ F & F & F & T & F & T \\ F & F & F & F & F & F \end{bmatrix} \\
b & & & & & & \\
c & & & & & & \\
d & & & & & & \\
e & & & & & & \\
f & & & & & &
\end{array}
$$

b. Rewrite the matrix with the rows and columns ordered by the new labels. What pattern can you see in the new version?

5. Give a complete inductive proof that every acyclic digraph contains a source and a sink.

6. Apply the *Topological Sort* algorithm to the digraph in exercise 1.

7. Write an algorithm to compute the antecedent sets $A(v)$, given a matrix representation of a digraph.

8. The following list shows a set of computer science courses and their prerequisites. How can the digraph model for the house-building tasks be modified to accommodate the need to represent *or*? What if we had a course that had two or more prerequisites?

Course	Prerequisites
CS101	None
CS102	CS101
CS150	None
CS201	CS102
CS202	CS201
CS203	CS201
CS204	CS203
CS301	CS202 or CS204

9. The table of house-building tasks is expanded in Table 6.4 to include estimates of the time it takes to complete each task.

TABLE 6.4
Estimated Time for House-Building Tasks

Task	Estimated completion time
1. Make foundation	10 days
2. Add subflooring	3 days
3. Frame walls/roof	12 days
4. Add siding	3 days
5. Install windows	2 days
6. Partition rooms	6 days
7. Shingle roof	2 days
8. Install wiring	4 days
9. Install plumbing	5 days
10. Install interior doors	6 days
11. Paint interior walls/ceiling	7 days
12. Lay carpeting	2 days

a. What is the minimum amount of time in which all the tasks can be completed?
b. Find a path through the digraph of Figure 6.4 whose total weight is the value found in (a). This path is called the **critical path** in a weighted digraph. What happens if the completion time for any of the tasks along this path increases?
c. What happens to the critical path if the estimated time to complete the wiring increases from 4 to 5 days? from 4 to 8 days?

6.2   Path Problems in Digraphs

The idea of a path in a digraph has many interpretations, depending on the application being modeled by the digraph. For instance, if the digraph is a model of a communications network, the paths in the graph show possible routings for messages between different distributed computers. There are three issues that arise in trying to apply the model to reality: establishing the *existence* of paths between two vertices, finding the *shortest* path between a pair of vertices, and counting the *number of paths* between two vertices. We can address all three issues best with algorithms. We present detailed discussions of the first two, and briefly indicate a method for solving the third in the case of an acyclic digraph.

Existence of Paths

A solution to the problem of determining which vertices are connected by paths to other vertices can be stated simply in terms of the logical matrix representing the edges of the digraph.

Let $G = (V, E)$ be a digraph. E is a specification of paths of length 1. Recalling how we defined composition of relations, we can conclude that the composition of E with itself will give us paths of length 2, and the composition of this result with E again will give us paths of length 3, and so on. Then the obvious thing to do is to proceed with an induction argument to find all paths. Observe that the process we have described is identical to the process of computing the transitive closure of E.

Define the kth powers of E as follows: Let $E^1 = E$, and for $k = 2, 3, \ldots$.

$$E^k = E^{k-1} \circ E$$

Theorem 6.3 summarizes this idea.

Theorem 6.3

Let $G = (V, E)$ be a digraph. Let $v_1, v_2, \ldots, v_n$ be the vertices. For each $i, j \in S_n$, there is a path from v_i to v_j in G if and only if there exists $k = 1, 2, 3, \ldots$ such that

$$v_i E^k v_j$$

Hence, $E^* = E \cup E^2 \cup E^3 \cup \ldots$.

$\bullet \qquad \bullet \qquad \bullet$

The first step in making the computation more tractable is to realize that not all powers of E need to be computed. In fact, if G has n vertices, then any path of length greater than n must contain repeated vertices (remember the Pigeonhole Principle?), which means that the digraph must contain a cycle. Any cycles in the path can be deleted until a path of length less than or equal to n can be obtained. Figure 6.8 shows how this reduction works. Hence, we obtain the result of Theorem 6.4.

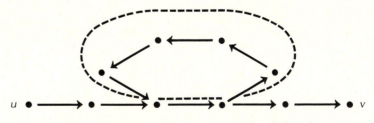

Figure 6.8. Deleting cycles to reduce path length

Theorem 6.4

Let $G = (V, E)$ be a digraph, and let $n = |V|$.

$$E^* = E^1 \cup E^2 \cup \ldots \cup E^n$$

$\bullet \qquad \bullet \qquad \bullet$

Given that E has adjacency matrix M and that E^* has adjacency matrix M^*, we can see how to compute M^* from M by taking successive products of M with itself. By recognizing that the powers of the adjacency matrix should be combined with the **or** operation, we obtain the formula

$$M^* = M^1 \text{ or } M^2 \text{ or } \ldots \text{ or } M^n \qquad (6.1)$$

The matrix M^* is called the **reachability matrix** of the digraph.

EXAMPLE 6.6

Find the reachability matrix for the digraph in Figure 6.9.

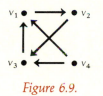

Figure 6.9.

Solution

M^1 is the matrix for the digraph.

$$M^1 = \begin{bmatrix} F & T & F & F \\ F & F & T & F \\ T & F & F & F \\ T & F & T & F \end{bmatrix}$$

The successive powers of M, and the digraph for each, are pictured in Figures 6.10 through 6.13. You should check the matrix multiplications against the digraphs in each case to be sure that each M^k gives all paths of length k.

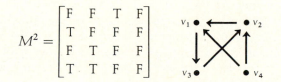

$$M^2 = \begin{bmatrix} F & F & T & F \\ T & F & F & F \\ F & T & F & F \\ T & T & F & F \end{bmatrix}$$

Figure 6.10. M^2 and its digraph

$$M^3 = \begin{bmatrix} T & F & F & F \\ F & T & F & F \\ F & F & T & F \\ F & T & T & F \end{bmatrix}$$

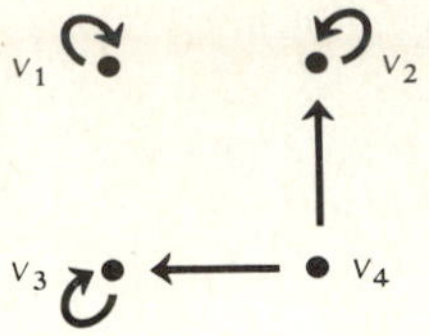

Figure 6.11. M^3 and its digraph

$$M^4 = \begin{bmatrix} F & T & F & F \\ F & F & T & F \\ T & F & F & F \\ T & F & T & F \end{bmatrix}$$

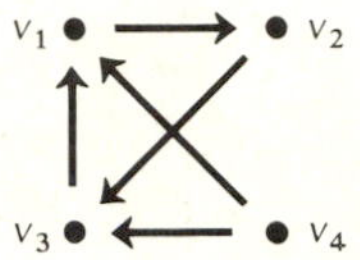

Figure 6.12. M^4 and its digraph

The reachability matrix is obtained by combining, in each row and column entry, the corresponding entries of M^1 through M^4. The result is given in the matrix M^*, shown in Figure 6.13. This result agrees with paths found by an inspection of the digraph in the figure.

$$M^* = \begin{bmatrix} T & T & T & F \\ T & T & T & F \\ T & T & T & F \\ T & T & T & F \end{bmatrix}$$

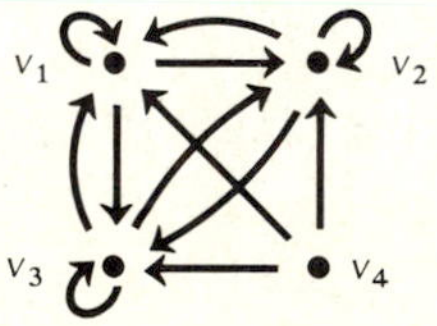

Figure 6.13. M^* and its digraph

The computation of M^* is fairly laborious, even for a 4-vertex graph. We can make an estimate of the complexity of the calculation. In Equation 6.1, there are $n - 1$ matrix products to be computed. Each matrix has n^2 entries, and for each entry there are n **and** operations to be performed. By the Fundamental Principle of Counting, there are approximately n^4 **and** operations to be performed in calculating M^*. We say that the calculation is of order n^4. This means that if the number of vertices doubles, the magnitude of the computation grows by a factor of 16. This type of complexity calculation is an important consideration in comparing two algorithms that perform the same task.

Warshall's Algorithm

There are other algorithms for computing the reachability matrix. The best known of these algorithms is **Warshall's algorithm**. As we shall see, Warshall's algorithm is a power of n more efficient than the calculation based on Equation (6.1). The approach taken in the algorithm is somewhat more difficult to understand at first reading, but more efficient algorithms are often more complex.

Let $G = (V, E)$ be a digraph. Define the family of relations $W^{(k)}$ as follows: Let $v_1, v_2, \ldots, v_n$ be a labeling of the vertices.

1. For $u, v \in V$, we say $uW^{(1)}v$ if and only if uEv or there exists a path from u to v using v_1 as an intermediate vertex.
2. For $u, v \in V$, we say $uW^{(k+1)}v$ if and only if $uW^{(k)}v$ or there exists a path from u to v using some subset of the vertices $v_1, \ldots, v_{k+1}$ as intermediate vertices.

It follows from the definition that $E^* = W^{(n)}$. Thus, if we can compute $W^{(n)}$ effectively, we will have an alternate to the tedious calculation above. If we consider the relations $W^{(k)}$ as sets, we see that $W^{(1)} \subseteq W^{(2)} \subseteq \ldots W^{(n)}$, so that each successive relation contains the preceding relations.

EXAMPLE 6.7

Use the definitions to compute $W^{(1)}$ and $W^{(2)}$ for the digraph in Figure 6.9.

Solution

There are 16 entries to be computed in the matrix for $W^{(1)}$. At the outset, the logical matrix contains the same entries as the matrix for E. Once an entry becomes *true* (T), it will remain *true*. Since v_3, v_1, v_2 is a path in G, we conclude that $v_3 W^{(1)} v_2$. Similarly, $v_4 W^{(1)} v_2$. There are no other additions to the matrix. Thus the matrix and its pictorial representation are as shown in Figure 6.14.

Figure 6.14. W^1 and its digraph

Notice that $W^{(1)}$ is not the same as E^1. For $W^{(2)}$, we start with all the paths found in $W^{(1)}$, and look for paths that use v_1 and v_2 as intermediate vertices. The only such paths that have not been found are v_1, v_2, v_3; and v_3, v_1, v_2, v_3; and v_4, v_1, v_2, v_3. Hence, the matrix and its pictorial representation are as shown in Figure 6.15.

$$W^{(2)} = \begin{bmatrix} F & T & T & F \\ F & F & T & F \\ T & T & T & F \\ T & T & T & F \end{bmatrix}$$

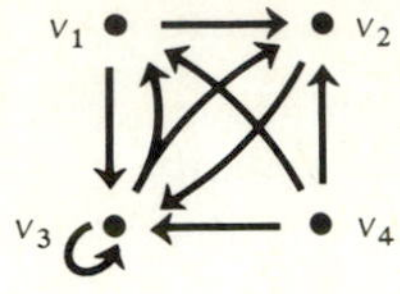

Figure 6.15. W^2 and its digraph

Observe the differences between $W^{(1)}$ and $W^{(2)}$ and between E^1 and E^2. In particular, observe that the only positions that can change in going from $W^{(1)}$ to $W^{(2)}$ are those that contain F; this means that all the edges appearing in the picture of $W^{(1)}$ must also appear in the picture of $W^{(2)}$.

Warshall's Algorithm

{Warshall's algorithm computes the reachability}
{matrix $M*$ of a digraph with adjacency matrix }
{M. At the end of the calculation, the result is   }
{contained in the matrix W.   }

begin
    $W \leftarrow M$
    for $k = 1$ **to** n **do**
        for $i = 1$ **to** n **do**
            for $j = 1$ **to** n **do**
                $W(i, j) = W(i, j)$ **or** $(W(i, k)$ **and** $W(k, j))$
end.

The algorithm is quite simple to state, as you can see. Showing that it computes the reachability matrix is another matter. We prove the correctness of the algorithm by induction on k. The first thing to notice about the algorithm is that the matrix W is built *in place*; that is, W is set to M initially, and each pass through the outer loop (indexed by k) generates $W^{(k+1)}$ from $W^{(k)}$. The two inner loops compute the elements $W(i, j)$ for $i, j = 1$ to n.

When $k = 1$, the value left in $W(i, j)$ will be either the value of $W(i, j)$ or the value of $(W(i, 1)$ **and** $W(1, j))$. The latter term is *true* if and only if there is a path from v_i to v_j through v_1. Hence, after the first pass through the loop W equals $W^{(1)}$.

Assume that the kth pass through the outer loop has generated $W^{(k)}$. The next time through the outer loop, the value left in $W(i, j)$ is either the old value or it is $(W(i, k + 1)$ **and** $W(k + 1, j))$. If the latter value is *true*, there is a path from v_i to v_{k+1} and from v_{k+1} to v_j using only v_1 through v_k as intermediate vertices. Hence, there is a path from v_i to v_j using the first $k + 1$ vertices. Therefore, after pass $k + 1$, W will equal $W^{(k+1)}$.

The assignment statement within the innermost loop will be executed n^3 times. (Why?) Thus Warshall's algorithm is much more efficient, especially for large n, than our previous method for computing M^*. Moreover, Warshall's algorithm builds the reachability matrix in place, so it uses less storage than the algorithm using conventional matrix multiplication, which requires at least one working matrix. Such improvements become important when dealing with large matrices in a programming context.

EXAMPLE 6.8

Trace Warshall's algorithm on the graph in Figure 6.9.

Solution

There are two ways to approach the trace. First, we could simply use the pictorial representation and show the successive digraphs that are constructed by each pass through the main loop. Second, we could simulate the way a computer might perform the logical operations required. We choose the former because it allows us to capitalize on the geometric form of the graph. Of course, we have cheated a little since Example 6.7 gives us the values of $W^{(1)}$ and $W^{(2)}$. For completeness, we repeat those here.

$$
W^{(1)} = \begin{array}{c} v_1 \\ v_2 \\ v_3 \\ v_4 \end{array}
\begin{bmatrix}
F & T & F & F \\
F & F & T & F \\
T & T & F & F \\
T & T & T & F
\end{bmatrix}
\qquad
W^{(2)} = \begin{array}{c} v_1 \\ v_2 \\ v_3 \\ v_4 \end{array}
\begin{bmatrix}
F & T & T & F \\
F & F & T & F \\
T & T & T & F \\
T & T & T & F
\end{bmatrix}
$$

Picking up the calculation at this point, we find it is possible to get from any of the vertices v_1, v_2, and v_3 to any of the other vertices through the cycle involving these three vertices. Thus $W^{(3)}$ has the matrix shown in Figure 6.16.

$$
W^{(3)} = \begin{bmatrix}
T & T & T & F \\
T & T & T & F \\
T & T & T & F \\
T & T & T & F
\end{bmatrix}
$$

Figure 6.16. $W^{(3)}$ and its digraph

At the final step (that is, $k = 4$) there are no further paths to be found, since vertex 4 is a source vertex. Hence, $W^{(4)}$ is as shown in Figure 6.17.

$$W^{(4)} = \begin{bmatrix} T & T & T & F \\ T & T & T & F \\ T & T & T & F \\ T & T & T & F \end{bmatrix}$$

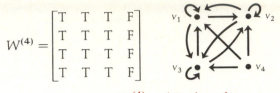

Figure 6.17. $W^{(4)}$ and its digraph

SELF TEST 3

1. Find the reachability matrix for the digraph whose adjacency matrix is given below.

$$\begin{bmatrix} T & F & F & F & T \\ T & T & T & F & F \\ T & F & F & T & F \\ F & F & T & T & T \\ F & T & F & F & F \end{bmatrix}$$

Shortest Paths

The second issue relating to paths in digraphs is finding the *best* path between two vertices. The simplest form of the problem is to compute the path between two vertices that uses the fewest edges (which is sometimes called a *minimum hop path*). We generalize somewhat and allow the edges of the digraph to be assigned **weights** that have nonnegative values. We call such a digraph a **weighted digraph**. The **weight of a path** is then the sum of the weights of the edges in the path. The **shortest path** between two vertices is the path with least total weight. The length of the shortest path between two vertices u and v is called the **distance** from u to v. If there is no path from u to v, the distance is said to be *infinity* (∞). To obtain the minimum hop path, all we have to do is assign each edge a weight of 1.

Weighted digraphs play a significant role in modeling situations such as transportation networks (where the vertices represent terminals and the weights represent distances or costs) and work scheduling algorithms (where the weights may represent time).

There are many algorithms for finding shortest paths in digraphs. The one we present here was developed by Dijkstra. We begin with a thorough analysis of a particular case, and then give the general algorithm. If you look at the weighted digraph in Figure 6.18, you will discover without difficulty that the shortest path from A to F

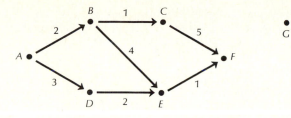

Figure 6.18. A weighted digraph

has weight 6, and it consists of the vertices A, D, E, F. In general, the digraphs are more complicated, and one does not always get such a global view of them. The digraph in Figure 6.18 could be represented by a 7×7 matrix w whose entries have one of three values:

$$w(x, y) = \begin{cases} 0 & \text{if } x = y \\ \infty & \text{if } (x, y) \text{ is not an edge} \\ \text{the weight of the edge } (x, y), \text{ otherwise} \end{cases}$$

the symbol ∞ is read "infinity," and represents a value greater than any finite real value. This matrix w is called the **weight matrix** for the weighted digraph. The weight matrix for the graph in Figure 6.13 is given below.

$$
\begin{array}{c|ccccccc}
 & A & B & C & D & E & F & G \\
\hline
A & 0 & 2 & \infty & 3 & \infty & \infty & \infty \\
B & \infty & 0 & 1 & \infty & 4 & \infty & \infty \\
C & \infty & \infty & 0 & \infty & \infty & 5 & \infty \\
D & \infty & \infty & \infty & 0 & 2 & \infty & \infty \\
E & \infty & \infty & \infty & \infty & 0 & 1 & \infty \\
F & \infty & \infty & \infty & \infty & \infty & 0 & \infty \\
G & \infty & \infty & \infty & \infty & \infty & \infty & 0
\end{array}
$$

With this representation, it is no longer clear what the paths are, much less which one is the best. For large digraphs, however, the representation will likely be such a matrix, or its equivalent as lists of edges and weights.

We will describe the sequence of steps necessary to find the lengths of the shortest paths from A to each of the other vertices in Figure 6.18. Then we will present the full algorithm.

For each vertex x in the digraph, let $d[x]$ represent the distance from A to x. First we will assign appropriate initial values to each $d[x]$. The fundamental idea behind the procedure is to traverse the vertices of the graph, perhaps visiting some vertices more than once, and to improve the computed value of $d[x]$ as we go. When we are sure we have found the best path to a vertex, we say it is *marked* and remove it from further consideration. For each step described below, a row of Table 6.5 shows the vertex to be marked, the current values of $d[x]$; and the remaining unmarked vertices.

TABLE 6.5

Summary of Steps in Finding Distances from A

| Step | Vertex to be marked | Distance to vertex | | | | | | | Unmarked vertices |
		A	B	C	D	E	F	G	
0	A	0	2	∞	3	∞	∞	∞	B, C, D, E, F, G
1	B	0	2	3	3	6	∞	∞	C, D, E, F, G
2	D	0	2	3	3	5	∞	∞	C, E, F, G
3	C	0	2	3	3	5	8	∞	E, F, G
4	E	0	2	3	3	5	6	∞	F, G
5	F	0	2	3	3	5	6	∞	G

Step 0 (Initialization).   We set the value of $d[x]$ to 0 when $x = A$, to ∞ when (A, x) is not an edge, and to the weight of the edge, $w(A, x)$, when (A, x) is an edge. These values are shown in the first row of Table 6.5. Since A is closest to itself, we mark it. All other vertices are unmarked at this point.

Step 1.   The next vertex to be marked is B because it is the unmarked vertex closest to A. Then we remove B from the list of unmarked vertices. For each of the other unmarked vertices y such that (B, y) is an edge, we check to see if the new path to y through B is shorter than the previously computed path. In this instance, we are able to reach C and E for the first time, so we reduce their distances from ∞ to 3 and from ∞ to 6, respectively.

Step 2.   Of the remaining unmarked vertices, D and C are tied for closest. We arbitrarily pick D to mark first. Because E can be reached through D with a path of length 5, we reduce the entry under E from 6 to 5.

Step 3. C is still the closest vertex to A, so we will mark it next. Since F is reachable through C for the first time, we reduce its distance to 8.

Step 4.   We mark E next, which causes us to reduce the distance to F from 8 to 6.

Step 5.   Then we mark F, which leaves no more vertices that can be reached. The process must end with G left unmarked.

In the sample calculation, we have found only the lengths of the shortest paths, not the paths themselves. This step will be added in the main algorithm. At each step in the above procedure we identified the unmarked vertex x closest to A. This is equivalent to saying that x satisfies the condition

$$d[x] = \min\{d[z] \text{ such that } z \text{ is unmarked and } d[z] < \infty\}$$

We mark the vertex x, and for each y for which (x, y) is an edge, we determine whether the path to y through x is shorter than any previously computed path. This means that we let

$$d[y] = \min\{d[y], d[x] + w(x, y)\}$$

At this point, we can update the actual path to y as well. The procedure must terminate when the set of unmarked vertices at a finite distance from A is empty. We are now ready to state Dijkstra's algorithm.

Dijkstra's Algorithm

{Let (V, E) be a weighted digraph. Let A be a vertex.}
{The objective is to find, for each vertex x, the   }
{distance, $d[x]$, from A to x and the shortest path }
{from A to x. We assume that for y and z in V,   }
{$w(y, z)$ is defined by   }

$$\{w(y, z) = \begin{cases} 0, \text{ if } y = z, \\ \infty, \text{ when } (y, z) \notin E \\ \text{weight of the edge from } y \text{ to } z, \text{ otherwise} \end{cases}$$

{Define $pathto(z)$ to be the list of vertices in the   }
{shortest current path from A to z.   }

```
begin
    for each x ∈ V do
        begin
            d[x] ← w(A, x)
            pathto(x) ← A
        end
    Mark vertex A
    while unmarked vertices remain that
            are a finite distance from A do
        begin
            x ← one of the unmarked vertices whose distance
                        from A is minimal
            Mark vertex x
            for each unmarked y ∈ V such that (x, y) ∈ E do
                begin
                    d' ← d[x] + w(x, y)
                    if d' < d[y] then
                        begin
                            d[y] ← d'
                            pathto(y) ← pathto(x), x
                        end
                end
        end
end.
```

SELF TEST 4

1. Given the weight matrix below, use Dijkstra's algorithm to find for each vertex the shortest path from A to the vertex.

$$
\begin{array}{c@{\quad}ccccccc}
 & A & B & C & D & E & F & G \\
A & 0 & \infty & \infty & 5 & 3 & 2 & \infty \\
B & \infty & 0 & \infty & \infty & \infty & \infty & \infty \\
C & 4 & 2 & 0 & \infty & 3 & 4 & 6 \\
D & \infty & \infty & 7 & 0 & 2 & 5 & 9 \\
E & 8 & 2 & 1 & \infty & 0 & 3 & \infty \\
F & 1 & 7 & \infty & 9 & 2 & 0 & \infty \\
G & \infty & \infty & \infty & \infty & \infty & \infty & 0
\end{array}
$$

Number of Paths

So far we have two variations on path problems in digraphs: Warshall's algorithm gives an efficient means for computing the reachability matrix of a digraph, which specifies whether two vertices are connected by a path; and Dijkstra's algorithm determines the *best* path between two vertices in a weighted digraph. A third problem is to determine *how many* paths there are between two vertices. We restrict our attention to acyclic digraphs, because vertices in a cycle will be connected by an infinite number of paths. To make the computation, we will make use of one more matrix representing the digraph.

Let $G = (V, E)$ be a digraph. Let $V = \{v_1, v_2, \ldots, v_n\}$. Define the $n \times n$ matrix N by

$$
N(i, j) = \begin{cases} 0, & \text{if} \quad (v_i, v_j) \notin E \\ 1, & \text{if} \quad (v_i, v_j) \in E \end{cases}
$$

$N(i, j)$ contains the number of paths of length 1 from vertex v_i to v_j. As you might suspect, the next step is an inductive step that claims the number of paths of length k between v_i and v_j is given by the ijth entry of the kth "power" of N. The only hitch is that we haven't defined the *power* of a matrix such as N. The pattern used here has already been used in defining the logical matrix product. Suppose N_1 is an $n \times m$ matrix of 0's and 1's, and N_2 is an $m \times p$ matrix of 0's and 1's. These correspond to relations R_1 and R_2, as shown in Figure 6.19.

Finding the composition of these two relations is like finding two-hop paths in a digraph, but these are spread out laterally. Let $i \in S_n$ and $j \in S_p$. There is a two-hop path from v_i to v_j if and only if there exists $k \in S_m$ such that $N(i, k) = 1$ and $N(k, j) = 1$. The number of two-hop paths will be the number of all such instances. Based on

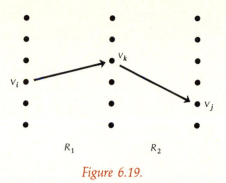

Figure 6.19.

this observation, we define $N_1 \times N_2$ to be the $n \times p$ matrix N_3 given by

$$N_3(i, j) = \sum_{k=1}^{m} N_1(i, k) \cdot N_2(k, j) \qquad (6.2)$$

Now we can define powers of an $n \times n$ real matrix N inductively. We define

$$N^1 = N$$

$$N^k = N \times N^{k-1}, \text{ for } k \geq 2$$

EXAMPLE 6.9

Compute the first two powers of the matrix N.

$$N = \begin{bmatrix} 0 & 0 & 1 & 0 \\ 1 & 0 & 1 & 0 \\ 0 & 0 & 0 & 0 \\ 1 & 1 & 0 & 0 \end{bmatrix}$$

Solution

We show the digraph corresponding to N in Figure 6.20. N^1 is just N, so we compute N^2. For instance, to compute $N^2(2, 3)$, we take the second row and the third column

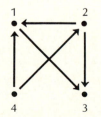

Figure 6.20. The digraph for the matrix N

of N and arrange them as shown below:

$$1 \quad 0 \quad 1 \quad 0$$
$$1 \quad 1 \quad 0 \quad 0$$

Multiplying each of the four pairs of entries and summing the results yields 1. If this is true, then there should be one two-hop path from v_2 to v_3. We see by inspection of the digraph that there is one such path, namely v_2, v_1, v_3. Fifteen more times through the process yields N^2:

$$N^2 = \begin{bmatrix} 0 & 0 & 0 & 0 \\ 0 & 0 & 1 & 0 \\ 0 & 0 & 0 & 0 \\ 0 & 0 & 2 & 0 \end{bmatrix}$$

The kth power of the matrix N gives the number of paths of length k. To complete the original task of finding how many paths there are between two vertices, we introduce the real $n \times n$ matrix NP. $NP(i, j)$ should be the number of paths of any length between the ith and jth vertices of the acyclic digraph. This situation is similar to that with the powers of the logical matrix and the reachability matrix. Continuing the analogy, it should not surprise you that NP can be computed as

$$NP(i, j) = \sum_{k=1}^{n} N^k(i, j) \tag{6.3}$$

Exercises 6.2

1. Trace Warshall's algorithm on the following graphs. In each case, draw the graph and verify your results by inspection.

a. $\begin{bmatrix} F & T & F & F \\ F & F & T & F \\ F & F & F & T \\ T & F & F & F \end{bmatrix}$

b. $\begin{bmatrix} F & F & T & T & F \\ F & F & T & T & T \\ F & F & F & T & F \\ T & T & T & F & T \\ F & F & F & F & F \end{bmatrix}$

2. Modify Warshall's algorithm by inserting an **if** test so the innermost loop (indexed by j) can be omitted when $W(i, k) = F$. Why does this work?

3. Suppose you want to compute the reachability matrix for a digraph with $n = 10$. How many operations are required for the conventional algorithm? For Warshall's algorithm? How do these numbers change when $n = 100$ and when $n = 1000$?

4. *Variations on connectivity for digraphs.* A digraph $G = (V, E)$ is said to be *unilaterally connected* if for all distinct vertices $u, v \in E$ there is either a path from u to v or a path from v to u. G is said to be *strongly connected* if for all distinct vertices $u, v \in V$ there is a path from u to v and a path from v to u. How can you use the reachability matrix to determine whether a graph is unilaterally or strongly connected?

5. How can you use the reachability matrix of a digraph to detect cycles?

6. Trace Dijkstra's algorithm for the graph in Figure 6.21, finding the shortest path from each of the following vertices.

   a. vertex A
   b. vertex C

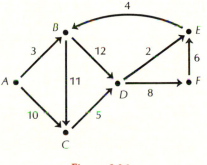

Figure 6.21.

7. Compute the real matrix products N^1, N^2, N^3, and N^4 for the graph in exercise 1(a).

Routing in Communications Networks

The advent of small, relatively inexpensive computers has enabled computer users to work with systems that solve problems at a local level. There is still a need, however, to share information and programs on a larger geographic scale, requiring interconnections among systems within a building, throughout an industrial plant, or across a nation. Computer systems interconnected by communications media are called **computer networks**.

Uses of a Computer Network

A computer network has many uses, including:

- Providing remote access to large data banks, which cannot be maintained easily on small, individual systems
- Providing mail services
- Providing access to large programs on a time-sharing basis

Most interconnections within a narrow geographical range are attained through a *local area network* (LAN). A typical arrangement of a LAN is shown in Figure 6.22. The various computers and/or terminals communicate with one another over a shared *link*. Networks covering broader ranges typically use telephone lines or satellite channels to transfer messages from place to place.

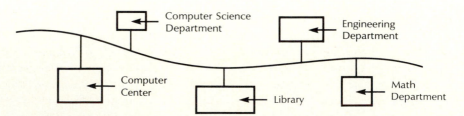

Figure 6.22. A local area network (LAN)

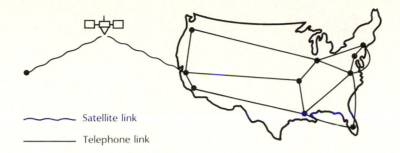

Satellite link

Telephone link

Figure 6.23. Graph of a communications network

Figure 6.23 illustrates the interconnection of computers in a nationwide network as a graph. The computers are the vertices, called *nodes,* and the communications channels are the edges, called *links.* The connections in this graph are said to be *point-to-point* because each computer can communicate directly only with its neighbors, relying on a *store-and-forward* message service to exchange messages with computers not directly connected to it. This delivery service can transfer a message from one place to another by giving the message to one of its neighbors, which in turn gives it to one of its neighbors, and so forth until the message eventually arrives at its destination.

Networks and Graph Models

The design and implementation of a new computer network is a nontrivial hardware and software engineering task. There are many problems in the design of point-to-point computer networks that can be solved by using models from graph theory [1]. To show the degree of connectivity, a network can be modeled as an undirected graph. The reliability of a point-to-point network is increased if for every pair of nodes there are multiple paths between them. Obviously, the most reliable point-to-point network would have a completely connected graph, but the cost of this solution is generally prohibitive. Thus, a network designer wants to obtain maximum redundancy of paths and minimum path lengths within a certain cost/performance range.

Modeling a point-to-point network as a digraph can help with the problems of *capacity assignment, message routing,* and *flow control.* Here we will concentrate on the problem of message routing.

The Routing Problem

Once a point-to-point network has been designed and installed, the problem of how to route messages between nonadjacent nodes remains. There are two strategies that we can employ. A *static routing* strategy uses information on link capacities to determine a set of fixed routes between nodes. This is a simple strategy to implement because the initial configuration of the network can be modeled as a weighted digraph (with the link capacities as the weights) and then a best-path algorithm such as

Dijkstra's algorithm can be used. However, a static routing procedure tends to break down when there are failures in links or network nodes. Moreover, it does not take into account changing patterns of message traffic. Over a period of time, changes in traffic patterns can lead to increased loads on particular links, causing severe delays in message services when the traffic demand on a link is close to its capacity. For these reasons, a *dynamic routing* procedure is more desirable, although it is more difficult to implement. A dynamic routing procedure allows periodic recomputation of best paths between nodes based on changing network connections and changing conditions.

Facets of a Dynamic Routing Procedure

There are three major subtasks in designing a dynamic routing procedure:

1. Decide on appropriate performance statistics and make provisions for measuring them during network operation. Typically, the most relevant performance statistic is the *average link delay*; that is, the amount of time (usually measured in milliseconds) a message must wait before it can be sent over a given link. Because of changing traffic patterns, this calculation must be run periodically. When the delay changes, the weight assigned to the link in the digraph is either reduced or increased, leading to possible changes in the best paths.

2. Devise the protocol for distributing changes in link information. The protocol is the set of rules used by individual nodes to decide when to transmit new information and how to respond to information received. It must be designed so that transmitting the link change data doesn't use too much of the network's link capacity.

3. Design the algorithm to generate new routes. Typically, each node runs its own route calculation rather than relying on one centralized computer. Thus, the algorithm is distributed, with all nodes cooperating in the calculation. Each node also maintains its own set of routing tables for determining how to send messages to a nonadjacent node.

In the simple 7-node network shown in Figure 6.24, each link is represented by an arrow and a link delay number for each direction of message traffic on the link. For instance, the link delay from 1 to 2 is greater than from 2 to 1, presumably because there is more message traffic in the direction from 1 to 2.

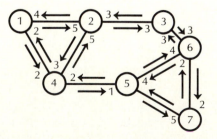

Figure 6.24. A simple 7-node network

The routing calculation can be performed for each node separately using a best-path algorithm such as Dijkstra's algorithm. One major nationwide network, the ARPANET[2], uses a variation of Dijkstra's algorithm that allows incremental changes to the best-path information without having to recompute paths to nodes not affected by network changes. Once the calculation is run, the node has information about best paths from itself to all other nodes. These can be stored in a tree, with the "home" node as the root. The tree of best paths for node 1 in Figure 6.24 is shown in Figure 6.25.

To actually forward messages, a given node needs only to have a table that indicates the neighbor to which to pass the message for each destination. Then it can rely on the neighbor to do the same thing. Here is the routing table for node 1:

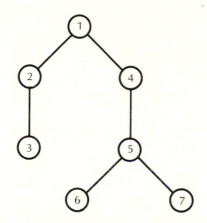

Figure 6.25. Tree of shortest paths from node 1

Destination	2	3	4	5	6	7
Next node	2	2	4	4	4	4

Thus, when forwarding a message, node 1 in Figure 6.19 can forward the message either to node 2 or to node 4. The above table tells it which node to choose. Thereafter, the neighbor has the responsibility for forwarding the message, if necessary.

Here is the routing table for node 2:

Destination	1	3	4	5	6	7
Next node	1	3	4	4	3	3

Let's look at how nodes 1 and 2 cooperate to get a message from node 1 to node 3: When node 1 has a message whose destination is node 3, it looks in its routing table, sees that the best way to get to node 3 is through node 2, and then forwards the message to node 2. Node 2 receives the message and sees that node 2 is not the destination, so it looks at its own routing table and sees that it can forward the message to node 3. Node 3 receives the message, sees that the message is destined for itself, and does not forward it further.

In the exercises below, we give you a chance to examine how changing the delay information changes the shortest path tree and routing tables.

Exercises 6.3

1. How do the shortest path trees and routing tables in nodes 1 and 2 change if the delay from node 2 to node 4 increases from 3 to 6? What if it decreases from 3 to 1?

2. How do the shortest path trees and routing tables in nodes 1 and 2 change if the link between nodes 5 and 6 is removed from the network?

REFERENCES

[1] A. S. Tanenbaum, *Computer Networks* (Englewood Cliffs, N.J.: Prentice-Hall, 1981).

[2] J. M. McQuillan et. al., "The New Routing Algorithm for the ARPANET," *IEEE Transactions for Communications*, Vol. Com−28, May 1980, 711–719.

SUMMARY

1. A **directed graph**, or **digraph**, is a pair (V, E), where V is a finite set of elements called **vertices**, and E is a relation on V. The elements of E are called the **edges** of the digraph. For any pair of vertices $u, v \in V$, the set of edges E will contain at most one edge (u, v) from u to v, and at most one edge (v, u) from v to u.

2. If $(u, v) \in E$, we will say that u **precedes** v, or is an **antecedent** of v.

3. Given a graph $G = (V, E)$ and a vertex $v \in V$, the **indegree** of v is the number of edges arriving at v. That is, the indegree of v is given by

$$\text{indegree}(v) = \left|\{u \in V : (u, v) \in E\}\right|$$

4. The **outdegree** of a vertex $v \in V$ is the number of edges leaving v. Thus the outdegree of v is given by

$$\text{outdegree}(v) = \left|\{u \in V : (v, u) \in E\}\right|$$

5. A vertex with indegree 0 is called a **source**, and a vertex with outdegree 0 is called a **sink**.

6. **Theorem 6.1**. Let $G = (V, E)$ be a digraph. Let M be the $n \times n$ adjacency matrix for E.
   1. $\sum \{\text{indegree}(v) : v \in V\} = \sum \{\text{outdegree}(v) : v \in V\}$.
   2. For all $v_i \in V$, $\text{indegree}(v_i) =$ the number of T's in the ith column of M.
   3. For all $v_i \in V$, $\text{outdegree}(v_i) =$ number of T's in the ith row of M.

7. A **path** from vertex u to vertex v is a sequence of vertices $v_0, v_1, \ldots, v_k$ such that $v_0 = u$, $v_k = v$, and for $i = 1, 2, \ldots, k$, the edge $e_i = (v_{i-1}, v_i) \in E$. Alternatively, the path can be defined as the sequence of edges $e_1, e_2, \ldots, e_k$. k is called the **length** of the path.

8. A **cycle** (circuit) is a path from a vertex to itself. When a digraph does not contain a cycle, it is said to be **acyclic**.

9. Let $G = (V, E)$ be a digraph. Let $\{v_1, v_2, \ldots, v_n\}$ be a labeling of the vertices. The labeling is said to be a consistent labeling if

$$(v_i, v_j) \in E \Rightarrow i < j$$

10. An important property of an acyclic digraph is that it must have at least one source vertex.

11. **Theorem 6.2**. Let $G = (V, E)$ be a digraph. G has a consistent labeling if and only if G is acyclic.

12. **Theorem 6.3**. Let $G = (V, E)$ be a digraph. Let $v_1, v_2, \ldots, v_n$ be the vertices. For each $i, j \in S_n$, there is a path from v_i to v_j in G if and only if there exists $k = 1, 2, 3, \ldots$ such that

$$v_i E^k v_j$$

Hence, $E^* = E \cup E^2 \cup E^3 \cup \ldots$.

13. **Theorem 6.4**. Let $G = (V, E)$ be a digraph and let $n = |V|$.

$$E^* = E^1 \cup E^2 \cup \ldots \cup E^n$$

14. The matrix M^*, defined below, is called the **reachability matrix** of the digraph.

$$M^* = M^1 \text{ or } M^2 \text{ or } \ldots \text{ or } M^n$$

15. A **weighted digraph** is a digraph with nonnegative values assigned to its edges.

16. The **weight of a path** is the sum of the weights of the edges in the path.

17. The **shortest path** between two vertices is the path with least total weight.

18. The length of the shortest path between vertices u and v is called the **distance** from u to v. If there is no path from u to v, the distance is said to be infinity (∞).

19. Let $G = (V, E)$ be an acyclic digraph. Let $V = \{v_1, v_2, \ldots, v_n\}$. Define the $n \times n$ matrix N by

$$N(i, j) = \begin{cases} 0, & \text{if } (v_i, v_j) \notin E \\ 1, & \text{if } (v_i, v_j) \in E \end{cases}$$

$N(i, j)$ contains the number of paths of length 1 from vertex v_i to v_j.

20. The number of paths of length k between v_i and v_j is given by the ijth entry of the kth "power" of N.

21. Modeling a point-to-point network as a digraph can help with the problems of capacity assignment, message routing, and flow control.

REVIEW PROBLEMS

1. Let $G = (V, E)$ be the digraph pictured in Figure 6.26.
   a. Find the adjacency matrix for G.
   b. Find the indegree and outdegree of each vertex.
   c. Does G contain any sources or sinks? If not, find a cycle in G.

2. Let G be the digraph with adjacency matrix given below.

	a	b	c	d	e	f	g	h
a	T	F	F	F	T	T	F	F
b	T	F	F	F	F	F	F	T
c	F	F	F	F	F	F	F	F
d	T	F	T	F	T	F	T	F
e	F	F	F	F	F	T	T	T
f	F	F	T	T	F	F	F	F
g	F	F	T	T	F	F	F	F
h	F	F	F	F	T	T	F	F

  a. Sketch the digraph.

  b. How can you determine the sources and sinks in the digraph just by looking at the adjacency matrix?

3. a. Apply the *Topological Sort* algorithm to the digraph in problem 2. Does the digraph have a consistent labeling?

4. A digraph $G = (V, E)$ is said to be *circular* if for all $u, v, w \in V$, $(u, v) \in E$ and $(v, w) \in E$ implies that $(w, u) \in E$.

  a. Give an example of a circular digraph with four edges on the vertices a, b, c, d.

  b. Is the digraph in Figure 6.27 circular? Why or why not?

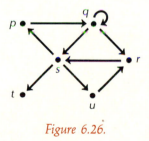

Figure 6.26.

  c. Show that a symmetric and circular digraph must be transitive.

5. Compute the reachability matrix for the digraph in Figure 6.28 by each of the following means.

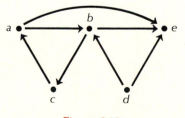

Figure 6.27.

    a. inspection
    b. Equation (6.1)
    c. Warshall's algorithm

*6. The reachability matrix is sometimes defined so that it is always reflexive; that is, the matrix is the reflexive and transitive closure of the original edge relation. Thus we would write

$$E^* = I \cup E \cup E^2 \cup \ldots$$

How does this alternate definition affect each of the following?
    a. The formula derived in Theorem 6.4
    b. The initialization step of Warshall's algorithm
    c. Our ability to use the reachability matrix to detect the existence of cycles

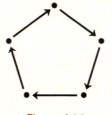

Figure 6.28.

7. Find the weight matrix for the digraph in Figure 6.29.

8. Given the digraph in Figure 6.29, use Dijkstra's algorithm to find, for each vertex, the shortest path from A to the vertex.

9. Repeat problem 8, but use vertex D as the starting point.

10. A digraph is said to be *unilaterally connected* if for each pair of vertices u and v, either there is a path from u to v or there is a path from v to u.

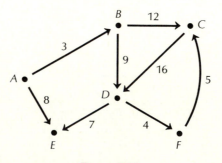

Figure 6.29.

a. Is the digraph of Figure 6.30 unilaterally connected?

*b. Show that a digraph is unilaterally connected if and only if there is a path in the graph that uses all vertices.

11. A digraph is said to be *strongly connected* if for each pair of vertices u and v there is a path from u to v and a path from v to u.

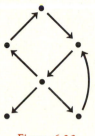

Figure 6.30.

a. Is the digraph in Figure 6.30 strongly connected?

b. Show that a digraph is strongly connected if and only if there is a path in the digraph that uses all vertices and starts and ends at the same vertex.

12. Show that an acyclic digraph with at least two vertices cannot be strongly connected. Can an acyclic digraph be unilaterally connected?

13. Another matrix used to represent a digraph $G = (V, E)$ is called the *incidence matrix*. Let $|V| = n$, and let $|E| = m$. The incidence matrix is the $n \times m$ matrix $(a_{i,j})$ given by

$$a_{i,j} = \begin{cases} 1, & \text{if the jth edge} = (v_i, v_k), \text{ for some } k \\ -1, & \text{if the jth edge} = (v_k, v_i), \text{ for some } k \\ 0, & \text{otherwise} \end{cases}$$

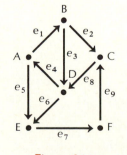

Figure 6.31.

a. Construct the incidence matrix for the digraph of Figure 6.31.

b. Show that the columns of an incidence matrix must sum to 0, and that the ith row sums to the difference of the indegree and outdegree of the ith vertex.

c. How can the incidence matrix be used to determine the sources and sinks of a digraph?

*14. Let $G = (V, E)$ be a digraph, and suppose that E is irreflexive. Suppose $\min\{\text{indegree}(v),\ \text{outdegree}(v): v \in V\} = k$. Show that G contains a cycle of length greater than or equal to k.

15. Find three distinct consistent labelings of the digraph with adjacency matrix given below.

	a	b	c	d	e	f	g	h
a	F	F	F	F	T	T	F	F
b	F	F	T	F	F	F	F	T
c	F	F	F	T	F	F	F	F
d	F	F	F	F	T	F	T	F
e	F	F	F	F	F	T	T	T
f	F	F	F	F	F	F	T	F
g	F	F	F	F	F	F	F	F
h	F	F	F	F	F	F	F	F

16. Find the number of paths of length 3 or less from vertex a to vertex f in the digraph of problem 15 by each of the following means.

a. inspection

b. computing N^1, N^2, and N^3

17. Given the weight matrix below, find the shortest paths from vertex a and vertex b.

	a	b	c	d	e	f	g	h
a	0	∞	∞	3	2	1	∞	∞
b	2	0	∞	∞	∞	∞	∞	5
c	∞	∞	0	∞	∞	∞	∞	∞
d	3	∞	3	0	8	∞	9	∞
e	∞	∞	∞	∞	0	3	5	2
f	∞	∞	2	2	∞	0	∞	∞
g	∞	∞	3	1	∞	∞	0	∞
h	∞	∞	∞	∞	7	6	∞	0

Programming Problems

18. Write a program that uses the adjacency matrix to compute the antecedent sets $A(v)$ for all vertices in a digraph.

19. Write a program that uses the adjacency matrix to compute the indegree and outdegree of all vertices, and that determines which vertices are sources and sinks.

20. Write a program to compute the powers of the adjacency matrix.

21. Rewrite the programs in problems 18 and 19 so that they use the incidence matrices instead of the adjacency matrices.

22. Write a program to convert an incidence matrix to the equivalent adjacency matrix.

23. Write a program to convert an adjacency matrix to the equivalent incidence matrix.

24. Write a program to convert the adjacency matrix to the real matrix N and to compute N^k, $k = 2, 3, \ldots, n$, where n is the number of vertices in the digraph.

SELF TEST ANSWERS

Self Test 1

1.

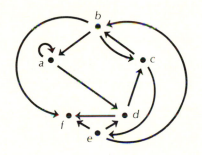

2. a, d, c, b is a path from a to b.

3. There are numerous cycles of length 5. Two particular ones are a, a, a, a, a, a and e, d, c, b, c, e.

4. The table below contains the indegrees and outdegrees of each vertex. The indegree of a vertex is the number of T's in the column corresponding to the vertex, and the outdegree is the number of T's in the row corresponding to the vertex. Observe that the sum of the indegrees equals the sum of the outdegrees.

vertex	a	b	c	d	e	f
indegree	2	2	2	2	1	3
outdegree	2	3	2	2	3	0

Self Test 2

Initially, the antecedent sets are as follows:

$$A(a) = \{\ \} \qquad A(b) = \{a, g\} \qquad A(c) = \{b, g\}$$
$$A(d) = \{c, e\} \qquad A(e) = \{g\} \qquad A(f) = \{\ \}$$
$$A(g) = \{a, f\}$$

Thus a and f are sources. Label a with 1. Then $A(b) = \{g\}$, and $A(c) = \{f\}$. f is still the only source, so label it with 2. Then $A(g) = \{\ \}$, and g is the only new source. Label g with 3. Then the antecedent sets for the remaining vertices are as follows:

$$A(b) = \{\ \} \qquad A(c) = \{b\} \qquad A(d) = \{c, e\} \qquad A(e) = \{\ \}$$

Then label either b or e. Choosing to label b with 4 reduces $A(c)$ to $\{\ \}$. Labeling e with 5 leaves $A(d) = \{c\}$. Labeling c with 6 reduces $A(d)$ to $\{\ \}$, so d gets label 7.

Self Test 3

1. The reachability matrix can be obtained either from Theorem 6.4 or by using Warshall's algorithm.

$$\begin{bmatrix} T & T & T & T & T \\ T & T & T & T & T \\ T & T & T & T & T \\ T & T & T & T & T \\ T & T & T & T & T \end{bmatrix}$$

Self Test 4

1. Mark A first and set

$$d[B] = \infty \qquad d[C] = \infty \qquad d[D] = 5$$
$$d[E] = 3 \qquad d[F] = 2 \qquad d[G] = \infty$$

where ∞ denotes infinity. Mark F. The remaining distances are as follows:

$$d[B] = 9 \qquad d[C] = \infty \qquad d[D] = 5$$
$$d[E] = 3 \qquad d[G] = \infty$$

Then mark E and recompute the distances:

$$d[B] = 5 \qquad d[C] = 4$$
$$d[D] = 5 \qquad d[G] = \infty$$

Mark C next. The remaining distances are as follows:

$$d[E] = 5 \qquad d[D] = 5 \qquad d[G] = 10$$

B or D can be chosen next. Choosing B reduces the remaining distances:

$$d[D] = 5 \qquad d[G] = 10$$

Then mark D, and mark G last.

7 Boolean Algebra

Boolean algebra is quite a different kind of mathematical tool from the graphs and digraphs we have discussed so far. Boolean algebra takes its name from the nineteenth century British mathematician George Boole, who sought to develop tools for a *logical calculus*, that is, a set of structures and rules that could be applied to logical symbols in the same way that ordinary algebra is applied to symbols representing numeric quantities. His success was such that Bertrand Russell once credited Boole with the discovery of pure mathematics. Our interest in Boolean algebra, however, is not limited to applied logic. Indeed, as with many areas of "pure" mathematics, Boolean algebra has been applied in significant areas that did not exist during Boole's time, including the design of logical circuits.

7.1   Boolean Expressions

There are many sets that can be called *Boolean algebras*. In our context, a Boolean algebra will be built on the set $B = \{0, 1\}$, equipped with the operations of **disjunction**, **conjunction**, and **negation**.

Disjunction ($\vee$) is the binary operation on B defined by

$\vee$	0	1
0	0	1
1	1	1

Conjunction ($\wedge$) is the binary operation on B defined by

$\wedge$	0	1
0	0	0
1	0	1

Negation (') is the unary operation on B defined by

$$0' = 1 \quad \text{and} \quad 1' = 0$$

There is an obvious similarity between the Boolean operations $\wedge$, $\vee$, and ' and the logical operations **or**, **and**, and **not**. The analogy between the Boolean operators and the logical operators is so strong that $\wedge$ is often referred to as **AND**, $\vee$ as **OR**, and ' as **NOT**. We can see the similarity more clearly by expressing the definitions of the Boolean operations with truth tables. To do so, we let p and q be Boolean variables; that is, let the values of p and q be elements of B. The truth tables equivalent to the binary tables above are shown in Figure 7.1.

p	p'
0	1
1	0

p	q	$p \vee q$	$p \wedge q$
0	0	0	0
0	1	1	0
1	0	1	0
1	1	1	1

Figure 7.1. Truth table definitions of Boolean operations

EXAMPLE 7.1

Use Boolean truth tables to verify the following identities involving the Boolean variable p.

1. $p \vee 0 = p$
2. $p \wedge 1 = p$
3. $p \vee p' = 1$

Solution

The combined truth tables are shown below. Since the first three columns agree, statements (1) and (2) are *true*. Since the last column is entirely 1, statement (3) is *true*.

p	$p \vee 0$	$p \wedge 1$	$p \vee p'$
0	0	0	1
1	1	1	1

Expressions

$p \vee p'$ is an example of a **Boolean expression** because it is made up of a combination of Boolean variables and Boolean operations. More precisely, Boolean expressions are defined recursively as follows:

1. The constants 0, 1 are Boolean expressions.
2. Boolean variables are Boolean expressions.
3. For Boolean expressions f, g, $f \vee g$, $f \wedge g$, f', and (f) are also Boolean expressions.

For example, if p, q, and r are Boolean variables, the following are examples of Boolean expressions:

$$p \vee (q \wedge r')' \vee (p \vee r)'$$
$$(p \vee q) \wedge (p' \wedge q')$$

Boolean expressions are similar in form to algebraic expressions. There are an infinite number of Boolean expressions involving only two Boolean variables. Each can be evaluated by substituting all possible values of the Boolean variables involved. For Boolean expressions, unlike for algebraic expressions, there will be a finite number of solutions, since each variable must take on one of two values. When two Boolean expressions have the same truth table, we say they are **equivalent**. If f and g are equivalent Boolean expressions, we write $f = g$.

There are many useful identities that can be derived for Boolean expressions. These are summarized, without proof, in Theorem 7.1.

Theorem 7.1

Let p, q, and r be Boolean variables. The following identities hold.

Commutative Law

1. a. $p \wedge q = q \wedge p$ b. $p \vee q = q \vee p$

Associative Law

2. a. $p \wedge (q \wedge r) = (p \wedge q) \wedge r$ b. $p \vee (q \vee r) = (p \vee q) \vee r$

Distributive Law

3. a. $p \wedge (q \vee r) = (p \wedge q) \vee (p \wedge r)$ b. $p \vee (q \wedge r) = (p \vee q) \wedge (p \vee r)$

Idempotent Law

4. a. $p \wedge p = p$ b. $p \vee p = p$

Absorption Law

5. a. $p \wedge (p \vee q) = p$ b. $p \vee (p \wedge q) = p$

EXAMPLE 7.2

Apply the identities of Theorem 7.1 to show that

$$(p \wedge (q \vee r')) \vee (p \wedge q) = p \wedge (q \vee r')$$

for all Boolean variables p, q, and r.

Solution

$$
\begin{aligned}
(p \wedge (q \vee r')) \vee (p \wedge q) &= ((p \wedge q) \vee (p \wedge r')) \vee (p \wedge q) && \{\text{by identity 3a}\} \\
&= ((p \wedge r') \vee (p \wedge q)) \vee (p \wedge q) && \{\text{by identity 1b}\} \\
&= (p \wedge r') \vee ((p \wedge q) \vee (p \wedge q)) && \{\text{by identity 2b}\} \\
&= (p \wedge q) \vee (p \wedge r') && \{\text{by identities 4b and 1b}\} \\
&= p \wedge (q \vee r') && \{\text{by identity 3a}\}
\end{aligned}
$$

EXAMPLE 7.3

Evaluate the following Boolean expressions:

$$p \vee (q \wedge r') \vee (p \vee r)'$$
$$(p \vee q) \wedge (p' \wedge q')$$

Solution

Each of the expressions is evaluated, in stages, in the following truth tables:

p	q	r	$q \wedge r'$	$(p \vee r)'$	$p \vee (q \wedge r') \vee (p \vee r)'$
0	0	0	0	1	1
0	0	1	0	0	0
0	1	0	1	1	1
0	1	1	0	0	0
1	0	0	0	0	1
1	0	1	0	0	1
1	1	0	1	0	1
1	1	1	0	0	1

p	q	$p \vee q$	$p' \wedge q'$	$(p \vee q) \wedge (p' \wedge q')$
0	0	0	1	0
0	1	1	0	0
1	0	1	0	0
1	1	1	0	0

Although it may appear in Example 7.3 that the only way to find the truth table for an expression is to laboriously evaluate each row, there is another, more efficient method. Look at the expression and ask, "When can this have a value of 1?" When two subexpressions are connected by $\vee$, the value of the whole expression will be 1 if either subexpression has a value of 1. When two subexpressions are connected by $\wedge$, the value of the whole expression will be 1 only if both subexpressions have a value of 1. In the first expression evaluated in Example 7.3, all three subexpressions are connected by $\vee$, so the whole expression has a value of 1 if any of the three has a value of 1. For instance, when $p = 1$, the value of the first subexpression is 1, so there is no need to evaluate the other two subexpressions. Thus we can put 1 in the last four rows of the truth table. Then we can consider the value of the second subexpression, $(q \wedge r')$. Its value is 1 only when $q = 1$ and $r = 0$, giving us a value of 1 for the whole expression in the third and seventh rows. Finally, we consider the subexpression $(p \vee r)'$, which has a value of 1 only if $p = 0$ and $r = 0$, giving us a value of 1 for the whole expression in the first and third rows. This method gives us a value of 1 for the whole expression in the first through third and fifth through eighth rows, which agrees with the first table shown in the solution for Example 7.3.

DeMorgan's Laws

Looking at the second table in the solution for Example 7.3, we see that the expression has constant value; that is, the expression has a value of 0 for any combination of p and q. This means we have found another identity, namely,

$$(p \vee q) \wedge (p' \wedge q') = 0$$

Even more interesting is the fact that the first subexpression has a value of 1 whenever the second subexpression has a value of 0, and vice versa. This means that each subexpression is the negation of the other, or

$$(p \vee q)' = p' \wedge q'$$

This result is yet another version of DeMorgan's laws, which we previously encountered in logic and in set theory. We state the Boolean form of DeMorgan's laws in Theorem 7.2.

Theorem 7.2
DeMorgan's Laws

For all Boolean expressions p and q:

1. $(p \vee q)' = p' \wedge q'$
2. $(p \wedge q)' = p' \vee q'$

• • •

Theorem 7.3 says that any expression that behaves like the negation of a variable is the negation of the variable. In the proof we will use properties proved previously.

Theorem 7.3

If $p \vee q = 1$ and $p \wedge q = 0$, then $q = p'$.

Proof

Observe that $p \vee p' = 1$ and $p \wedge p' = 0$. Hence

$$q = q \wedge 1 = q \wedge (p \vee p') = (q \wedge p) \vee (q \wedge p')$$
$$= 0 \vee (q \wedge p') = (q \wedge p') \vee (p \wedge p')$$
$$= (q \vee p) \wedge p' = 1 \wedge p' = p'$$

EXAMPLE 7.4

Give a direct symbolic proof of the first of DeMorgan's laws.

Solution

In order to show that $(p \vee q)' = p' \wedge q'$, it is enough to show that $p' \wedge q'$ behaves like the negation of $(p \vee q)$; that is,

$$(p \vee q) \wedge (p' \wedge q') = 0 \quad \text{and} \quad (p \vee q) \vee (p' \wedge q') = 1$$

and then apply Theorem 7.3. But

$$(p \vee q) \wedge (p' \wedge q') = (p \wedge (p' \wedge q')) \vee (q \wedge (p' \wedge q'))$$
$$= ((p \wedge p') \wedge q') \vee (p' \wedge (q \wedge q'))$$
$$= (0 \wedge q') \vee (p' \wedge 0)$$
$$= 0 \vee 0 = 0$$

Similarly,

$$(p \vee q) \vee (p' \wedge q') = ((p \vee q) \vee p') \wedge ((p \vee q) \vee q')$$
$$= (1 \vee q) \wedge (p \vee 1) = 1$$

SELF TEST 1

1. Find the truth tables for each of the following Boolean expressions.
   a. $p \wedge (q \vee r)'$ $\Rightarrow p(q+r)'$
   b. $(x \wedge y')' \vee (z \wedge (u \vee z'))$
2. Verify each of the following identities.
   a. $p \vee p' = 1$
   b. $p \wedge 1 = p$
3. Apply DeMorgan's laws to verify that

$$((p \wedge q)' \vee r)' = p \wedge q \wedge r'$$

Exercises 7.1

1. Use truth tables to verify each of the following identities.
   a. $p \vee 1 = 1$
   b. $p \wedge p' = 0$
   c. $p \wedge 0 = 0$
2. Find a simpler Boolean expression that is equivalent to the first expression in Example 7.3.
3. Prove the idempotent and absorption identities of Theorem 7.1.
4. Given the first DeMorgan's law in Theorem 7.2, derive the second.
5. Using any of the identities or theorems, verify the following equalities.
   a. $(p \vee q')' \wedge r' = p' \wedge q \wedge r'$
   b. $(z \wedge w) \vee (z' \wedge w) \vee (z \wedge w') = z \vee w$
   c. $((p \vee q') \vee (r \wedge (p \vee q')))' = p' \wedge q$
   d. $x \wedge y \vee x' \wedge y \vee x' \wedge y' \vee x \wedge y' = 1$

7.2 Representation of Expressions

Although there are an infinite number of Boolean expressions that are equivalent to each other, the number of truth tables involving a fixed number of variables is always finite. In this section, we begin the process of finding a "best" way to write a Boolean expression. We begin by considering the distinct functions of Boolean variables.

For each $n \geq 1$, let $F^n = \{f | f : B^n \to B\}$. It is easy to define binary operations on F^n in terms of the binary operations on B. For $f, g \in F^n$, define $f \vee g$, $f \wedge g$, and f' as follows:

$$(f \vee g)(x) = f(x) \vee g(x)$$
$$(f \wedge g)(x) = f(x) \wedge g(x)$$
$$(f)'(x) = (f(x))'$$

for all $x = (b_1, b_2, \ldots, b_n) \in B^n$.

It is not hard to check that F^n, equipped with the operations defined above, satisfies the five conditions of Theorem 7.1. Moreover, if we define 1_n and $0_n \in F^n$ by

$$1_n(x) = 1 \quad \text{and} \quad 0_n(x) = 0$$

for all $x \in B^n$, then we can obtain identities similar to the identities for Boolean expressions. For example, for all $f \in F^n$,

$$f \vee f' = 1_n$$

and

$$f \wedge f' = 0_n$$

EXAMPLE 7.5

List all the elements of F^1 and F^2.

Solution

We list the elements in truth table form. Here are the elements of F^1:

b_1	f_0	f_1	f_2	f_3
0	0	0	1	1
1	0	1	0	1

Here are the elements of F^2:

b_1	b_2	f_0	f_1	f_2	f_3	f_4	f_5	f_6	f_7	f_8	f_9	f_{10}	f_{11}	f_{12}	f_{13}	f_{14}	f_{15}
0	0	0	0	0	0	0	0	0	0	1	1	1	1	1	1	1	1
0	1	0	0	0	0	1	1	1	1	0	0	0	0	1	1	1	1
1	0	0	0	1	1	0	0	1	1	0	0	1	1	0	0	1	1
1	1	0	1	0	1	0	1	0	1	0	1	0	1	0	1	0	1

To read the tables, pick a particular function and look in its column to find its values. For instance, using the table for F^2, $f_3(0, 1) = 0$, and $f_{14}(1, 0) = 1$.

EXAMPLE 7.6

Using the table for F^2, find $f_0 \vee f_4$, $f_3 \wedge f_{13}$, and f'_2. Which of the sixteen functions is 1_2 and which is 0_2?

   Solution

Looking at the appropriate columns from the table, we obtain the following result for $f_0 \vee f_4$.

b_1	b_2	f_0	f_4	$f_0 \vee f_4$
0	0	0	0	0
0	1	0	1	1
1	0	0	0	0
1	1	0	0	0

Comparing the third and fourth columns, it's easy to see that $f_0 \vee f_4 = f_4$. Similarly, we obtain the following table for $f_3 \wedge f_{13}$:

b_1	b_2	f_3	f_{13}	$f_3 \wedge f_{13}$
0	0	0	1	0
0	1	0	1	0
1	0	1	0	0
1	1	1	1	1

Comparing the fifth column with the definitions in the table for F^2, we see that $f_3 \wedge f_{13} = f_1$. Finally, we obtain the following table for f'_2:

b_1	b_2	f_2	f'_2
0	0	0	1
0	1	0	1
1	0	1	0
1	1	0	1

Comparing the fourth column with the table of elements of F^2, we see that $f'_2 = f_{13}$. To conclude the example, observe that f_0 has all 0's in its column, and that f_{15} has all 1's in its column, so $f_0 = 0_2$, and $f_{15} = 1_2$.

EXAMPLE 7.7

Determine the number of elements of F^n.

Solution

For each n, there are 2^n elements of B^n. Since each element of F^n assigns either 0 or 1 to every element of B^n, there are 2^{2^n} elements of F^n. The following table shows the rapid growth of 2^{2^n} for $n = 1$ through 5.

n	2^{2^n}
1	4
2	16
3	256
4	65,536
5	4,294,967,296

Although the number of functions of a fixed number of Boolean variables is finite, it tends to be very large for even a small number of variables.

Minterms

Because there is such a large number of possible functions for even a few Boolean variables, we seek a method for representing them in a standard, or **canonical**, form. The basis for the method we use is the set of minterms. A **minterm** is a Boolean function that has exactly one 1 in its truth table. For example, consider the following truth table for a minterm of three variables p, q, r:

p	q	r	$m(p, q, r)$
0	0	0	0
0	0	1	0
0	1	0	0
0	1	1	1
1	0	0	0
1	0	1	0
1	1	0	0
1	1	1	0

Since $m(p, q, r) = 1$ only if $p = 0$, $q = 1$, and $r = 1$, it must be the case that $m(p, q, r) = p' \wedge q \wedge r$. If you doubt that this is so, construct the truth table for the expression $p' \wedge q \wedge r$ and compare it with the truth table for m.

Representing a minterm as a conjunction of either the variables or their complements is typical. In general, look for the row in which m is 1. Suppose, for instance, that m is a minterm involving the n variables $b_1, b_2, \ldots, b_n$, and that $m(a_1, a_2, \ldots, a_n) = 1$. First we combine all the variables b_i with **AND** operations. Whenever $a_i = 0$, we take the complement of b_i. The resulting expression is the representation of the minterm. For the minterm of p, q, and r discussed above, we start with

$$p \wedge q \wedge r$$

Since $m(0, 1, 1) = 1$, the first variable, in this case p, is to be replaced by its complement. Hence,

$$m(p, q, r) = p' \wedge q \wedge r$$

as we saw before. The representation of a minterm in this form is called the **product representation** of the minterm.

EXAMPLE 7.8

Find the product representation of the minterm defined in the following table.

p	q	r	s	$m(p, q, r, s)$
0	0	0	0	0
0	0	0	1	0
0	0	1	0	0
0	0	1	1	0
0	1	0	0	0
0	1	0	1	0
0	1	1	0	0
0	1	1	1	0
1	0	0	0	0
1	0	0	1	1
1	0	1	0	0
1	0	1	1	0
1	1	0	0	0
1	1	0	1	0
1	1	1	0	0
1	1	1	1	0

Solution

First we observe that the minterm is 1 on the input (1, 0, 0, 1). Hence we should take the complements of the second and third variables (q and r):

$$m(p, q, r, s) = p \wedge q' \wedge r' \wedge s$$

Normal Form

So far we have shown that any minterm can be expressed using only the operations of conjunction and negation. The next step is to show that any Boolean function can be written as a disjunction of minterms. The representation we obtain is called the **disjunctive normal** form, or the **sum of products** form.

Theorem 7.4

Every Boolean function has a unique representation as a disjunction of minterms.

Proof

Let $f \in F^n$. There is a minterm for each 1 in the truth table of f. The disjunction of those minterms is f.

EXAMPLE 7.9

Find the sum of products form of the function $f(x, y, z)$ that has the following truth table:

x	y	z	$f(x, y, z)$
0	0	0	1
0	0	1	0
0	1	0	0
0	1	1	1
1	0	0	1
1	0	1	0
1	1	0	1
1	1	1	0

Solution

There are 1's in f's truth tables corresponding to the four minterms

$$x' \wedge y' \wedge z'$$
$$x' \wedge y \wedge z$$
$$x \wedge y' \wedge z'$$
$$x \wedge y \wedge z'$$

We can obtain the truth table for f by overlaying the truth tables of the four minterms. Hence,

$$f(x, y, z) = x' \wedge y' \wedge z' \vee x' \wedge y \wedge z \vee x \wedge y' \wedge z' \vee x \wedge y \wedge z'$$

EXAMPLE 7.10

Find the sum of products representation of the Boolean function $g(x, y, z, w)$ that has the following truth table:

x	y	z	w	g
0	0	0	0	0
0	0	0	1	0
0	0	1	0	1
0	0	1	1	1
0	1	0	0	0
0	1	0	1	0
0	1	1	0	0
0	1	1	1	1
1	0	0	0	0
1	0	0	1	0
1	0	1	0	0
1	0	1	1	1
1	1	0	0	1
1	1	0	1	0
1	1	1	0	0
1	1	1	1	0

Solution

The minterms in this example will involve the four variables x, y, z, and w. We find which of the sixteen minterms to use in the representation of f by looking at the rows of the truth table that are 1 and connecting these rows by disjunction. The result is the disjunction of the five minterms:

$$f(x, y, z, w) = x' \wedge y' \wedge z \wedge w' \vee x' \wedge y' \wedge z \wedge w \vee x' \wedge y \wedge z \wedge w$$
$$\vee x \wedge y' \wedge z \wedge w \vee x \wedge y \wedge z' \wedge w'$$

Complete Sets of Operators

The result of Theorem 7.4 is that any Boolean function can be represented in terms of the operators $\wedge$, $\vee$, and $'$. Any set of Boolean operators that is sufficient to represent all Boolean expressions is called a **complete set of operators**. Thus we can say that the set $\{\wedge, \vee, '\}$ is a complete set of operators. There are other complete sets of Boolean operators.

EXAMPLE 7.11

Show that the set $\{\wedge, '\}$ is a complete set of Boolean operators.

Solution

All we have to do is to show that the operator $\vee$ can be obtained by combining $\wedge$ and $'$. Recall DeMorgan's law that states that for all Boolean expressions p and q,

$$(p \vee q)' = p' \wedge q'$$

If we apply the negation operator $'$ to both sides of the equality, we obtain

$$p \vee q = (p \vee q)'' = (p' \wedge q')'$$

which establishes the desired result.

SELF TEST 2

1. Use the truth tables for F^2 in Example 7.5 to find $f_3 \wedge f_8$.
2. Use the product form to represent the minterm given by the following table.

p	q	r	$m(p, q, r)$
0	0	0	0
0	0	1	0
0	1	0	0
0	1	1	0
1	0	0	0
1	0	1	0
1	1	0	1
1	1	1	0

3. Find the sum of products form of the Boolean function given by the following table.

p	q	r	s	$m(p, q, r, s)$
0	0	0	0	1
0	0	0	1	0
0	0	1	0	0
0	0	1	1	0
0	1	0	0	0
0	1	0	1	1
0	1	1	0	0
0	1	1	1	0
1	0	0	0	0
1	0	0	1	1
1	0	1	0	0
1	0	1	1	0
1	1	0	0	1
1	1	0	1	0
1	1	1	0	0
1	1	1	1	0

Exercises 7.2

1. Use the truth tables for F^1 in Example 7.5 to compute $f_1 \wedge f_3$ and $f_2 \vee (f_3 \wedge f_2)'$.
2. Use the truth tables for F^2 in Example 7.5 to compute f'_{10} and $f_{11} \wedge (f_6 \vee f_8)$.
3. How many minterms are there in F^n, for $n \geq 1$?
4. Which of the functions in F^2 are minterms?
5. Find the sum of products form of the function with the following truth table.

x	y	z	w	g
0	0	0	0	0
0	0	0	1	0
0	0	1	0	1
0	0	1	1	1
0	1	0	0	0
0	1	0	1	0
0	1	1	0	0
0	1	1	1	1
1	0	0	0	0
1	0	0	1	0
1	0	1	0	0
1	0	1	1	1
1	1	0	0	1
1	1	0	1	0
1	1	1	0	0
1	1	1	1	0

6. Find the disjunctive normal form for the following Boolean expression. (*Hint*: First find the truth table for the expression.)

$$p \wedge (q' \vee r) \vee (p' \wedge q \vee r')$$

7. a. Show that the set of operators $\{\vee,'\}$ is a complete set of operators.
   b. Write the expression $x \wedge y \wedge z$ in terms of $\vee$ and $'$.
8. The Boolean operator **NAND** (for *not and*) is defined by

$$x\mathbf{NAND}y = (x \wedge y)'$$

Show that $\{\mathbf{NAND}\}$ is a complete set of operators.

7.3   Minimization of Boolean Expressions

In the previous section we showed that every Boolean expression corresponds to a unique truth table with a unique representation as a sum of products. Now we go one step further and find a simpler expression equivalent to the original one. The tool for the reduction is known as a **Karnaugh map**. Karnaugh maps were devised in the 1950's to aid in logical circuit design. They are rectangular arrays of boxes for which the top and sides of the arrays are labeled with combinations of Boolean variables and their complements. Each box is used to represent the presence or absence of a minterm in the sum of products representation of an expression.

For the rest of this chapter we will dispense with writing $\wedge$, so that an expression such as $x \wedge y \wedge z$ will be written as xyz. This is analogous to the shorthand in algebra in which multiplication is denoted by writing two variables next to each other without a multiplication sign.

Karnaugh Maps

To motivate the construction of Karnaugh maps for expressions with three variables, consider the expression

$$xyz \vee x'yz \vee xyz'$$

By the idempotent law, we can duplicate the term xyz, and the expression becomes

$$xyz \vee x'yz \vee xyz \vee xyz' = (x \vee x')yz \vee xy(z \vee z')$$
$$= yz \vee xy$$

The last expression is certainly simpler than the original one. (If you wonder why we chose to duplicate the term xyz, try duplicating the expression $x'yz$ and see what happens.) For the graphical form of this case we need an array of eight squares for the eight minterms with three variables. This is shown in the 2×4 array in Figure 7.2. The labels across the top cover all four combinations of the two variables x and y. The labels on the side cover both possibilities z and z'. Hence, each box has a row and column index that corresponds to a minterm in the variables x, y, and z.

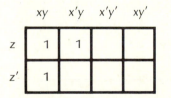

Figure 7.2. Karnaugh map for $xyz \vee x'yz \vee xyz'$

This particular choice of labeling is arbitrary; we could just as easily use y and z across the top and x on the side. The only restriction in laying out a Karnaugh map is that in moving from one column or row to the next, only one of the variables can be changed by applying or removing '. For instance, we could not have $x'y'$ adjacent to xy, because the two minterms have two variables changed. If a minterm is present in the sum of products representation, we mark its box with 1. Otherwise, we leave the box empty. The picture in Figure 7.2 is the **Karnaugh map** for $xyz \vee x'yz \vee xyz'$.

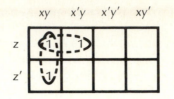

Figure 7.3. Clusters from the Karnaugh map of Figure 7.2

The primary rule of Karnaugh maps is that "good things come in powers of 2." Inspecting the map of Figure 7.2, we see two clusters of minterms. These are shown circled in Figure 7.3. Note that the box corresponding to xyz is shared by both clusters, and recall that xyz is the minterm we chose to duplicate. Look at the cluster in the first row. This corresponds to $xyz \vee x'yz$, from which we can factor yz, yielding $(x \vee x')yz$. Since $x \vee x' = 1$, the cluster reduces to yz. Similarly, the cluster in the first column shares xy with both z and z', so it reduces to xy. The net result is the reduced expression $yz \vee xy$, which was obtained before.

EXAMPLE 7.12

For the Karnaugh map in Figure 7.4, find the clusters, identify the minterms involved in each, and find the simplified expression that results.

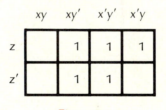

Figure 7.4.

Solution

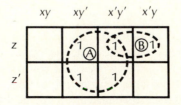

Figure 7.5.

The clusters are shown circled in Figure 7.5. The cluster labeled $\textcircled{A}$ corresponds to the expression

$$xy'z \vee x'y'z \vee xy'z' \vee x'y'z' = (x \vee x')y'z \vee (x \vee x')y'z'$$
$$= y'z \vee y'z'$$
$$= y'(z \vee z') = y'$$

The cluster labeled $\textcircled{B}$ corresponds to the expression

$$x'y'z \vee x'yz = x'z(y' \vee y) = x'z$$

Then the net result is that the sum of products form reduces to the simpler expression

$$y' \vee x'z$$

Observe that the cluster of four elements, $\textcircled{A}$, reduces to a single variable, and the cluster of two elements, $\textcircled{B}$, reduces to an expression with two variables.

There are many possibilities for relabeling Karnaugh maps. However, it is central to the use of maps that no matter how the labels are switched around, adjacent minterms remain adjacent, and minterms not adjacent to one another remain apart. The Karnaugh map in Figure 7.6 contains the same information as the map in Figure 7.4, but the labels are changed drastically. Note that there are still two clusters, one of four minterms and one of two minterms, and that the clusters still share the minterm $x'y'z$.

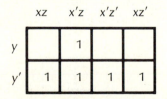

Figure 7.6. A relabeling of the Karnaugh map of Figure 7.4

EXAMPLE 7.13

Analyze each of the Karnaugh maps shown in Figure 7.7 to find a reduced form for the expression represented.

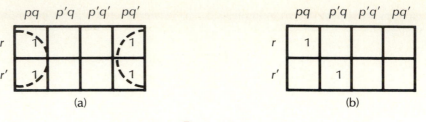

Figure 7.7.

Solution

a. The Karnaugh map shown in Figure 7.7(a) exhibits the wraparound effect. That is, the first and last columns, labeled pq and pq', can be considered adjacent because one can be obtained from the other by negating the variable q (or q'). Hence, there is a cluster of four minterms. The four minterms cover all combinations of q and r, so the expression reduces to p.

b. The Karnaugh map shown in Figure 7.7(b) contains two minterms that are not adjacent and that cannot be made adjacent by any relabeling of the Karnaugh map. (Try a few relabelings to satisfy yourself that this is so.) Hence, no reduction is possible, and the expression is $pqr \vee p'qr'$.

Karnaugh Maps for Four Variables

We have presented Karnaugh maps for three variables, and at this point can go on to maps for four variables or two variables. A typical 2-variable Karnaugh map is shown in Figure 7.8, and represents the expression q. A Karnaugh map for four variables requires sixteen boxes. In such maps we look for clusters of two, four, eight, or sixteen minterms. Of course, if all sixteen minterms are represented, the expression is the constant 1.

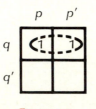

Figure 7.8.

EXAMPLE 7.14

Analyze the Karnaugh map in Figure 7.9.

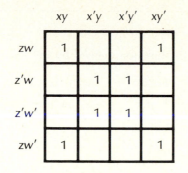

Figure 7.9.

Solution

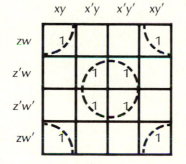

Figure 7.10. The clusters of Figure 7.9

There are two clusters in Figure 7.9, which are circled in Figure 7.10. Observe the wraparound, which indicates that the four corners are adjacent. The cluster of the four corners represents the reduced expression xz, because all combinations of z and w are accounted for. The second cluster represents the expression $x'z'$. Thus the expression reduces to $xz \lor x'z'$. Figure 7.11 shows how the wraparound cluster can be brought together by relabeling.

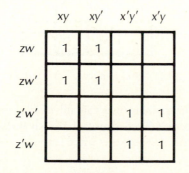

Figure 7.11. A relabeling of Figure 7.9

EXAMPLE 7.15

Find a simplified form for the function defined by the following truth table:

p	q	r	s	f
0	0	0	0	1
0	0	0	1	0
0	0	1	0	1
0	0	1	1	0
0	1	0	0	1
0	1	0	1	0
0	1	1	0	1
0	1	1	1	0
1	0	0	0	0
1	0	0	1	0
1	0	1	0	0
1	0	1	1	0
1	1	0	0	1
1	1	0	1	0
1	1	1	0	0
1	1	1	1	0

Solution

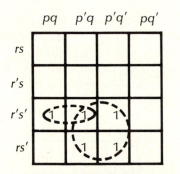

Figure 7.12. Karnaugh map for Example 7.15

Since we begin with the truth table, we can fill in the appropriate Karnaugh map. The result is shown in Figure 7.12. Because of the clustering shown, the expression simplifies to

$$p's' \vee qr's'$$

SELF TEST 3

1. Simplify the expression represented by the Karnaugh map in Figure 7.13.

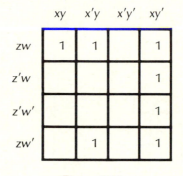

	xy	x'y	x'y'	xy'
zw	1	1		1
z'w				1
z'w'				1
zw'		1		1

Figure 7.13.

2. Find the Karnaugh map for the following expression. Then use the map to find a reduced form of the expression.

$$p' \vee (q' \wedge r)' \vee (p' \vee r)'$$

3. Use a Karnaugh map to simplify the Boolean function f given by the following table.

u	v	w	x	f
0	0	0	0	1
0	0	0	1	1
0	0	1	0	1
0	0	1	1	1
0	1	0	0	1
0	1	0	1	0
0	1	1	0	0
0	1	1	1	0
1	0	0	0	1
1	0	0	1	0
1	0	1	0	0
1	0	1	1	0
1	1	0	0	0
1	1	0	1	0
1	1	1	0	0
1	1	1	1	0

Exercises 7.3

1. Simplify the expressions represented by the Karnaugh maps in Figures 7.14 and 7.15.

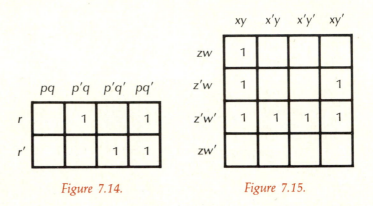

Figure 7.14.

Figure 7.15.

2. Use Karnaugh maps to find a simpler expression for the function defined by each of the following truth tables.

a.

x	y	z	$f(x, y, z)$
0	0	0	1
0	0	1	0
0	1	0	0
0	1	1	1
1	0	0	1
1	0	1	0
1	1	0	1
1	1	1	0

b.

x	y	z	w	$g(x, y, z, w)$
0	0	0	0	0
0	0	0	1	0
0	0	1	0	1
0	0	1	1	1
0	1	0	0	0
0	1	0	1	0
0	1	1	0	0
0	1	1	1	1
1	0	0	0	0
1	0	0	1	0
1	0	1	0	0
1	0	1	1	1
1	1	0	0	1
1	1	0	1	0
1	1	1	0	0
1	1	1	1	0

3. Find a simplified version of each of the following Boolean expressions.

   a. $p \wedge (q' \vee r) \vee (p' \wedge q \vee r')$

   b. $rs \vee (q' \wedge r)' \vee (p' \vee r)'$

7.4   Switching Theory

In this section, we consider the application of Boolean algebra to binary devices and logic design, which are part of the field called switching theory. Logic design is important to designers of the electronic circuits that make up modern digital computers built from binary devices. A **binary device** is a device, usually electronic, that accepts a finite number of "inputs" and produces a finite number of "outputs," where each of the inputs and outputs have only two possible values.

Circuit Diagrams

Inputs and outputs of a binary device can be represented as a Boolean quantity. Thus binary devices can be built up out of even simpler devices, called **logic gates**, which perform basic Boolean operations such as disjunction, conjunction, and negation. Each of the gates has a graphical symbol. The graphical symbols for the common logic gates are shown with their definitions in Figure 7.16. A diagram showing the interconnection of the logic gates to implement a collection of expressions is called a **circuit diagram**, or just a **circuit**. The implementation of a binary device through logic gates is called a **combinational network**.

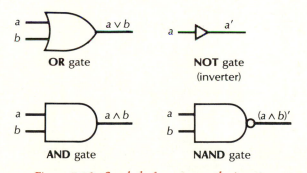

Figure 7.16. Symbols for common logic gates

As we have seen, there can be many different Boolean expressions that are equivalent. Hence, there is no unique implementation of a binary device in terms of logic gates. We seek a minimal representation for Boolean expressions so that the expressions can be implemented with fewer gates, resulting in concrete savings. Figure 7.17(a) shows two equivalent binary devices that represent one of DeMorgan's laws using logic gates. Figure 7.17(b) shows two logic gate representations of a distributive law.

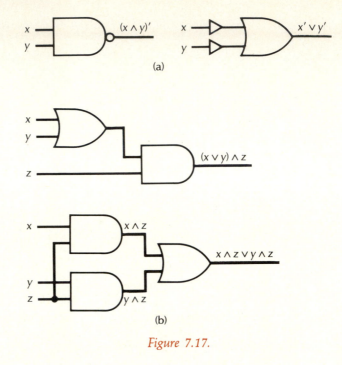

(a)

(b)

Figure 7.17.

EXAMPLE 7.16

1. Verify that the two logic circuits pictured in Figure 7.18 are equivalent; that is, they convert the same inputs to the same outputs.

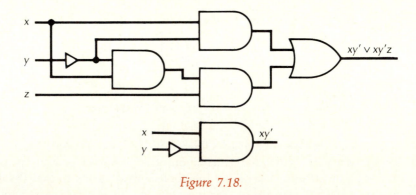

Figure 7.18.

2. Find a logic circuit that implements the Boolean expression $x'y \lor zy'$.

Solution

1. The first circuit implements the expression $xy' \vee xy'z$, and the second circuit implements the expression xy'. But

$$xy' = xy' \vee xy'z$$

by the absorption law, so the circuit diagrams are equivalent.

2. The expression $x'y$ is represented by the circuit in Figure 7.19(a). The expression zy' is represented by the circuit in Figure 7.19(b). Combining the two in one circuit, as depicted in Figure 7.20, yields the desired result.

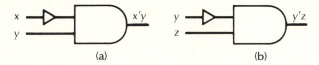

(a)   (b)

Figure 7.19.

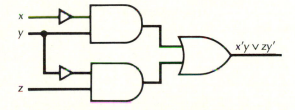

Figure 7.20. Circuit diagram for $x'y \vee zy'$

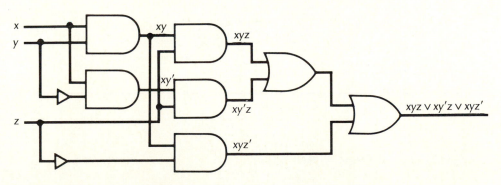

Figure 7.21. Circuit for $xyz \vee xy'z \vee xyz'$

Suppose we want to design a circuit that implements the function $xyz \vee xy'z \vee xyz'$. The straightforward circuit is shown in Figure 7.21. Each of the three term products requires an additional **AND** because of the restriction that allows only two inputs per gate. If we allow **AND** and **OR** gates to have more than two inputs, the diagram simplifies to the circuit in Figure 7.22.

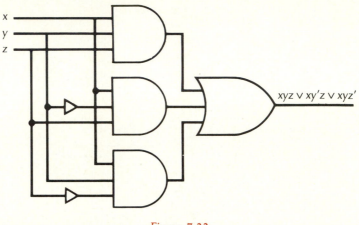

Figure 7.22.

We can obtain even more dramatic economies by applying Karnaugh maps to reduce the complexity of the expression before drawing the circuit. The Karnaugh map for the expression represented by Figure 7.22 is shown in Figure 7.23. The expression reduces to $xz \lor xy = x(y \lor z)$. The simplified circuit diagram is shown in Figure 7.24.

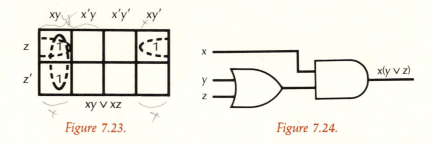

Figure 7.23.   *Figure 7.24.*

Complete Sets of Logic Gates

It is not necessary to use all types of available gates in constructing a circuit diagram. For instance, since $\{\land, '\}$ is a complete set of Boolean operators, it is theoretically possible to build any binary device using only **AND** and **NOT** gates. This is generally not practical, however, because it increases the number of components. Likewise, it is conceivable that a circuit designer could be restricted to using only **NAND** gates since {**NAND**} is a complete set of operators. This is not as difficult as it might seem, since the sum of products form for a Boolean expression easily reduces to a **NAND** implementation.

EXAMPLE 7.17

Find a logic circuit using **NAND**'s and **NOT**'s for the expression

$$xyz \vee xy'z' \vee x'yz'$$

Solution

By DeMorgan's law, the expression can be written as

$$((xyz)' \wedge (xy'z')' \wedge (x'yz'))'$$

How?

Each of the three inner terms is a 3-input **NAND**. The three inner terms are then combined by another **NAND**. This yields the circuit shown in Figure 7.25. It would be more difficult to represent this expression in terms of 2-input **NAND**'s.

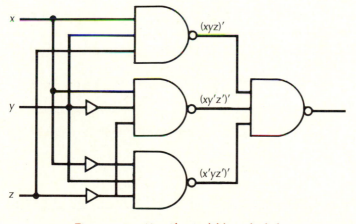

Figure 7.25. $((xyz)' \wedge (xy'z')' \wedge (x'yz'))'$

A Control Switch Example

We now consider a more complete example in which the problem is stated in less obvious form. Figure 7.26 shows a 5-position switch that controls three lights. The table in the figure defines the effects of the switch settings on the lights.

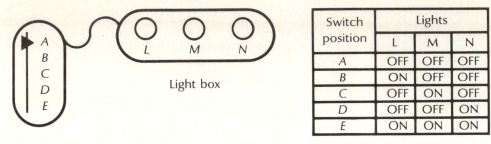

Switch

Light box

Switch position	Lights		
	L	M	N
A	OFF	OFF	OFF
B	ON	OFF	OFF
C	OFF	ON	OFF
D	OFF	OFF	ON
E	ON	ON	ON

Figure 7.26. A switch controlling three lights

To design a Boolean circuit that performs the same function as the switch, we need to find a suitable **encoding** of its inputs and outputs as Boolean variables. One simple approach is to represent the switch positions as integers in binary form. Since 2-digit binary numbers range only from 0 to 3, we need at least a 3-digit representation. Thus we can map A to 000, B to 001, and so on (although other mappings could be used). For the lights, the only outputs are ON and OFF, which we map to 1 and 0, respectively. Denoting the Boolean inputs by x, y, z, and the outputs as L, M, N, the encoding transforms the representation shown in Figure 7.26 to the representation shown in Figure 7.27.

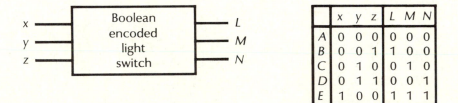

	x	y	z	L	M	N
A	0	0	0	0	0	0
B	0	0	1	1	0	0
C	0	1	0	0	1	0
D	0	1	1	0	0	1
E	1	0	0	1	1	1

Figure 7.27. Boolean representation of multilight switch

Having found an encoding of the switch inputs and outputs, we can now proceed with the logical circuit design as before. The Karnaugh maps for each of the functions L, M, and N are given in Figure 7.28. None of the three maps has a cluster to allow

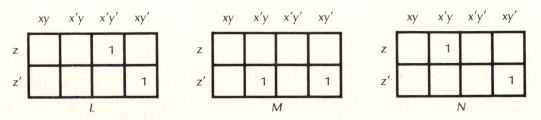

Figure 7.28. Karnaugh maps for the switch outputs

reduction of the expression, so

$$L = x'y'z \lor xy'z'$$
$$M = x'yz' \lor xy'z'$$
$$N = x'yz \lor xy'z'$$

A circuit diagram for these expressions is shown in Figure 7.29.

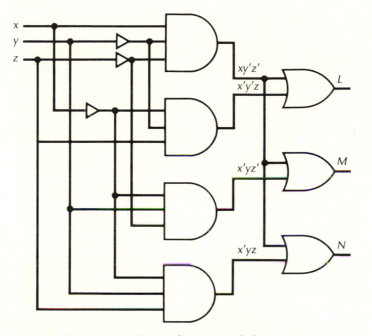

Figure 7.29. Circuit diagram for light switch

The circuit design for the switch is the result of a "natural" encoding of the inputs and outputs. However, there is nothing inherent in the encoding process that requires us to map the switch positions as we did. In fact, we could map A to any of the eight Boolean inputs, B to any of the remaining seven, and so on. In other words, we have looked at only 1 of 6720 possible encodings of the switch positions as 3-digit Boolean inputs. For each encoding, there will always be three unused input combinations. These are called "**don't care**" inputs, and we can either use them or not, as we choose.

The objective of choosing an alternative encoding is to make the resulting expressions simpler. If there were only one output, choosing a different encoding would be less difficult. As it is, there are three outputs, so we need to try to find a better encoding that simplifies all three expressions at once. Karnaugh maps can be helpful in this process, because we can show all three outputs on one map by using their names as place

	xy	x'y	x'y'	xy'
z	d	N	L	d
z'	d	M		L, M, N

Figure 7.30. A composite Karnaugh map for L, M, N

holders. Figure 7.30 is a composite Karnaugh map for the Karnaugh maps of Figure 7.28. The d is the place holder for the "don't care" conditions, in this case the three input combinations that are not used by the switch. The box with L, M, and N in it corresponds to the fact that for switch position E, all three lights are on.

Choosing a new encoding for the input switch positions has the same effect as moving the place holders L, M, N, and d around in the Karnaugh map. Our objective is to get clusters for each of the expressions. Then we can go back from the Karnaugh map to the table defining the switch action. It appears that the best we can do in this

	xy	x'y	x'y'	xy'
z	L, M, N	M	d_1	N
z'	L	d_2		d_3

	x	y	z	L	M	N
A	0	0	0	0	0	0
d_1	0	0	1	0	0	0
d_2	0	1	0	0	0	0
C	0	1	1	0	1	0
d_3	1	0	0	0	0	0
D	1	0	1	0	0	1
B	1	1	0	1	0	0
E	1	1	1	1	1	1

Figure 7.31. A new encoding for the multilight switch

case is the clustering shown in Figure 7.31. The Boolean expressions are just

$$L = xy$$
$$M = yz$$
$$N = xz$$

The corresponding circuit diagram is shown in Figure 7.32.

As a final variation on the problem, we consider the possibilities of using the "don't care" positions to improve the design. As illustrated in Figure 7.31, there is nothing to be gained by using the "don't care" positions in that encoding. However, we could

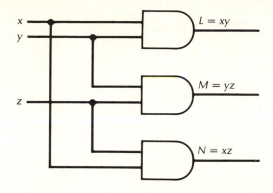

Figure 7.32. Circuit diagram for remodeled switch

try to enhance the expression for two of the outputs at the expense of the third. For instance, we could arrange the encoding as in Figure 7.33, in which the three "don't

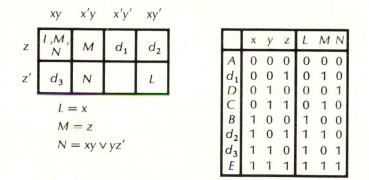

	xy	x'y	x'y'	xy'
z	l,M,N	M	d_1	d_2
z'	d_3	N		L

$L = x$

$M = z$

$N = xy \vee yz'$

	x	y	z	L	M	N
A	0	0	0	0	0	0
d_1	0	0	1	0	1	0
D	0	1	0	0	0	1
C	0	1	1	0	1	0
B	1	0	0	1	0	0
d_2	1	0	1	1	1	0
d_3	1	1	0	1	0	1
E	1	1	1	1	1	1

Figure 7.33. Another encoding of the multilight switch

care" positions can be of use. Observe that in the table for the encoding, the "don't care" positions appear to turn the lights ON or OFF, though of course they have no effect. It appears that we are getting something for nothing!

The simpler output expressions are

$$L = x$$
$$M = z$$
$$N = xy \vee yz'$$

These expressions lead to the circuit diagram shown in Figure 7.34. Despite the extra effort, this new circuit diagram is essentially no simpler than the circuit diagram obtained before.

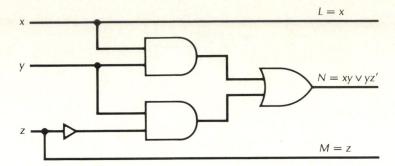

Figure 7.34. *Yet another circuit diagram for the multilight switch*

SELF TEST 4

1. Draw a circuit diagram for the expression

$$(p \vee q)'p'r$$

   a. using **AND**, **OR**, and **NOT** gates.
   b. using only **AND** and **NOT** gates.
   c. using only **NAND** gates.

2. Identify the expression corresponding to the circuit diagram in Figure 7.35.

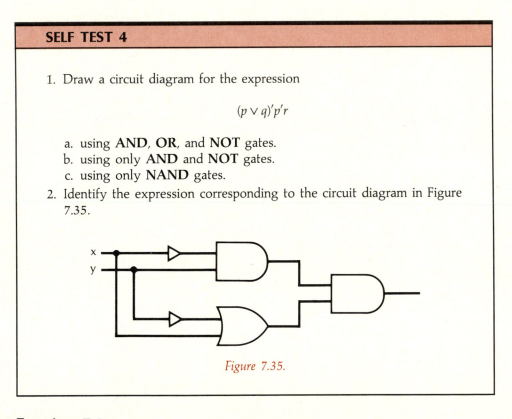

Figure 7.35.

Exercises 7.4

1. Draw the circuit diagram for the expression

$$(x \vee y)' \vee yx'z$$

   a. using any of the gates defined in Figure 7.16.

b. using only **NAND** gates.

c. using **AND** and **NOT** gates.

2. Design a circuit that implements an expression that equals 1 when exactly three of its four inputs are 1.

3. Simplify the expression in exercise 1, and then repeat parts (a), (b), and (c).

4. a. Identify the expression implemented by the circuit in Figure 7.36.

   b. Simplify the expression found in (a), and then find a circuit implementing the simpler expression.

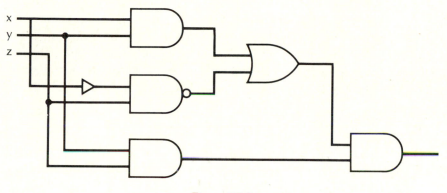

Figure 7.36.

5. Design a circuit for a *half-adder*; that is, a binary device that accepts two binary digits as inputs, and outputs their 2-digit sum.

6. The table below defines the action of a 6-position switch that governs the three lights L, M, N. Design a circuit to implement the switch as a binary device encoding the switch positions in each of the following ways.

   a. A as 0, B as 1, and so on

   b. So that it minimizes the output expressions, but does not take advantage of the "don't care" conditions

   c. So that it minimizes the output expressions, using "don't care" conditions where they help

Switch position	Light		
	L	M	N
A	ON	OFF	OFF
B	OFF	ON	ON
C	OFF	OFF	OFF
D	ON	ON	ON
E	OFF	OFF	OFF
F	OFF	ON	OFF

Designing a
2-Bit Adder

An important example of a binary device is a *2-bit adder*, which is not a cheap rubber snake but a device that adds two 2-bit binary numbers, producing a 3-bit binary number as the sum. Figure 7.37 shows the 2-bit adder as a "black box," which, when given the inputs a, b, c, d, generates the sum of the binary numbers ab and cd as the 3-bit number efg.

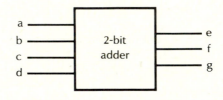

Figure 7.37. 2-bit adder

We show how to design a circuit for this device using a simpler device called a **half-adder**. A half-adder computes the sum of two single binary digits, producing two outputs, the *sum* bit and the *carry* bit. Let x and y be the two inputs. The sum of x and y is a 2-digit binary number, which we denote uv. The sum is thus expressed as follows:

$$\begin{array}{r} x \\ +y \\ \hline uv \end{array}$$

For instance, $0 + 1 = 1$, and $1 + 1 = 10$. All possible cases are covered in the following truth table for u and v:

x	y	u	v
0	0	0	0
0	1	0	1
1	0	0	1
1	1	1	0

Then $u = xy$, and $v = x'y \lor xy'$. Using these expressions, we can derive the circuit for the half-adder, which is depicted in Figure 7.38. For convenience, we refer to u as the *carry bit* and to v as the *sum bit*.

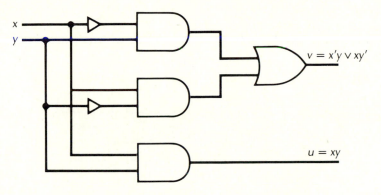

Figure 7.38. *Circuit diagram for the half-adder*

To finish the design of the 2-bit adder, let the first 2-bit number be denoted ab and the second cd. Then the result of adding ab and cd is a 3-bit number, which we denote efg. The sum has the form

$$\begin{array}{r} ab \\ + cd \\ \hline efg \end{array}$$

We call a and c the *high-order bits* and b and d the *low-order bits*. We could proceed by writing out the 16-row truth tables for e, f, and g, drawing their Karnaugh maps, and finding minimal expressions for the functions. An alternative is to use the half-adder as one of the building blocks for the 2-bit adder. First we add a to c and b to d, producing the diagram in Figure 7.39.

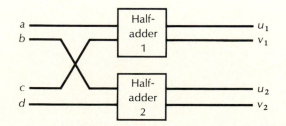

Figure 7.39. *Two half-adders to be used in the 2-bit adder*

The problem in putting the two half-adders together to form the 2-bit adder is to be sure to add the carry bit from the addition of the low-order bits to the sum bit of the addition of the two high-order bits. Since v_2 is the sum of the low order inputs b and d, it must be the same as g. Then u_2 is the bit to be carried over to the next higher term, in this case to the sum of a and c. Hence, we need to connect the outputs u_2 and v_1 as inputs to a third half-adder. The sum bit produced will then be the middle binary digit, f. Since the carry bit u_3 and u_1 cannot be 0, they can be combined with an **OR** gate to produce the third output bit, e. The final logic diagram is found in Figure 7.40.

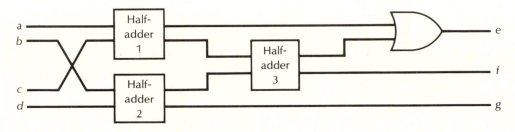

Figure 7.40. Circuit diagram for a 2-bit adder

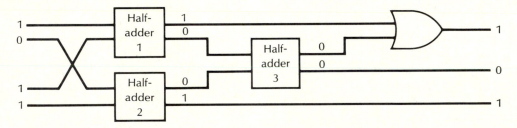

Figure 7.41. Addition of 10 and 11 by the 2-bit adder

To complete the study of the 2-bit adder, we trace its operation in adding 2 and 3. We represent 2 as the input $ab = 10$ and 3 as the input $cd = 11$. Figure 7.41 shows how the 2-bit adder finds the sum.

$$\begin{array}{r} 10 \\ +11 \\ \hline 101 \end{array}$$

Exercises 7.5

1. Extend the 2-bit adder to a 3-bit adder, using half-adders as components.

2. Derive a circuit for the 2-bit adder that uses only **AND**, **OR**, and **NOT** gates. (*Hint*: Find the truth tables for the output expressions e, f, g.)

SUMMARY

1. **Boolean expressions** are defined recursively as follows:
   1. The constants 0, 1 are Boolean expressions.
   2. Boolean variables are Boolean expressions.
   3. For Boolean expressions $f, g, f \vee g, f \wedge g, f'$ and (f) are also Boolean expressions.

2. **Theorem 7.1**. Let p, q, and r be Boolean variables. The following identities hold.
   1. a. $p \wedge q = q \wedge p$ b. $p \vee q = q \vee p$
   2. a. $p \wedge (q \wedge r) = (p \wedge q) \wedge r$ b. $p \vee (q \vee r) = (p \vee q) \vee r$
   3. a. $p \wedge (q \vee r) = (p \wedge q) \vee (p \wedge r)$ b. $p \vee (q \wedge r) = (p \vee q) \wedge (p \vee r)$
   4. a. $p \wedge p = p$ b. $p \vee p = p$
   5. a. $p \wedge (p \vee q) = p$ b. $p \vee (p \wedge q) = p$

3. **Theorem 7.2 (DeMorgan's Laws)**. For all Boolean expressions p and q:
   1. $(p \vee q)' = p' \wedge q'$
   2. $(p \wedge q)' = p' \vee q'$

4. **Theorem 7.3**. If $p \vee q = 1$ and $p \wedge q = 0$, then $q = p'$.

5. For each $n \geq 1$, let $F^n = \{f: f:B^n \to B\}$. We can define binary operations on F^n in terms of the binary operations on B. For $f, g \in F^n$, define $f \vee g, f \wedge g$, and f' as follows:

$$(f \vee g)(x) = f(x) \vee g(x)$$
$$(f \wedge g)(x) = f(x) \wedge g(x)$$
$$(f)'(x) = (f(x))'$$

   for all $x = (b_1, b_2, \ldots, b_n) \in B^n$.

6. **Theorem 7.4**. Every Boolean function has a unique representation as a disjunction of minterms.

7. Any Boolean function can be represented in terms of the operators $\wedge$, $\vee$, and $'$.

8. Any set of Boolean operators that is sufficient to represent all Boolean expressions is called a **complete set of operators**. Thus we can say that the set $\{\wedge, \vee, '\}$ is a complete set of operators.

9. **Karnaugh maps** are rectangular arrays of boxes for which the top and sides of the arrays are labeled with combinations of Boolean variables and their complements. Each box is used to represent the presence or absence of a minterm in the sum of products representation of an expression. Karnaugh maps are used to minimize Boolean expressions.

10. A **binary device** is a device, usually electronic, that accepts a finite number of "inputs" and produces a finite number of "outputs," where each of the inputs and outputs have only two possible values.

11. Inputs and outputs of a binary device can be represented as a Boolean quantity.

12. Binary devices can be built up out of even simpler devices, called **logic gates**, which perform basic Boolean operations such as disjunction, conjunction, and negation.

13. The implementation of a binary device through logic gates is called a **combinational network**.

14. Dramatic economies in number of components for a circuit can be obtained by applying Karnaugh maps to reduce the complexity of the expression before designing the circuit.

REVIEW PROBLEMS

1. Prove the first three identities of Theorem 7.1.

2. Find truth tables for each of the following expressions.
   a. $q' \wedge (r \vee p)'$
   b. $p \vee (q' \wedge r') \vee (p \vee r)'$

3. Verify the equality $p \vee q = (p' \wedge q) \vee (p \wedge q) \vee (p \wedge q')$.

4. Verify the equality $(p' \wedge q)' = p \vee q'$.

5. Write the product form of the minterm given by each of the following tables.

a.

p	q	r	$m(p, q, r)$
0	0	0	0
0	0	1	0
0	1	0	1
0	1	1	0
1	0	0	0
1	0	1	0
1	1	0	0
1	1	1	0

b.

p	q	r	s	$m(p, q, r, s)$
0	0	0	0	0
0	0	0	1	0
0	0	1	0	0
0	0	1	1	0
0	1	0	0	0
0	1	0	1	1
0	1	1	0	0
0	1	1	1	0
1	0	0	0	0
1	0	0	1	0
1	0	1	0	0
1	0	1	1	0
1	1	0	0	0
1	1	0	1	0
1	1	1	0	0
1	1	1	1	0

6. Find the sum of products form of the function $f(x, y, z)$ with the following truth table.

x	y	z	$f(x, y, z)$
0	0	0	1
0	0	1	0
0	1	0	0
0	1	1	1
1	0	0	1
1	0	1	0
1	1	0	1
1	1	1	0

7. Find a minimum expression for the Boolean function $g(x, y, z, w)$ with the following truth table.

x	y	z	w	g
0	0	0	0	0
0	0	0	1	1
0	0	1	0	0
0	0	1	1	1
0	1	0	0	1
0	1	0	1	0
0	1	1	0	1
0	1	1	1	0
1	0	0	0	1
1	0	0	1	0
1	0	1	0	0
1	0	1	1	1
1	1	0	0	1
1	1	0	1	0
1	1	1	0	0
1	1	1	1	0

8. Find the truth table for the function f.

$$f(x, y, z, w) = x' \wedge y' \wedge z \wedge w' \vee x' \wedge y' \wedge z \wedge w$$
$$\vee x' \wedge y \wedge z \wedge w \vee x \wedge y' \wedge z \wedge w \vee x \wedge y \wedge z' \wedge w'$$

9. Use the truth tables for F^2 in Example 7.5 to find $f_{12} \wedge f_7$.

10. Use the product form to represent the minterm given by the following table.

p	q	r	$m(p, q, r)$
0	0	0	1
0	0	1	0
0	1	0	0
0	1	1	0
1	0	0	0
1	0	1	0
1	1	0	0
1	1	1	0

11. Find the sum of products form of the Boolean function given by the following table.

p	q	r	s	$m(p, q, r, s)$
0	0	0	0	1
0	0	0	1	0
0	0	1	0	0
0	0	1	1	0
0	1	0	0	0
0	1	0	1	0
0	1	1	0	0
0	1	1	1	1
1	0	0	0	0
1	0	0	1	1
1	0	1	0	0
1	0	1	1	0
1	1	0	0	0
1	1	0	1	0
1	1	1	0	0
1	1	1	1	1

12. Use the truth tables for F^1 in Example 7.5 to compute each of the following.
   a. $f_2 \wedge f_3$
   b. $f_1 \vee (f_2 \wedge f_3)'$

13. Using the truth tables for F^2 in Example 7.5, compute each of the following.
    a. f'_8
    b. $f_{12} \wedge (f_2 \vee f_5)$

14. Find the sum of products form of the function with the following truth table.

x	y	z	w	g
0	0	0	0	0
0	0	0	1	1
0	0	1	0	0
0	0	1	1	0
0	1	0	0	1
0	1	0	1	0
0	1	1	0	1
0	1	1	1	1
1	0	0	0	0
1	0	0	1	0
1	0	1	0	0
1	0	1	1	1
1	1	0	0	1
1	1	0	1	0
1	1	1	0	0
1	1	1	1	0

15. Find the disjunctive normal form for the Boolean expression

$$x \wedge (y \vee z)' \vee (x' \wedge y \vee z)$$

16. Write the expression $(x \wedge y)' \wedge z$ in terms of each of the following.
    a. $\vee$ and $'$
    b. the **NAND** operator

17. In some of the circuits involving **NAND**'s, we simplified the figures by using a 3-input **NAND**. Write the **NAND** of the three variables x, y, and z in terms of **NAND**'s involving only two variables at a time. Translate this into a circuit diagram for the 3-input **NAND** in terms of 2-input **NAND**'s.

18. The Boolean operator **NOR** (for *not or*) is defined by

$$x\mathbf{NOR}y = (x \vee y)'$$

Show that $\{\mathbf{NOR}\}$ is a complete set of operators.

19. Draw the Karnaugh map for the function with the following truth table. Then simplify the expression.

p	q	r	h
0	0	0	0
0	0	1	0
0	1	0	1
0	1	1	1
1	0	0	1
1	0	1	0
1	1	0	1
1	1	1	0

20. Find a simplified form for the function defined by the following truth table.

p	q	r	s	f
0	0	0	0	1
0	0	0	1	0
0	0	1	0	1
0	0	1	1	0
0	1	0	0	1
0	1	0	1	0
0	1	1	0	1
0	1	1	1	0
1	0	0	0	0
1	0	0	1	0
1	0	1	0	0
1	0	1	1	0
1	1	0	0	1
1	1	0	1	0
1	1	1	0	0
1	1	1	1	0

21. Simplify the expression represented by the Karnaugh map in Figure 7.42.

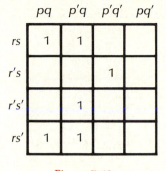

	pq	$p'q$	$p'q'$	pq'
rs	1	1		
$r's$			1	
$r's'$		1		
rs'	1	1		

Figure 7.42.

22. Find the Karnaugh map for the expression

$$(p \vee (q' \wedge r)')'$$

23. Find the Karnaugh map for the expression

$$x \vee (y' \vee z)'$$

Use the map to find a reduced form of the expression. Then simplify the complement of the expression. Which has a simpler form?

24. Use a Karnaugh map to simplify the Boolean function f given by the following table.

u	v	w	x	f
0	0	0	0	1
0	0	0	1	0
0	0	1	0	0
0	0	1	1	1
0	1	0	0	1
0	1	0	1	1
0	1	1	0	1
0	1	1	1	0
1	0	0	0	0
1	0	0	1	1
1	0	1	0	0
1	0	1	1	0
1	1	0	0	1
1	1	0	1	0
1	1	1	0	0
1	1	1	1	0

25. Simplify the expression represented by the Karnaugh map in Figure 7.43.

Figure 7.43.

26. Use Karnaugh maps to find a simpler expression for the function defined by each of the following truth tables.

a.

x	y	z	$f(x, y, z)$
0	0	0	1
0	0	1	1
0	1	0	0
0	1	1	1
1	0	0	1
1	0	1	1
1	1	0	0
1	1	1	0

b.

x	y	z	w	g
0	0	0	0	0
0	0	0	1	0
0	0	1	0	1
0	0	1	1	1
0	1	0	0	0
0	1	0	1	1
0	1	1	0	1
0	1	1	1	0
1	0	0	0	0
1	0	0	1	1
1	0	1	0	0
1	0	1	1	1
1	1	0	0	1
1	1	0	1	0
1	1	1	0	1
1	1	1	1	0

27. How many Boolean functions of four variables take on the value 1 exactly twice? Three times?

28. Find the Karnaugh map for the Boolean function of four variables that is 1 when exactly two of its inputs are 1.

29. Find a simplified version of each of the following Boolean expressions.
    a. $x' \land (q' \lor z) \lor (x' \land q \lor z')$
    b. $rs \lor (q' \land r)' \lor (p' \lor r)'$

30. a. Verify that the two logic circuits pictured in Figure 7.44 are equivalent; that is, they convert the same inputs to the same outputs.

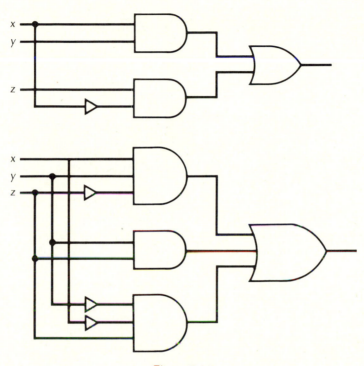

Figure 7.44.

  b. Find a logic circuit that implements the Boolean expression

$$zy' \vee z'y'$$

31. Draw a logic circuit using **OR**'s and **NOT**'s for the expression

$$xyz \vee xy'z' \vee x'yz'$$

32. Trace the addition of 1 and 3 by the 2-bit adder.

33. Draw a circuit diagram for the expression

$$(pqr' \vee pq'r)'p'r$$

  a. using **AND**, **OR**, and **NOT** gates.
  b. using only **AND** and **NOT** gates.
  c. using only **NAND** gates.

34. Identify the expression corresponding to the circuit diagram in Figure 7.45.

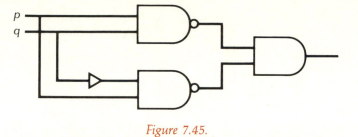

Figure 7.45.

35. Design a circuit that implements an expression with five inputs that is 1 when exactly three of the inputs are 1.

36. Simplify the expression in problem 35, and then repeat parts (a), (b), and (c).

37. a. Identify the expression implemented by the circuit in Figure 7.46.
    b. Simplify the expression found in (a), and then find a circuit implementing the simpler expression.

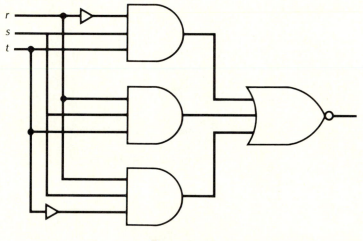

Figure 7.46.

38. Extend the 2-bit adder to a 3-bit adder, using half-adders as components. Show how the new adder works to add 6 and 7.

*39. Derive a circuit for the 2-bit adder that uses only **NAND** gates.

40. The table below defines the action of a 7-position switch that governs the three lights L, M, N. Design a circuit to implement the switch as a binary device encoding the switch positions in each of the following ways.

a. *A* as 0, *B* as 1, and so on
b. So that it minimizes the output expressions, but does not take advantage of the "don't care" conditions
c. So that it minimizes the output expressions, using "don't care" conditions where they help

Switch position	Light		
	L	*M*	*N*
A	ON	OFF	OFF
B	OFF	ON	ON
C	ON	OFF	ON
D	OFF	OFF	OFF
E	OFF	ON	OFF
F	OFF	OFF	ON
G	ON	ON	OFF

41. A 7-element digital display can display each of the digits 0 through 9, as shown in Figure 7.47. Encode the input digits in their standard binary representation, and then find Boolean functions for each of the seven elements in the display. Use "don't care" conditions, if possible.

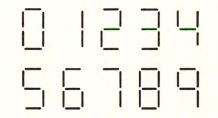

Figure 7.47. *Display of the digits 0–9 with a 7-element display*

Programming Problems

42. Write a program that accepts as input a truth table for a Boolean function of four variables and prints as output the disjunctive normal form of the function.

43. Using the Boolean expressions found in problem 41, write a program that accepts digits in the range 0 through 9 and, for each, displays the digit using the seven elements shown in Figure 7.47. Use bars (|) for the vertical elements and dashes (—) for the horizontal elements.

Self Test 1

1. a.

p	q	r	$p \wedge (q \vee r)'$
0	0	0	0
0	0	1	0
0	1	0	0
0	1	1	0
1	0	0	1
1	0	1	0
1	1	0	0
1	1	1	0

b.

x	y	z	u	$(x \wedge y')' \vee (z \wedge (u \vee z'))$
0	0	0	0	1
0	0	0	1	1
0	0	1	0	1
0	0	1	1	1
0	1	0	0	1
0	1	0	1	1
0	1	1	0	1
0	1	1	1	1
1	0	0	0	0
1	0	0	1	0
1	0	1	0	0
1	0	1	1	1
1	1	0	0	1
1	1	0	1	1
1	1	1	0	1
1	1	1	1	1

2. a. The truth table for $p \vee p'$ is

p	$p \vee p'$
0	1
1	1

so $p \vee p' = 1$.

b. The truth table for $p \wedge 1$ is

p	$p \wedge 1$
0	0
1	1

so $p \wedge 1 = p$.

3. Applying DeMorgan's laws to the outermost ' on the left yields

$$((p \wedge q)' \vee r))' = (p \wedge q)'' \wedge r' = p \wedge q \wedge r'$$

Self Test 2

1. Isolating f_3 and f_8, we get the reduced truth table given below.

b_1	b_2	f_3	f_8	$f_3 \wedge f_8$
0	0	0	1	0
0	1	0	0	0
1	0	1	0	0
1	1	1	0	0

2. The row of the truth table in which m is 1 has $p = 1$, $q = 1$, and $r = 0$. Hence, $m(p, q, r) = p \wedge q \wedge r'$.

3. $f(p, q, r, s) = p' \wedge q' \wedge r' \wedge s' \vee p' \wedge q \wedge r' \wedge s \vee p \wedge q' \wedge r' \wedge s \vee p \wedge q \wedge r' \wedge s'$

Self Test 3

1. Using the clusterings shown in the figure, the expression simplifies to

$$yzw \vee x'yz \vee xy'$$

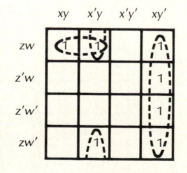

Figure 7.48.

2. The map can be derived easily from the following truth table.

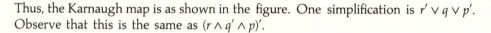

p	q	r	$p' \lor (q' \land r)' \lor (p' \lor r)'$
0	0	0	1
0	0	1	1
0	1	0	1
0	1	1	1
1	0	0	1
1	0	1	0
1	1	0	1
1	1	1	1

Thus, the Karnaugh map is as shown in the figure. One simplification is $r' \lor q \lor p'$. Observe that this is the same as $(r \land q' \land p)'$.

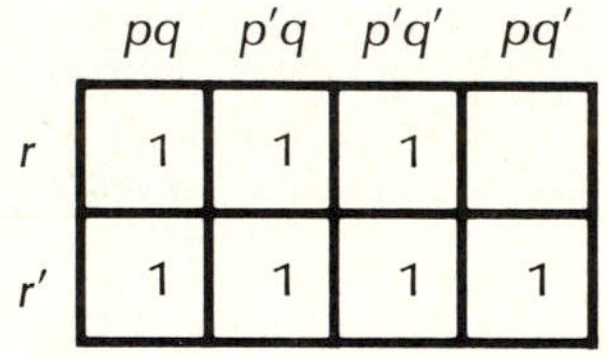

	pq	$p'q$	$p'q'$	pq'
r	1	1	1	
r'	1	1	1	1

Figure 7.49.

3. The map for the function is as shown in the figure. Using the three clusters shown, the expression simplifies to

$$u'v' \lor u'w'x' \lor v'w'x'$$

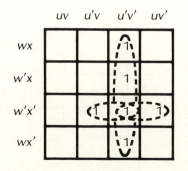

Figure 7.50.

Self Test 4

1. a.
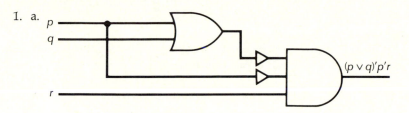

b. To use **AND** and **NOT** gates, we rewrite the expression using DeMorgan's laws, obtaining

$$(p \vee q)'p'r = p'q'p'r = p'q'r$$

The circuit diagram is

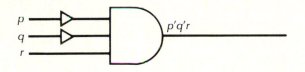

c.
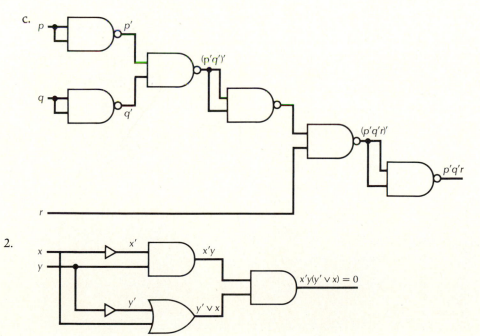

2.

8 Algebraic Systems

When we described the modeling process in Chapter 1, we pointed out that one of our goals was to study mathematical structures that can be used to model something in the real world. To be useful, the structure of a mathematical model must reflect the structure of the objects in the real world that it is supposed to represent.

In previous chapters, we studied numerous examples of models consisting of sets of objects for which there is at least one binary operation defined. We took a basic set of objects and placed them into abstract structures by defining unary and binary operations on the objects in the set and adding a set of fundamental propositions, called *identities*, which codified the important features of the operations. From these identities we derived other useful propositions about the operations. In studying Boolean algebra, for instance, we began with constants and variables, and then defined the two binary operations **AND** and **OR** and the unary operator **NOT**. Then we derived identities such as the associative law and the commutative law. Further work led to additional rules such as DeMorgan's laws.

In this chapter, we discuss the actual mathematical structures. The study of these structures should facilitate our understanding of abstract characterizations of models and provide a basis for creating new models. The mathematical structures we discuss in this chapter are **algebras**, or **algebraic structures**, and the study of them is often referred to as abstract algebra. These algebras have been used in computer science to characterize abstract data structures, to design error-correcting codes for data transmission, to describe the functions computable by classes of machines, and to serve as a basis for programming-language semantics.

We begin with simple algebraic systems known as semigroups, monoids, and groups. The class of monoids includes most of the algebraic structures we have seen in earlier chapters. Groups are more complex structures than monoids and therefore have a richer body of results. Finally, we move on to the study of homomorphisms and isomorphisms. Morphisms are the mechanism by which we study likenesses between algebraic systems.

8.1   Semigroups, Monoids, and Groups

Defining an algebraic system requires specifying a set of objects, one or more operations on the set, and a collection of rules, or *axioms*, that the operations must satisfy. We begin with a simple structure, a semigroup, in which there is a set of objects, one binary operation on these objects, and only two rules the operation must satisfy.

Semigroups

A **semigroup** is a pair $(S, \cdot)$, where S is a nonempty set and $\cdot$ is a binary operation on S that is **associative**. That is, for all $x, y, z \in S$,

$$x \cdot (y \cdot z) = (x \cdot y) \cdot z$$

We have encountered this structure many times before. In order to discuss examples in this section and in later sections, we summarize many of the sets we have studied and their operations in Table 8.1. Several of the sets have more than one operation defined on them, so it is natural to ask which, if any, of the operations are associative.

TABLE 8.1
Some Sets and Operations on Them

Set	Name	Binary operation	Unary operation
$\mathscr{P}(S)$	Power Set	Union, intersection, symmetric difference	Complement
N	Natural numbers	Addition, multiplication	
Z	Integers	Addition, multiplication	Additive inverse
R	Real numbers	Addition, multiplication	Additive inverse
$\mathbf{R}^+$	Positive real numbers	Addition, multiplication	Reciprocal
$R(S)$	Relations on a set S	Composition	Inverse
$F(S)$	Functions on a set S	Composition	
M_n	$n \times n$ logical matrices	Multiplication	
B	Boolean algebra	Conjunction, disjunction	Negation

EXAMPLE 8.1

Verify that each of the following is a semigroup:

1. $(\mathbf{R}^+, *)$, where $*$ represents multiplication
2. $(R(S), \circ)$, where $\circ$ represents composition
3. $(B, \vee)$, where $\vee$ represents disjunction.

Solution

The first of these should be familiar to you from algebra, and we have discussed the latter two in previous parts of this book. For each, we know that the set is closed under the operation. That is, we know that the product of two real numbers is a real number, that the composition of two relations on a set is a relation on the set, and that disjunction applied to two Boolean values yields a Boolean value. We write out the associative law in each case:

1. For all $x, y, z \in \mathbf{R}^+$, associativity of multiplication means that

$$x * (y * z) = (x * y) * z$$

which is well known.

2. Associativity of composition means that for all relations $R_1, R_2, R_3 \in R(S)$, and for all $s \in S$,

$$R_1 \circ (R_2 \circ R_3)(s) = (R_1 \circ R_2) \circ R_3(s)$$

The equality follows directly from the definition of composition.

3. We need to know that for all $p, q, r \in B$,

$$p \vee (q \vee r) = (p \vee q) \vee r$$

which we verified in the previous chapter.

EXAMPLE 8.2

1. Show that the pair $(\{a, b, c\}, \cdot)$ is a semigroup, where $\cdot$ is defined by the following table:

·	a	b	c
a	a	a	a
b	b	b	b
c	c	c	c

2. Show that the pair $(\{a, b, c\}, *)$ is not a semigroup, where $*$ is defined by the following table:

*	a	b	c
a	a	b	c
b	b	a	b
c	c	b	a

Solution

1. We have to show that for all $x, y, z \in \{a, b, c\}$,

$$x \cdot y \in \{a, b, c\} \quad \text{and} \quad x \cdot (y \cdot z) = (x \cdot y) \cdot z$$

The set $\{a, b, c\}$ is closed with respect to the operation $\cdot$ since there are no missing entries in the table and all entries are either a, b, or c. In the absence of a summary organizing principle for the table given, we have to work out each of the 27 possible cases to determine if $\cdot$ is associative. For instance, if $x = a$, $y = a$, and $z = b$, then $x \cdot (y \cdot z) = a \cdot (a \cdot b) = a \cdot a = a$, and $(x \cdot y) \cdot z = (a \cdot a) \cdot b = a \cdot b = a$. To shorten the work, we construct the following table, and we include only a sampling of the 27 results, leaving the rest as an exercise. We expect the last two columns to agree with each other.

x	y	z	$x \cdot (y \cdot z)$	$(x \cdot y) \cdot z$
a	a	a	$a \cdot a = a$	$a \cdot a = a$
a	a	b	$a \cdot a = a$	$a \cdot b = a$
a	b	a	$a \cdot a = a$	$a \cdot a = a$
a	b	b	$a \cdot b = a$	$a \cdot b = a$
b	a	a	$b \cdot a = b$	$b \cdot a = b$
b	a	c	$b \cdot a = b$	$b \cdot c = b$
c	a	b	$c \cdot a = c$	$c \cdot b = c$
c	b	a	$c \cdot b = c$	$c \cdot a = c$
c	c	c	$c \cdot c = c$	$c \cdot c = c$

An alternative solution for this example, which is much quicker, uses the fact that for any x, y, the product $x \cdot y = x$. Hence, each instance of the associative rule becomes trivial.

2. The set $\{a, b, c\}$ is closed with respect to the operation $*$. To show that $*$ is not associative, all we need to do is to find one instance for which the equality $x * (y * z) = (x * y) * z$ fails:

$$(b * b) * c = a * c = c$$

but

$$b * (b * c) = b * b = a$$

Thus the associative property fails.

Consider the set $\{a, b\}$ of two elements called "letters." We are interested in the set of all "words" that can be generated from $\{a, b\}$. We can define a *word* recursively as a letter or a letter followed by a word. Here are the first elements of the set of words, starting with the words of one letter and progressing to longer ones:

$$a, b, aa, ab, ba, bb, aaa, aab, aba, abb,$$
$$baa, bba, bbb, aaaa, aaab, aaba, aabb, \ldots$$

EXAMPLE 8.3

Show that the set of words generated from $\{a, b\}$, together with the operation of **concatenation**, is a semigroup. To concatenate two words, we merely connect the letters of the second word with those of the first word.

Solution

We can always concatenate any two words. Here we concatenate *aabb* and *abbaba*.

$$(aabb)(abbaba) = aabbabbaba$$

Since concatenation of two words using the letters $\{a, b\}$ results in a word using the letters $\{a, b\}$, the set of words over $\{a, b\}$ is closed under concatenation. The associativity of concatenation is obvious from its definition. Consequently, we have a semigroup.

Monoids

A slightly more complex system than a semigroup is a monoid. A **monoid** is a semigroup $(S, \cdot)$ such that S contains a special element e, called the **identity** relative to $\cdot$, satisfying the condition

$$x \cdot e = e \cdot x = x$$

for all $x \in S$.

EXAMPLE 8.4

Show that each of the following semigroups is a monoid by finding its identity element.

1. $(\mathbf{R}^+, *)$
2. $(\mathscr{P}(S), \cup)$
3. $(\mathscr{P}(S), \cap)$

Solution

1. Since multiplying any number by 1 yields the number itself, 1 is the identity in $(\mathbf{R}^+, *)$.
2. $(\mathscr{P}(S), \cup)$ has identity $\varnothing$, since for any subset A of S,

$$A \cup \varnothing = A = \varnothing \cup A$$

3. $(\mathscr{P}(S), \cap)$ has identity S, since for any subset A of S,

$$A \cap S = A = S \cap A$$

EXAMPLE 8.5

1. Consider $(\mathbf{Z}, *)$, where the operation $*$ is defined as $x * y = x + y - 1, x, y \in \mathbf{Z}$. Show $(\mathbf{Z}, *)$ is a monoid, and that 1 is the identity.
2. Consider $(\mathbf{Z}, \cdot)$, where the operation $\cdot$ is defined as $x \cdot y = xy + 1, x, y \in \mathbf{Z}$. Show that $(\mathbf{Z}, \cdot)$ is not a monoid.

Solution

1. First we show that $*$ is associative. Consider $(x * y) * z$. We have $(x * y) * z = (x + y - 1) * z = (x + y - 1) + z - 1 = x - 1 + (y + z - 1) = x * (y * z)$. Since $x * 1 = x + 1 - 1 = x$, and $1 * x = 1 + x - 1 = x$, the identity element is 1 and therefore $(\mathbf{Z}, *)$ is a monoid.
2. We will show that $\cdot$ has no identity element. If $\cdot$ had an identity element z, then for any x, $x \cdot z = z \cdot x = xz + 1 = x$. But $xz + 1 = x$ is equivalent to $1 = x(1 - z)$, which has no solution in the set of integers. Hence, there is no identity for $\cdot$. Therefore, $(\mathbf{Z}, \cdot)$ is not a monoid.

It is not hard to see that the semigroup $(\mathbf{N}, +)$ has no identity, since 0 is not a member of $\mathbf{N}$. Thus, the collection of semigroups is larger than the collection of monoids. Theorem 8.1 states that when a semigroup has an identity element, it can have only one.

Theorem 8.1

Let $(S, \cdot)$ be a monoid with identity element e. Let x be an element of S such that for all $y \in S$, $x \cdot y = y \cdot x = y$. Then $e = x$.

Proof

Since x "behaves like an identity," we can show it is the identity element. We know that $x = e \cdot x$ since e is the identity element. But since x is an identity element, we have $e \cdot x = e$. Hence, $e = x$.

Submonoids

Let $(S, \cdot)$ be a monoid with identity e. A subset S' of S is a **submonoid** of S if $e \in S'$, and if $x \cdot y \in S'$ for all $x, y \in S'$. (That is, S' is **closed** under the operation $\cdot$). A monoid always has at least two (*trivial*) submonoids: the singleton set $\{e\}$ and the entire set S.

EXAMPLE 8.6

Let $\mathbf{Z}^+$ be the set of nonnegative integers $\{0, 1, 2, \ldots ,\}$. $(\mathbf{Z}^+, +)$ is a monoid with identity element 0. Find a submonoid that is not one of the trivial ones.

Solution

We must have a set containing 0 and at least one other element. Since adding two positive integers gives a larger positive integer, maintaining closure under addition will require an infinite set. Arbitrarily, let's start with 0 and 10 in the subset. Then we must have $10 + 10 = 20$ in the submonoid. But then we need 30 and 40, and so on, so we end up with a set consisting of 0 and all positive multiples of 10. This subset defines a submonoid of $(\mathbf{Z}, +)$.

One feature of the submonoid we defined above is that it consists of the identity element and all multiples of an element (10) of $\mathbf{Z}$. In general, for a monoid $(M, \cdot)$ and an element $x \in M$, we can define the **powers** of x as follows:

$$x^0 = e$$

and for all $n \geq 0$,

$$x^{n+1} = x^n \cdot x$$

With $x = 10$ in $\mathbf{Z}$ as above, $x^0 = 0$, $x^1 = 10$, $x^2 = 10 + 10 = 20$, and for arbitrary n, $x^n = 10n$. Observe that we use the term *power of x* in the generic sense. For the case of addition, the "powers" of 10 are the multiples of 10, not the usual (multiplicative) powers.

Several of the monoids we have encountered have a binary operation that exhibits symmetry; that is, $x \cdot y = y \cdot x$ for every element of the monoid. This property is called **commutativity**. When the monoid is defined by a table, a commutative operation can be detected by symmetry of the table with respect to the main diagonal.

Groups

In solving equations, the simplest kind of equation to solve is a linear equation of the form $ax = b$, which has the solution $x = b/a$ whenever a is not 0. Two mathematicians of the nineteenth century, Galois and Abel, made significant contributions to the theory of solvability of equations and in the process helped to create the foundations of modern group theory.

A **group** is a monoid $(G, \cdot)$ with identity e such that for all $x \in G$ there exists an element $y \in G$ such that $x \cdot y = y \cdot x = e$. We call y the **inverse** of x with respect to $\cdot$. It is usually written as x^{-1}. If the group operation is commutative, we call the group a **commutative** group or an **abelian** group, after the mathematician Abel. We will often refer to a generic group operation as a "multiplication," although in particular instances it will have nothing to do with arithmetic multiplication.

EXAMPLE 8.7

Show that $(\mathbf{Z}, *)$ is a group, where $*$ is the binary operation $x * y = x + y - 1$, x, $y \in \mathbf{Z}$.

Solution

From Example 8.5 we know that $(\mathbf{Z}, *)$ is a monoid with identity 1. To show that $(\mathbf{Z}, *)$ is a group, we need to show that for each $x \in \mathbf{Z}$ there is an element $x^{-1} \in \mathbf{Z}$ such that $x * x^{-1} = x^{-1} * x = 1$. Since $x * x^{-1} = x + x^{-1} - 1$, we see that the inverse of x is $(-x + 2)$. Note that $x * (-x + 2) = (-x + 2) * x = -x + 2 + x - 1 = 1$.

Example 8.8 exhibits two monoids that are groups and one monoid that is not a group.

EXAMPLE 8.8

1. Show that the monoid $(\mathbf{Z}, +)$ with identity 0 is an abelian group.
2. Show that the monoid $(B(S), \circ)$, where $B(S) = \{f: S \to S: f \text{ is a bijection}\}$ and $\circ$ indicates composition, is a group. $B(S)$ is abelian if and only if $|S| \leq 2$.
3. Show that the monoid $(\mathbf{N}, \cdot)$ with identity 1 is not a group.

Solution

1. For $m \in \mathbf{Z}$, we need to find $n \in \mathbf{Z}$ such that $m + n = n + m = 0$. The solution is $n = -m$. Thus the (additive) inverse of m is $m^{-1} = -m$. Then $(\mathbf{Z}, +)$ is an abelian group since $m + n = n + m$ for all $m, n \in \mathbf{Z}$.

2. The identity is i_S. Our characterization of the inverse, g, of a function, f, is that $f \circ g = g \circ f = i_S$, which says that the set of bijections on S is a group. In fact, we wrote $g = f^{-1}$ for the compositional inverse. $B(S)$ is called the **permutation group** of the set S.

   We leave it as an exercise for you to show that if S has only two elements, then $B(S)$ is abelian. On the other hand, suppose that S has at least three elements, say a, b, c. We want to define two functions f and g such that $f \circ g$ is not the same as $g \circ f$. To make life simpler, we define $f(x) = g(x) = x$ when $x \in S - \{a, b, c\}$. Thus, the pathological behavior will come from twisting the elements a, b, and c. Define f and g on a, b, c according to the following table:

	a	b	c
f	a	c	b
g	c	b	a

   Then $f \circ g(b) = f(b) = c$, but $g \circ f(b) = g(c) = a$, so $f \circ g$ and $g \circ f$ are distinct.

3. For $m \in \mathbf{N}$, there is no element $n \in \mathbf{N}$ such that $m \cdot n = 1$, unless $m = 1$. Hence $(\mathbf{N}, \cdot)$ is not a group.

Theorem 8.2

Theorem 8.2 summarizes some properties of inverses. Let $(G, \cdot)$ be a group with identity e.

1. For all $x \in G$, the inverse is unique.
2. For all $x \in G$, $(x^{-1})^{-1} = x$.
3. For all $x, y, z \in G$, the **cancellation laws** hold:

$$x \cdot y = x \cdot z \Rightarrow y = z$$

and

$$y \cdot x = z \cdot x \Rightarrow y = z$$

4. If G is finite, then for all x in G there exists a positive integer k such that $0 < k \leq |G|$ and $x^k = e$.

Proof

1. Suppose $\tilde{x}$ behaves like the inverse of x; that is, $x \cdot \tilde{x} = \tilde{x} \cdot x = e$. Then $\tilde{x} = \tilde{x} \cdot e = \tilde{x} \cdot (x \cdot x^{-1}) = (\tilde{x} \cdot x) \cdot x^{-1} = e \cdot x^{-1} = x^{-1}$. Hence, x has only one inverse element.

2. The second property follows from (1) and the fact that by definition of x^{-1}, x behaves like the inverse of x^{-1}.

3. Consider the first equality, $x \cdot y = x \cdot z$. Multiplying both sides by x^{-1} on the left, we get $x^{-1} \cdot (x \cdot y) = x^{-1} \cdot (x \cdot z)$. Applying associativity yields $(x^{-1} \cdot x) \cdot y = (x^{-1} \cdot x) \cdot z$. But $(x^{-1} \cdot x) = e$, so $y = z$.

4. Let $x \in G$. The powers of x form a sequence $x, x^2, x^3, \ldots$. Since G is finite, there must be a repetition at some point. Thus there exist i and j such that $i < j$ and $x^i = x^j$. We can choose the smallest values of i and j for which this is true. In fact, if $|G| = n$, then we can be sure that $i \le n$, and at most $j = n + 1$. Thus $j - i \le n$. Let $k = j - i$. We also know that

$$x^i = x^j = x^i \cdot x^k$$

so multiplying x^i and $x^i \cdot x^k$ on the left by the inverse of x^i yields $e = x^k$.

Subgroups

Let G be a group with respect to the binary operation $\cdot$. A subset H of G is called a **subgroup** of a group $(G, \cdot)$ if H forms a group with respect to the binary operation $\cdot$ that is defined in G.

EXAMPLE 8.9

Show that the set $\mathbf{Z}$ of all integers is a group with respect to addition, and that the set E of all even integers is a nontrivial subgroup of $\mathbf{Z}$.

Solution

We already know that $\mathbf{Z}$ is closed with respect to addition, and that addition is associative. The identity element is 0, and the inverse for $x \in \mathbf{Z}$ is $-x$. Thus $(\mathbf{Z}, +)$ is a group. To show that E is a subgroup, we first show that E is closed with respect to addition. If x and y are any two elements of E, then $x = 2k_1$ and $y = 2k_2$ for some k_1 and k_2. It follows that $x + y = 2k_1 + 2k_2 = 2(k_1 + k_2)$. Thus $x + y \in E$, and E is closed with respect to addition. The fact that addition is associative on E follows directly from the fact that it is associative on $\mathbf{Z}$. The identity element, 0, is an element of E and for $x \in E$, we have $-x \in E$. Therefore, $(E, +)$ is a subgroup of $(\mathbf{Z}, +)$.

The **order** of a group is the number of elements in the group. For an element x in the group G, the **order of x** is the smallest positive integer k such that $x^k = e$.

The order of x is infinite if $x^k \neq e$ for all $k \geq 1$. The **subgroup generated by** an element x of finite order k is the set $\langle x \rangle = \{x, x^2, \ldots, x^k\}$. The *subgroup generated* by an element x of infinite order is $\langle x \rangle = \{e, x, x^{-1}, x^2, (x^{-1})^2, \ldots\}$. (We leave it as an exercise to verify that $\langle x \rangle$ is really a subgroup of G.) The set $\langle x \rangle$ is called a **cyclic** subgroup. If $G = \langle x \rangle$, we say that G is a **cyclic group.**

EXAMPLE 8.10

Let $(G, \cdot)$ be the monoid with the following multiplication table:

$\cdot$	e	p	q	r	s	t
e	e	p	q	r	s	t
p	p	e	s	t	q	r
q	q	s	r	e	t	p
r	r	t	e	q	p	s
s	s	q	t	p	r	e
t	t	r	p	s	e	q

1. Show that $(G, \cdot)$ is a group by finding the inverse of each element of G with respect to $\cdot$.
2. Find the cyclic subgroups of G.

   Solution

1. To find the inverse of an element, we look across the row labeled by the element and find the column in which e appears. In this example, each row and column are permutations of the elements of G. We find that $p^{-1} = p$, $q^{-1} = r$, $r^{-1} = q$, $s^{-1} = t$, and $t^{-1} = s$. Hence, each element has an inverse, so G is a group.
2. The cyclic subgroups of G are the groups generated by the elements of G. For instance, the distinct powers of q are q, $q^2 = r$, and $q^3 = e$. Thus, $\langle q \rangle = \{e, q, r\}$. For the other elements, we find

$$\langle e \rangle = \{e\}$$
$$\langle p \rangle = \{e, p\}$$
$$\langle r \rangle = \{e, q, r\}$$
$$\langle s \rangle = G$$
$$\langle t \rangle = G$$

Since $\langle s \rangle = G$, G is a cyclic group.

We summarize the properties of semigroups, monoids, and groups in the table of Figure 8.1. *S* is a set, and · is a binary operation.

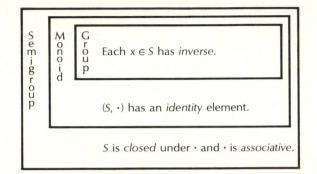

Figure 8.1. *Properties of semigroups, monoids, and groups*

SELF TEST 1

1. Consider the set $S = \{A, B, C, D\}$ and the binary operation · defined by the following table:

·	A	B	C	D
A	D	D	A	A
B	D	C	B	A
C	A	B	C	D
D	A	A	D	D

   a. Show that $(S, \cdot)$ is a semigroup.
   b. Show that $(S, \cdot)$ is a monoid.
   c. Show that $(S, \cdot)$ is not a group.

2. In each of the following parts, the operation · is defined on **Z** by the given rule. Determine in each case if **Z** is a group with respect to ·. State which, if any, conditions fail to hold. When all conditions hold, identify the identity and the inverse of an element x.

   a. $x \cdot y = x + y + 1$
   b. $x \cdot y = x - y$

3. Let E denote the even natural numbers; that is, $E = \{0, 2, 4, \ldots\}$. Show that $(\mathbf{E}, +)$ is a submonoid of $(\mathbf{Z}^+, +)$.

Exercises 8.1

1. Define the binary operation Δ on $\mathbf{N}$ by $n \Delta m = \min(n, m)$, for all n and m in $\mathbf{N}$. Show that $(\mathbf{N}, \Delta)$ is a semigroup but not a monoid.

2. Show that the set $\{a, b, c\}$ with binary operation given by

*	a	b	c
a	c	a	b
b	a	b	c
c	b	c	a

   is a monoid. What is the identity element? What are the submonoids of this monoid?

3. For each set and each binary operation on the set given in Table 8.1, determine if the set is a group. Which of the groups are abelian?

4. Find two binary operations on the set $G = \{x, y, z\}$ such that one is associative but does not have an identity element, and the other has an identity element but is not associative.

5. How many binary operations are there on a set with three elements?

6. Show that the table

·	p	q	r	s
p	p	q	r	s
q	q	p	s	r
r	r	r	r	r
s	s	s	s	s

   defines a monoid. What is the identity element? What are the submonoids of the monoid?

7. For each of the following monoids, find the successive powers of the given element.
   a. 2, in the monoid $(\mathbf{Z}_6, +_6)$
   b. f in the monoid $(F(S), \cdot)$, where $S = \mathbf{Z}$ and $f(x) = 2x$ for all x in $\mathbf{Z}$

*8. Let $(M, \cdot)$ be a monoid with identity e, and let $x \in M$. Verify that the additive power rule holds; that is, for all nonnegative integers n and m,

$$x^n \cdot x^m = x^{n+m}$$

9. Find two distinct groups of order 4. Show that both are abelian.

10. Let $(G, \cdot)$ be a group, and let $x \in G$ be an element of finite order k. Verify the claim that $\langle x \rangle$ is a group.

*11. Let G be a nonempty set that is closed under an associative binary operation $\cdot$. Prove that G is a group if and only if the equations $a \cdot x = b$ and $y \cdot a = b$ have solutions x and y in G for all choices of a and b in G.

12. Write the multiplication table for the set $G = \{1, -1, i, -i, j, -j, k, -k\}$ with multiplication as follows:

$$i^2 = j^2 = k^2 = -1$$
$$ij = -ji = k$$
$$jk = -kj = i$$
$$ki = -ik = j$$

13. Most computers represent numbers with binary sequences of fixed length. Only a finite set of numbers can be represented exactly, and *overflow* occurs when the result of a computation is greater than the largest number that can be represented. For simplicity, consider a computer that uses three bits to represent natural numbers. Let A be the binary sequences of length 3, and let $+$ represent binary addition on the sequences. For example, $011 + 010 = 101$. For each of the following algebras, construct the operation table for $+$ and determine if the algebra is a semigroup, monoid, or group.

   a. The three bits (binary digits) represent the least significant digits of each natural number. The operation $+$ is the usual binary addition, except if overflow occurs, the leading digits are lost. For example, $001 + 010 = 011$, and $101 + 110 = 011$.

   b. The leftmost bit is reserved as an overflow indicator. Its value is 0 if no overflow has occurred and it is 1 if there is overflow. The remaining two bits are used to represent the least significant digits of each natural number. The operation $+$ is the usual addition, except that overflow is set accordingly. For example, $001 + 010 = 011$, $011 + 011 = 110$, and $101 + 111 = 100$.

8.2   Building New Algebras

There are two common ways of building new instances of an algebraic system from existing instances: products and quotients. We considered both methods from a set-theoretical perspective when we defined product sets and partitions of sets in Chapter 2. Here we assume that the sets are monoids and groups, and determine what is necessary to obtain a monoid or group structure on the product set and on certain kinds of partitions.

Product Algebras

Let $A_1, A_2, \ldots, A_n$ be semigroups, and let $A = A_1 \times A_2 \times \cdots \times A_n$ be the product set. We define an operation on elements of A by using each of the component operations. Let $a = (a_1, a_2, \ldots, a_n)$ and $b = (b_1, b_2, \ldots, b_n) \in A$. Define $a \cdot b =$

$(a_1 \cdot b_1, a_2 \cdot b_2, \ldots, a_n \cdot b_n)$. In each coordinate, the operation indicated is taken from the corresponding semigroup. Although tedious, it is not too hard to see that A is closed under the product operation $\cdot$ and that $\cdot$ is associative because each coordinate operation is associative. Hence A is a semigroup, called the **product semigroup**. If each of the A_i are monoids and $e_1, e_2, \ldots, e_n$ are the identity elements in $A_1, A_2, \ldots, A_n$, respectively, then the element $e = (e_1, e_2, \ldots, e_n)$ is the identity element in A, and A is a monoid, called the **product monoid**.

Finally, if each of the coordinate sets is a group, it is not hard to see that the inverse of an element $a = (a_1, a_2, \ldots, a_n)$ relative to the product operation is the element $a^{-1} = (a_1^{-1}, a_2^{-1}, \ldots, a_n^{-1})$. Thus the product of a sequence of groups is a group. Observe also that if each factor is abelian, the product is abelian.

EXAMPLE 8.11

Let $A_1 = (\mathbf{Z}, +)$ and $A_2 = (\mathbf{Z}, *)$. Describe the product monoid $A_1 \times A_2$. What is the identity element? Compute $(1, 3) \cdot (5, -1)$.

Solution

The product set consists of pairs of integers (x, y). The operation on the product set finds the sum of the first coordinates and the product of the second coordinates. For instance $(1, 3) \cdot (5, -1) = (6, -3)$. The identity element is $(0, 1)$.

EXAMPLE 8.12

Consider B^n, the set of binary strings n bits long. For $(a_1, a_2, \ldots, a_n), (b_1, b_2, \ldots, b_n) \in B^n$, define $(a_1, a_2, \ldots, a_n) + (b_1, b_2, \ldots, b_n) = (a_1 + b_1, a_2 + b_2, \ldots, a_n + b_n)$ with $+$ defined on $\{0, 1\}$ by the following table:

+	0	1
0	0	1
1	1	0

Show that $(B^n, +)$ is a group.

Solution

It is easy to show that $(\{0, 1\}, +)$ is an abelian group. The set $\{0, 1\}$ is closed with respect to the binary operation of addition; the operation of addition is both associative and commutative; the identity element is 0; and each element is its own inverse. Therefore, $(B^n, +)$ is an abelian group. Note that the identity element is $(0, 0, \ldots, 0)$, and that the inverse of $(a_1, a_2, \ldots, a_n)$ is $(a_1, a_2, \ldots, a_n)$.

Quotient Algebras

Let $(S, \cdot)$ be a monoid with identity e. Let $\pi = \{A_1, A_2, \ldots, A_n\}$ be a partition of S. We seek a "natural" way to make π a monoid; that is, for each i and j, we need k so that we can write a "block-by-block" operation of the form $A_i \cdot A_j = A_k$.

Suppose we partition $(\mathbf{Z}, *)$ into $\pi = \{E, O\}$, where E is the set of even integers and O is the set of odd integers. What would it mean to "multiply" E times O (or E times E, and so on)? From elementary arithmetic, we know that any even number times any odd number yields an even number, and so on. Hence, we can say that the product of the blocks E and O ought to be E. We arrange these facts in a table:

$\cdot$	E	O
E	E	E
O	E	O

It is easy to check that the partition $\{E, O\}$ becomes a monoid with this "block" multiplication. The identity element is O.

From the example, we see that it makes sense to say that the product of two blocks A and B is a third block, C, only if any element of A times any element of B yields an element of C. This concept of "block arithmetic" is illustrated in Figure 8.2.

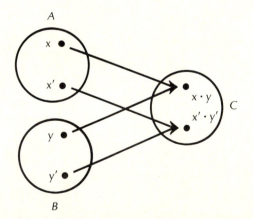

Figure 8.2. "Block-by-block" multiplication on a partition

The procedure for defining block arithmetic does not work for an arbitrary partition of a monoid. Consider the set $S = \{0, 1, 2, 3\}$ with operation $+$ given by the following table:

+	0	1	2	3
0	0	1	2	3
1	1	2	3	0
2	2	3	0	1
3	3	0	1	2

Let S be partitioned into the sets $A = \{0, 1\}$ and $B = \{2, 3\}$. Then it is not possible to define $A + B$ as above. If we choose $x = 1$ and $y = 3$, then $x + y = 0 \in A$, so perhaps we should say $A + B = A$. But $x = 1$ and $y = 2$ yields $x + y = 3 \in B$, suggesting that $A + B = B$.

What we need is a precise criterion that determines which partitions can be given valid binary operations. Recall that any partition of S corresponds to an equivalence relation R on S. For each element $x \in S$, we have an equivalence class $[x]$ consisting of all $y \in S$ that are related to x by R. The "natural" definition of block arithmetic that we tried above says that $[x] \cdot [y] = [x \cdot y]$. This process broke down where pairs of equivalent elements did not combine to give equivalent results. In terms of the relation R, the criterion we need is that if xRy and uRv, then $x \cdot uRy \cdot v$. An equivalence relation on a monoid that satisfies this criterion is called a **congruence relation**. We now state the following theorem. Since we have informally derived the criterion stated, we leave the formal proof as an exercise.

Theorem 8.3

Let $(S, \cdot)$ be a monoid with identity element e. Let R be an equivalence relation on S. Then the partition of S induced by R is a monoid under the operation $[x] \cdot [y] = [x \cdot y]$ if and only if R is a congruence relation. In this case, the identity element is $[e]$.

· · ·

The monoid induced on the partition of a congruence relation is called a **quotient monoid**. The word *quotient* is used because the partition factors the set into blocks of similar elements. If R is the congruence relation on $(M, \cdot)$, we denote the resulting quotient monoid by M/R. A quotient semigroup is similarly defined.

EXAMPLE 8.13

Define the relation R_5 on the set $\mathbf{Z}$ by mR_5n if and only if $m - n$ is a multiple of 5. Show that R_5 is a congruence relation on the monoid $(\mathbf{Z}, +)$. Find the partition of $\mathbf{Z}$ induced by R_5, and write the table for the corresponding quotient operation.

Solution

First we need to show that R_5 is an equivalence relation. It is not hard to see that it is reflexive and symmetric. Suppose, for some m, n, $p \in \mathbf{Z}$, that $m R_5 n$ and $n R_5 p$. Now, $m R_5 n$ means that $m - n = 5k$ for some k, and $n R_5 p$ means that $n - p = 5r$ for some r. Hence, $m - p = (m - n) + (n - p) = 5(k + r)$, so R_5 is transitive. Finally, suppose $m R_5 n$ and $p R_5 q$. We need to show that $(m + p) R_5 (n + q)$. As before, we know that $m - n = 5k$ and that $p - q = 5r$. Hence, $(m + p) - (n + q) = (m - n) + (p - q) = 5k + 5r = 5(k + r)$, so $(m + p) R_5 (n + q)$. This shows that R_5 is a congruence relation.

The next task is to find the equivalence classes. We start with $[0]$.

$$[0] = \{m \in \mathbf{Z}: m - 0 \text{ is a multiple of } 5\}$$
$$= \{0, 5, -5, 10, -10, \ldots\}$$

Similarly,

$$[1] = \{1, 6, -4, \ldots\}$$
$$[2] = \{2, 7, -3, \ldots\}$$
$$[3] = \{3, 8, -2, \ldots\}$$
$$[4] = \{4, 9, -1, \ldots\}$$

There is no need to go further, since every element of $\mathbf{Z}$ is related by R_5 to one of the elements 0, 1, 2, 3, or 4. The quotient operation on the partition $\{[0], [1], [2], [3], [4]\}$ is determined by the equation $[m] + [n] = [m + n]$. For example, $[3] + [4] = [7] = [2]$. The operation, $+_5$, which is called *addition modulo 5*, is summarized in the following table:

$+_5$	$[0]$	$[1]$	$[2]$	$[3]$	$[4]$
$[0]$	$[0]$	$[1]$	$[2]$	$[3]$	$[4]$
$[1]$	$[1]$	$[2]$	$[3]$	$[4]$	$[0]$
$[2]$	$[2]$	$[3]$	$[4]$	$[0]$	$[1]$
$[3]$	$[3]$	$[4]$	$[0]$	$[1]$	$[2]$
$[4]$	$[4]$	$[0]$	$[1]$	$[2]$	$[3]$

The monoid defined in Example 8.13 is denoted by $(\mathbf{Z}_5, +_5)$ and is called the **integers modulo 5**. The relation R_5 can be generalized by replacing the 5 in its definition by an arbitrary integer n. The resulting quotient monoid is called the integers modulo n, and is denoted by $(\mathbf{Z}_n, +_n)$ (see exercise 1 on page 328).

If $(G, \cdot)$ is a group, then for any $x \in G$, we know that $[x] \cdot [x^{-1}] = [x \cdot x^{-1}] = [e]$, so we can say that the inverse of block $[x]$ is the block $[x^{-1}]$. Thus the quotient monoid is a group when $(G, \cdot)$ is a group. The next theorem captures some properties of the quotient when $(G, \cdot)$ is a group.

Theorem 8.4

Let $(G, \cdot)$ be a group with identity element e. Let R be a congruence relation on G.

1. If xRy, then $x^{-1}Ry^{-1}$.
2. $[e]$ is a subgroup of G.
3. For $x, y \in G$, xRy if and only if $x \cdot y^{-1} \in [e]$ (or $x^{-1} \cdot y \in [e]$).
4. For all x in G, $\{x \cdot y : y \in [e]\} = \{y \cdot x : y \in [e]\} = [x]$.

Proof

1. Suppose xRy. By the reflexivity of R, $x^{-1}Rx^{-1}$, so that $x \cdot x^{-1}Ry \cdot x^{-1}$. Thus $eRy \cdot x^{-1}$. But we also know that $y^{-1}Ry^{-1}$, which implies that $y^{-1} \cdot eRy^{-1} \cdot (y \cdot x^{-1})$, and hence $y^{-1}Rx^{-1}$. By symmetry of R, $x^{-1}Ry^{-1}$.

2. $[e]$ contains e. If x and $y \in [e]$, then xRe and yRe, so $x \cdot yRe$. Hence, $[e]$ is closed under the operation $\cdot$. Finally, if $x \in [e]$, we can apply (1) to conclude that $x^{-1}Re^{-1}$. But $e^{-1} = e$, so $x^{-1} \in [e]$.

3. The proof of (3) is left as an exercise.

4. Let $h \in \{x \cdot y : y \in [e]\}$. We show that $h \in \{y \cdot x : y \in [e]\}$ and leave the other half of the argument as an exercise. By assumption, there is some y such that yRe and $h = x \cdot y$. Since R is a congruence on $(G, \cdot)$, we know that hRx. Applying (3), we get $h \cdot x^{-1} \in [e]$, so there exists $y' \in [e]$ with $h \cdot x^{-1} = y'$. The final step is that $h = y' \cdot x$, so that $h \in \{y \cdot x : y \in [e]\}$.

EXAMPLE 8.14

Consider the group $(S, \cdot)$, where $S = \{\alpha, \beta, \gamma, \delta\}$ and the operation $\cdot$ is defined by the following table:

$\cdot$	α	β	γ	δ
α	α	β	γ	δ
β	β	α	δ	γ
γ	γ	δ	α	β
δ	δ	γ	β	α

Consider the congruence relation $R = \{(\alpha, \alpha), (\alpha, \beta), (\beta, \alpha), (\beta, \beta), (\gamma, \gamma), (\gamma, \delta), (\delta, \gamma), (\delta, \delta)\}$ on S. Determine the multiplication table for the quotient group S/R.

Solution

We first compute the equivalence classes determined by the congruence relation R:

$$[\alpha] = \{\alpha, \beta\} = [\beta]$$
$$[\gamma] = \{\gamma, \delta\} = [\delta]$$

Next we construct the multiplication table:

$\cdot_R$	$[\alpha]$	$[\gamma]$
$[\alpha]$	$[\alpha]$	$[\gamma]$
$[\gamma]$	$[\gamma]$	$[\alpha]$

Cosets

We will now examine a special form of congruence relations on groups. Let H be any subgroup of the group $(G, \cdot)$. We define the **left coset**, xH, of H by $xH = \{x \cdot h : h \in H\}$ and the **right coset**, Hx, of H by $Hx = \{h \cdot x : h \in H\}$. A subgroup H of a group $(G, \cdot)$ is said to be a **normal subgroup** if for all x in G, $xH = Hx$.

Part (4) of Theorem 8.4 asserts that the left and right cosets of $[e]$ coincide when $[e]$ is the equivalence class of the identity determined by a congruence relation. The equivalence class $[e]$ of a congruence relation is evidently a normal subgroup. For arbitrary subgroups, the left and right cosets may not be equal. However, we can derive some useful results about them.

Theorem 8.5

Let $(G, \cdot)$ be a group with identity element e. Let H be a subgroup of G.

1. The left cosets (or the right cosets) of H form a partition of G.
2. For all $x, y \in G$, $|xH| = |Hy| = |H|$.

Proof

We prove the results for the left cosets, since the proof for the right cosets is similar.

1. Since H is a subgroup of G and $e \in H$, it follows that for all $x \in G$, $x = x \cdot e \in xH$. Hence, the left cosets are not empty, and their union is G. To finish the proof, we show that if the intersection of xH and yH is not empty, then $xH = yH$. Suppose that $z = x \cdot h = y \cdot h'$ for $h, h' \in H$. Then $x = y \cdot h' \cdot h^{-1}$, so for all $k \in H$, $x \cdot k = y \cdot h' \cdot h^{-1} \cdot k$. Since the last three elements in the product belong to H, we know

that their product also belongs to H, so $x \cdot k \in yH$. Hence, the left coset xH is a subset of the left coset yH. Reversing the roles of x and y shows the opposite containment, so $xH = yH$, as desired.

2. Define the function $f: H \to xH$ by $f(h) = xh$. If f is a bijection, then the sets have the same number of elements. The function f is surjective by its definition, so we need to show only that f is injective. But $f(h) = f(h')$ implies that $xh = xh'$. Applying the cancellation law yields the result $h = h'$. Thus f is injective, and H and xH have the same number of elements.

•   •   •

Since the left (or right) cosets form a partition, they correspond to equivalence classes of an equivalence relation. In particular, $H = [e]$. When the equivalence relation is a congruence on $(G, \cdot)$, then Theorem 8.4 applies, and we know that the left and right cosets coincide. The converse is also true, in that when H is a normal subgroup of $(G, \cdot)$, the equivalence relation corresponding to the cosets of H is a congruence relation. In this instance, we denote the quotient group by G/H. In general, the equivalence relation is not a congruence, and the left and right cosets of a subgroup do not coincide, as Example 8.15 illustrates.

EXAMPLE 8.15

Let $(G, \cdot)$ be the group defined in the following table:

·	e	p	q	r	s	t
e	e	p	q	r	s	t
p	p	q	e	s	t	r
q	q	e	p	t	s	r
r	r	t	s	e	q	p
s	s	r	t	p	e	q
t	t	s	r	q	p	e

Find all the left and right cosets of the subgroup $H = \{e, r\}$.

Solution

$$eH = \{e, r\} \qquad He = \{e, r\}$$
$$pH = \{p, s\} \qquad Hp = \{p, t\}$$
$$qH = \{q, t\} \qquad Hq = \{q, s\}$$
$$rH = \{r, e\} \qquad Hr = \{r, e\}$$
$$sH = \{s, p\} \qquad Hs = \{s, q\}$$
$$tH = \{t, q\} \qquad Ht = \{t, p\}$$

The distinct left cosets of H are eH, pH, and qH. The distinct right cosets of H are He, Hp, and Hq. All have two elements, but the left and right cosets do not coincide. Hence, $\{e, r\}$ is not a normal subgroup.

Lagrange's Theorem

For any subgroup H of a finite group $(G, \cdot)$, we know that the left cosets of H form a partition of G, and we know that the left cosets all have the same number of elements as H. Let k be the number of distinct left cosets. A simple counting argument shows that $|G| = k|H|$, so that $|G|$ is an integer multiple of $|H|$. This result is known as **Lagrange's Theorem**.

Theorem 8.6
Lagrange's Theorem

Let $(G, \cdot)$ be a finite group, and let H be a subgroup of G. Then the order of H divides the order of G.

$$\bullet \qquad \bullet \qquad \bullet$$

Based on the result in Lagrange's Theorem, we define the **index** of a subgroup H in group G by $(G:H) = |G|/|H|$. The index of H in G is the same as the number of left (or right) cosets of H in G.

SELF TEST 2

1. Describe the product monoid $(B^2, \vee) \times (B^3, \vee)$.
2. Consider the group $(\mathbf{Z}, +)$, where $+$ is ordinary addition. Define a relation R on $\mathbf{Z}$ as

$$aRb \text{ if and only if } a^2 - a - 2 = b^2 - b - 2$$

   a. Show that R is an equivalence relation on $\mathbf{Z}$.
   b. Show that R is not a congruence relation on $(\mathbf{Z}, +)$.
3. Show that all subgroups of $(\mathbf{Z}_5, +_5)$ have index 1 or 5.

Exercises 8.2

1. Consider the monoid $(\mathbf{Z}, +)$. Let R_n be the relation on $\mathbf{Z}$ given by xR_ny if and only if $x = y \pmod{n}$. (This means that $x - y$ is a multiple of n.) Show that R is a congruence relation. Find the partitions of $\mathbf{Z}$ induced by R, and write out a partial table showing the induced quotient operation. The resulting quotient

monoid is called the integers modulo n, and is denoted $(\mathbf{Z}_n, +_n)$. Write out tables for $(\mathbf{Z}_2, +_2)$ and $(\mathbf{Z}_3, +_3)$.

2. Replace $(\mathbf{Z}, +)$ by $(\mathbf{Z}, *)$ in exercise 1. Is R_n still a congruence relation? If so, what is the identity element in the quotient monoid? Write out a table for the resulting quotient multiplication operation.

3. Describe the product monoid $(\mathbf{Z}_2 \times \mathbf{Z}_5, +)$.

4. Find the multiplication table for the product of the groups given in the following tables:

·	e	a
e	e	a
a	a	e

·	e	x	y
e	e	x	y
x	x	y	e
y	y	e	x

Then find all cyclic subgroups of the product group. Show that the product group is cyclic.

5. Prove that if $(G_1, \circ)$ and $(G_2, \cdot)$ are cyclic, then $G_1 \times G_2$ need not be cyclic.

6. Consider the monoid $(S, *)$ where $S = \{a, b, c, d\}$ and $*$ is given by the following multiplication table:

*	a	b	c	d
a	a	b	c	d
b	b	a	c	d
c	c	d	c	d
d	d	c	c	d

Consider the congruence relation $R = \{(a, a), (a, b), (b, a), (b, b), (c, c), (c, d), (d, c), (d, d)\}$ on S. Determine the multiplication table of the quotient monoid S/R.

7. Show that the multiplication table for a finite group must consist of rows and columns which are distinct permutations of the elements of the group.

8. Let $(G, \cdot)$ be a group with identity e. Show that $\{x \in G : x^2 \neq e\}$ has an even number of elements.

9. Find a binary operation on $\{x, y, z\}$ that makes it a monoid but not a group.

10. Let $(G, \cdot)$ be a monoid with identity e. Show that if there exists $x \in G - \{e\}$ such that $x^2 = x$, then G is not a group.

11. Show that the monoids $(\mathbf{Z}_n, +_n)$ are cyclic groups.

12. Let $(G, \cdot)$ be a group, and let $x \in G$. Show that $\langle x \rangle$ is the smallest subgroup of G containing x.

*13. Let $(G, \cdot)$ be an abelian group with identity e. Show that $\{x \in G : x^2 = e\}$ is a subgroup of G.

14. Show that if H_1 and H_2 are subgroups of a group, then so is $H_1 \cap H_2$.

15. Show that every cyclic group is abelian.

*16. Show that every subgroup of a cyclic group is cyclic.

17. Let $(G, \cdot)$ be a finite group with more than one element. If the only subgroups of $(G, \cdot)$ are the trivial ones, then $(G, \cdot)$ must be a cyclic group, and the order of G is a prime.

18. Verify part (3) of Theorem 8.4.

19. Show that $H = \{e, p, q\}$ is a normal subgroup of the group in Example 8.15. What is its index? Write out the multiplication table for the quotient group G/H.

20. The group $(S, \cdot_{13})$ where $S = \{1, 5, 8, 12\}$ is a subgroup of the group $(\mathbf{Z}_{13}, \cdot_{13})$. Compute each of the following.

   a. the left coset $3S$
   b. the right coset $S3$

8.3   Morphisms of Algebraic Structures

In this section of the chapter, we examine morphisms between algebraic structures. Morphisms are the mechanism with which we describe likenesses between two structures. Loosely speaking, a morphism is a function from one set to another that preserves the algebraic operations and properties of the underlying algebraic systems. Structurally, a monoid is determined by its associative binary operation and the identity element relative to the operation. For monoids, then, we expect morphisms to map the identity element of the domain monoid onto the identity element in the range, and to preserve the binary operation. For groups, a morphism should also map the inverse of an element to the inverse of the image of the element. We will consider homomorphisms and isomorphisms of monoids and groups. We also show how homomorphisms are closely related to congruence relations.

Homomorphisms Between Monoids

Let $(M, \cdot)$ and $(M', \circ)$ be monoids with identities e and e', respectively. Following the intuitive idea that a morphism is a structure-preserving function, we define a **homomorphism** between $(M, \cdot)$ and $(M', \circ)$ to be a function $f\colon M \to M'$ such that:

1. $f(e) = e'$
2. For all $x, y \in M$, $f(x \cdot y) = f(x) \circ f(y)$.

Condition (2) can be interpreted as "the image of a multiplication of two elements of M is the multiplication of their respective images in M'." This is what we mean when we say that the homomorphism preserves the monoid structure. Figure 8.3 illustrates the meaning of condition (2). In the figure, the result of taking the path through A is the same as the result of taking the path through $B \times B$. When this happens, the diagram is said to commute.

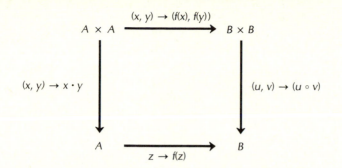

$$A \times A \xrightarrow{\quad (x, y) \to (f(x), f(y)) \quad} B \times B$$

$(x, y) \to x \cdot y$ $(u, v) \to (u \circ v)$

$$A \xrightarrow{\quad z \to f(z) \quad} B$$

Figure 8.3. Commuting diagram for monoid homomorphisms

EXAMPLE 8.16

1. Let $(M, \cdot)$ be the monoid $(\mathbf{Z}^+, +)$, and let $(M', \circ)$ be the monoid $(\mathbf{N}, *)$. Let f be the function $f: \mathbf{Z}^+ \to \mathbf{N}$ given by $f(m) = 2^m$, $m = 0, 1, 2, \dots$. Show that f is a monoid homomorphism.

2. Let $(M, \cdot)$ and $(M', \circ)$ be the monoids given in the following tables. Define $g: M \to M'$ as shown in Figure 8.4. Show that g is a monoid homomorphism.

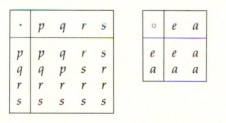

·	p	q	r	s
p	p	q	r	s
q	q	p	s	r
r	r	r	r	r
s	s	s	s	s

∘	e	a
e	e	a
a	a	a

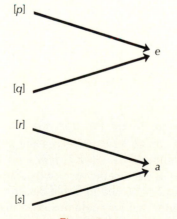

[p]

[q]   e

[r]

[s]   a

Figure 8.4.

3. Let $(M, \cdot)$ be the monoid $(\mathbf{Z}, +)$, and let $(M', \circ)$ be the monoid $(B, \vee)$. Define the function $h\colon \mathbf{Z} \to B$ by $h(m) = 0$ if m is even and $h(m) = 1$ if m is odd. Show that h is not a monoid homomorphism.

Solution

1. The identity element in $(\mathbf{Z}^+, +)$ is 0, and the identity element in $(\mathbf{N}, *)$ is 1. The first thing we need to check is that $f(0) = 1$. Since $2^0 = 1$, this is the case. Next we need to check that for all m, n in $\mathbf{Z}^+$, $f(m + n) = f(m) * f(n)$ in $\mathbf{N}$. This translates into the equality $2^{m+n} = 2^m * 2^n$, which is true because of the additive rule for exponents. Thus f is a homomorphism.

2. The identity element of $(M, \cdot)$ is p, and the identity element in $(M', \circ)$ is e. Since $g(p) = e$, g preserves the identity. To show that g preserves the binary operation, we check each of the 16 occurrences of the equality $g(x \cdot y) = g(x) \circ g(y)$ in Table 8.2.

TABLE 8.2

x	y	$x \cdot y$	$f(x)$	$f(y)$	$f(x \cdot y)$	$f(x) \circ f(y)$
p	p	p	e	e	e	e
p	q	q	e	e	e	e
p	r	r	e	a	a	a
p	s	s	e	a	a	a
q	p	q	e	e	e	e
q	q	p	e	e	e	e
q	r	s	e	a	a	a
q	s	r	e	a	a	a
r	p	r	a	e	a	a
r	q	r	a	e	a	a
r	r	r	a	a	a	a
r	s	r	a	a	a	a
s	p	s	a	e	a	a
s	q	s	a	e	a	a
s	r	s	a	a	a	a
s	s	s	a	a	a	a

The last two columns agree, so the function g is a monoid homomorphism.

3. Since $h(0) = 0$, the function at least preserves the identity element. The second condition to check is that for all m, n in $\mathbf{Z}^+$, $h(m + n) = h(m) \vee h(n)$ in B. Since the definition of h depends only on whether or not its argument is even or odd, we can arrange the conditions to be checked in Table 8.3. In order to be a monoid homomorphism, the results in the last two columns must match. The first three rows do match, but in the last row the match fails, so the function is not a homomorphism.

TABLE 8.3

m	n	$m + n$	$h(m) \vee h(n)$	$h(m + n)$
even	even	even	$0 \vee 0 = 0$	0
even	odd	odd	$0 \vee 1 = 1$	1
odd	even	odd	$1 \vee 0 = 1$	1
odd	odd	even	$1 \vee 1 = 1$	0

In the functions in (1) and (2) of Example 8.16, observe that the image of f is the set $f(\mathbf{Z}^+) = \{2^m : m \in \mathbf{Z}^+\}$, which is a submonoid of the codomain, $(\mathbf{N}, *)$. Similarly, the image of g is $\{e, a\} = M'$, which is a (trivial) submonoid of $(M', \circ)$. Also observe that $f^{-1}(1) = \{m \in \mathbf{Z}^+ : f(m) = 1\} = \{0\}$, which is a submonoid of the domain, $\mathbf{Z}^+$. In (2), $g^{-1}(e) = \{p, q\}$, which is also a submonoid of the domain of g. We generalize these observations about properties of homomorphisms in Theorem 8.7. We also include a result relating monoid homomorphisms to congruence relations, which we'll use later in The Fundamental Theorem of Monoid Homomorphisms.

Theorem 8.7

Let $(M, \cdot)$ and $(M', \circ)$ be monoids with identity elements e and e', respectively. Let $f: M \to M'$ be a monoid homomorphism.

1. The image of M by f is a submonoid of $(M', \circ)$.
2. $f^{-1}(e')$ is a submonoid of $(M, \cdot)$.
3. Let R_f be the relation on M defined by xR_fy if and only if $f(x) = f(y)$. Then R_f is a congruence relation on $(M, \cdot)$.

Proof

1. The image of M by f is the set $f(M) = \{f(x) : x \in M\}$. Since $f(e) = e'$, $e' \in f(M)$. We need to show that $f(M)$ is closed under the operation $\circ$. Let $f(x)$ and $f(y) \in f(M)$. To see that $f(x) \circ f(y) \in f(M)$, observe that $f(x) \circ f(y) = f(x \cdot y)$, since f is a monoid homomorphism.
2. Again, the fact that $f(e) = e'$ implies that $e \in f^{-1}(e')$. To see that $f^{-1}(e')$ is closed under the operation $\cdot$, let x and $y \in f^{-1}(e')$. Then $f(x) = f(y) = e'$, so $f(x \cdot y) = f(x) \circ f(y) = e' \circ e' = e'$. Hence, $x \cdot y \in f^{-1}(e')$.
3. We leave the fact that R_f is an equivalence relation on M as an exercise. We need to show that if xR_fy and uR_fv, then $x \cdot uR_fy \cdot v$. Now xR_fy means that $f(x) = f(y)$, and uR_fv means that $f(u) = f(v)$. Hence, $f(x \cdot u) = f(x) \circ f(u) = f(y) \circ f(v) = f(y \cdot v)$, so $x \cdot uR_fy \cdot v$, as desired. Thus R_f is a congruence relation on M.

$\bullet \qquad \bullet \qquad \bullet$

We have already commented on parts (1) and (2) of Theorem 8.7 in connection with the homomorphisms in Example 8.16. We know that congruence relations on a monoid G generate a quotient monoid. The elements of the quotient are equivalence classes $[x]$, $x \in G$. The multiplication on the quotient is defined by the formula $[x] \cdot [y] = [x \cdot y]$. Another way of looking at the equality is that it says that the map $x \to [x]$ is a monoid homomorphism!

Isomorphisms of Monoids

We now apply part (3) of Theorem 8.7 to the homomorphism g in the second part of Example 8.16. To begin with, R_g partitions the set $M = \{p, q, r, s\}$ into the two blocks $[p] = \{p, q\}$ and $[r] = \{r, s\}$. Since R_g is a congruence relation, we can write a table for the quotient monoid operation and compare the quotient with the monoid $(M', \circ)$.

$\cdot$	$[p]$	$[r]$
$[p]$	$[p]$	$[r]$
$[r]$	$[r]$	$[r]$

$\circ$	e	a
e	e	a
a	a	a

The interesting thing about the two tables is that they really define the *same* monoid. The only differences are superficial; the objects $[p]$ and $[r]$ play exactly the same role under operation $\cdot$ as e and a do under operation $\circ$. When two apparently different monoids turn out to be the same in the sense we have described, we say that they are **isomorphic**. To make the notion of isomorphism precise, we note that the mapping

$$[p] \to e \quad \text{and} \quad [r] \to a$$

is a monoid homomorphism that is also a bijection. Formally, a **monoid isomorphism** is a bijective monoid homomorphism. When monoid $(M, \cdot)$ is isomorphic to monoid $(M', \circ)$, we write $(M, \cdot) \cong (M', \circ)$ (and sometimes just $M \cong M'$).

Fundamental Theorem of Homomorphisms for Monoids

There are two important points to discuss in relation to the example on page 331. The first is that whenever f is a monoid homomorphism, the relation R_f is a congruence on the domain of f. Secondly, if f is surjective, then the quotient monoid M/R_f is isomorphic to the range of f. We capture these results in the following theorem, which is known as the **Fundamental Theorem of Homomorphisms for Monoids**. The theorem states that the only homomorphic images of a monoid are its quotient monoids. By *only* we mean up to an isomorphism. The significance of the theorem, then, is that knowledge of the homomorphisms of a monoid is equivalent to knowledge of the congruence relations on the monoid.

Theorem 8.8
Fundamental Theorem of Homomorphisms for Monoids

Let f be a surjective homomorphism from the monoid $(M, \cdot)$ to the monoid $(M', \circ)$. Then R_f is a congruence relation on $(M, \cdot)$ and $M/R_f \cong M'$. Conversely, if R is a congruence on the monoid $(M, \cdot)$, there exists a surjective monoid homomorphism $h \colon M \to M/R$ such that $R = R_h$.

Proof

Let f be a surjective homomorphism from the monoid $(M, \cdot)$ onto the monoid $(M', \circ)$. We know that R_f is a congruence relation on $(M, \cdot)$. We need an isomorphism $\tilde{f}$ from M/R_f to M'. Define $\tilde{f}([x]) = f(x)$. If $y \in [x]$, then $f(y) = f(x)$, so the definition of $\tilde{f}$ is unambiguous. Also, $\tilde{f}([x] \cdot [z]) = \tilde{f}([x \cdot z]) = f(x \cdot z) = f(x) \circ f(z) = \tilde{f}([x]) \circ \tilde{f}([z])$ for all x and z in M. Thus $\tilde{f}$ is a monoid homomorphism. The final thing to check is that $\tilde{f}$ is a bijection. But f is assumed to be surjective, so $\tilde{f}$ is also surjective. To see that $\tilde{f}$ is injective, assume that $\tilde{f}([x]) = \tilde{f}([z])$. Then $f(x) = f(z)$. We did not assume that f is injective, so we cannot conclude that $x = z$. However, we do know that $f(x) = f(z)$ implies that $x R_f z$, so $[x] = [z]$. This is enough to conclude that $\tilde{f}$ is injective. Putting all these facts together, we see that $M/R_f \cong M'$. The (commuting) diagram in Figure 8.5 shows the relationship between f and $\tilde{f}$.

The proof of the converse is left as an exercise.

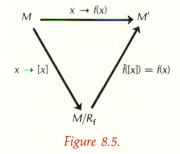

Figure 8.5.

EXAMPLE 8.17

As we saw in an earlier problem, most computers represent numbers as binary sequences of fixed length, n. Consider a computer that uses three bits to represent nonnegative integers; that is, as elements of B^3. For each of the following functions f, determine whether f is a homomorphism from $(\mathbf{Z}^+, +)$ to the specified algebra $(B^3, +)$. In each case, the operation $+$ is based on binary addition that includes carrying.

1. The three bits represent the least significant digits of each nonnegative integer. The operation $+$ is the usual binary addition except if overflow occurs, the leading

digits are lost. The function $f: \mathbf{Z}^+ \to B^3$ is defined as follows:

n	$f(n)$
0	000
1	001
2	010
3	011
4	100
5	101
6	110
7	111
8	000
9	001
⋮	⋮

We can describe the function f as the composition of two other functions h and g, where $h: \mathbf{Z}^+ \to \mathbf{Z}_8$ is defined as $f(n) = [n]$ modulo 8, and $g: \mathbf{Z} \to B^3$ is defined as $g([n]) = [n_{\text{binary}}]$. The function f can be defined as $f(n) = g(h(n))$. That is, f maps a nonnegative integer n into the binary representation of n modulo 8. For example, $f(13) = 101$, since the remainder of dividing 13 by 8 is 5, and the binary representation of 5 is 101.

2. The leftmost bit is reserved as an overflow indicator. Its value is 0 if no overflow has occurred, and its value is 1 if there is overflow. The remaining two bits are used to represent the least significant digits of each natural number. The function $f: \mathbf{Z}^+ \to B^3$ is defined as follows:

n	$f(n)$
0	000
1	001
2	010
3	011
4	100
5	101
6	110
7	111
8	100
9	101
⋮	⋮

Here, the function f maps nonnegative integers less than 4 to their exact binary values. For integers greater than or equal to 4, the remainder of division by 4 is first obtained. This is expressed in binary notation as the two rightmost bits. The third bit (the leading bit) is set to 1 to indicate that overflow has occurred. For example, $f(13) = 101$. This is because the remainder of 13 divided by 4 is 1, and the leading bit is set to 1.

Solution

1. Since the relation R that partitions $\mathbf{Z}^+$ into equivalence classes modulo 8 is a congruence relation, we conclude from the Fundamental Theorem that $h: \mathbf{Z}^+ \to \mathbf{Z}^+/R$ is a surjective monoid homomorphism such that $R = R_h$. The function g from $\mathbf{Z}^+/R$ to B^3 is an isomorphism. It maps the equivalence classes $[0], [1], \ldots, [7]$ to the corresponding binary value. Thus $g([0]) = 000, g([1]) = 001, \ldots, g([7]) = 111$. Since $f = g \circ h$, f is a homomorphism from $(\mathbf{Z}^+, +)$ to $(B^3, +)$.

2. We will argue this case directly. First, note that $f(0) = 000$, and 000 is the identity in $(B^3, +)$. Thus f preserves the identity. Next we need to show that $f(x + y) = f(x) + f(y)$. This is true since, aside from the overflow bit, we have essentially the situation as in part (1). As far as overflow is concerned, if x, y, and $x + y$ are all less than 4, or 2^{3-1}, then the overflow bit will not be set for either $f(x + y)$ or $f(x) + f(y)$. If either x or $y \geq 4$, then the overflow bit will be set for $f(x + y)$ and for $f(x) + f(y)$. Thus f is a homomorphism from $(\mathbf{Z}^+, +)$ to $(B^3, +)$.

Group Homomorphisms

Since groups are monoids, any result that applies to monoid homomorphisms will apply to groups. But groups have an additional property, namely that each element has an inverse. We will incorporate this property in defining a homomorphism between groups. Let $(G, \cdot)$ and $(G', \circ)$ be groups with identity elements e and e', respectively. Let $f: G \to G'$ be a function. We say that f is a **group homomorphism** if the following conditions are met:

1. $f(e) = e'$
2. $f(x \cdot y) = f(x) \circ f(y)$ for all $x, y \in G$
3. $f(x^{-1}) = (f(x))^{-1}$ for all $x \in G$

A **group isomorphism** is a bijective group homomorphism. We use the same symbol to denote isomorphic groups as we did for monoids ($\cong$).

EXAMPLE 8.18

Let $f: \mathbf{Z}_4 \to \mathbf{Z}_4$ be defined by $f([x]) = [2 * x]$ for $x = 0, 1, 2, 3$.

1. Show that f is a group homomorphism.
2. Find $f^{-1}(0)$ and $f(\mathbf{Z}_4)$.

Solution

1. f acts by multiplying the integers modulo 4 by 2. Since the set is small, we can write f in tabular form:

$[x]$	$[0]$	$[1]$	$[2]$	$[3]$
$f([x])$	$[0]$	$[2]$	$[0]$	$[2]$

Certainly $f([0]) = [0]$, so the first condition for homomorphisms is true. To verify condition (2), we must show that $[2 * (x + y)] = [2 * x] + [2 * y]$, which is straightforward to check from the table. To verify condition (3), we must show that $f([x]^{-1}) = (f([y]^{-1})$. To make it easier, we observe that for $x = 0, 1, 2, 3, [x]^{-1} = [4 - x]$. Using this fact and the table, you should be able to check that the function preserves inverses.

2. From the table, it is clear that $f^{-1}([0]) = \{[0], [2]\}$ and that $f(\mathbf{Z}_4) = \{[0], [2]\}$.

Our objective is to find a Fundamental Theorem of Homomorphisms for groups analogous to the theorem we found for monoids. There is a twist in the theorem for groups, which arises because of the correspondence between congruence relations on a group and normal subgroups. Anticipating this, we define, for any group homomorphism $f: G \rightarrow G'$, the **kernel** of f to be the set $K_f = f^{-1}(e')$, where e' is the identity in G'. Theorem 8.9 contains some useful properties of the kernel of a group homomorphism.

Theorem 8.9

Let $(G, \cdot)$ and $(G', \circ)$ be groups with identity elements e and e', respectively. Let $f: G \rightarrow G'$ be a group homomorphism.

1. K_f is a normal subgroup of G.
2. f is an injection if and only if $K_f = \{e\}$.

Proof

1. First of all, we know, by the second part of Theorem 8.2 that K_f is a submonoid of G. To show that K_f is a subgroup, it is enough to show that $x^{-1} \in K_f$ when-

ever $x \in K_f$. But if $f(x) = e'$, then $f(x^{-1}) = (f(x))^{-1} = e'^{-1} = e'$. Hence, K_f is closed under the inverse operation, and K_f is a subgroup of G. To see that K_f is a normal subgroup, let $x \in G$. We must show that $xK_f = K_f x$. We will show that $y \in K_f$ if and only if $x \cdot y \cdot x^{-1} \in K_f$, which means that there exists $z \in K_f$ such that $x \cdot y \cdot x^{-1} = z$. Consequently, $y \in K_f$ if and only if there exists $z \in K_f$ such that $x \cdot y = z \cdot x$. But $y \in K_f$ if and only if $f(y) = e'$, which is equivalent to the statement

$$f(x \cdot y \cdot x^{-1}) = f(x) \circ f(y) \circ f(x^{-1}) = f(x) \circ f(x^{-1}) = e'$$

The latter equality establishes the claim.

2. Certainly if f is injective, K_f can contain only one element, and since $f(e) = e'$, we must have $K_f = \{e\}$. Conversely, suppose that $K_f = \{e\}$ and that $f(x) = f(y)$. Then $f(x \cdot y^{-1}) = f(x) \circ f(y^{-1}) = f(x) \circ f(x)^{-1} = e'$. Hence, $x \cdot y^{-1} = e$. Multiplying both sides on the right by y^{-1} yields $x = y$. Thus f is injective.

$$\bullet \qquad \bullet \qquad \bullet$$

The second part of Theorem 8.9 makes it easy to decide whether a group homomorphism, f, is an isomorphism. A function f is surjective if its image group is the codomain, and f is injective if its kernel is $\{e\}$. The utility of the Fundamental Theorem of Homomorphisms for Groups is that it characterizes group homomorphisms according to their images and their kernels.

Fundamental Theorem of Homomorphisms for Groups

We are now ready to state the Fundamental Theorem of Homomorphisms for Groups. Its form is very similar to the form of the theorem for monoids. The main difference is that rather than writing G/R_f, we write G/K_f, where K_f is the kernel of the map f. Otherwise, the proof is similar to the version for monoids, and we omit the details.

Theorem 8.10
Fundamental Theorem of Homomorphisms for Groups

Let f be a surjective homomorphism from the group $(M, \cdot)$ to the group $(M', \circ)$. Then K_f is a normal subgroup of $(G, \cdot)$, and $G/K_f \cong G'$. Conversely, if K is a normal subgroup of the group $(G, \cdot)$, there exists a surjective group homomorphism $h: G \to G/K$ such that $K = K_h$.

SELF TEST 3

1. Let $(M, \cdot)$ be the monoid $(\mathbf{Z}^+, +)$, and let $(M', \circ)$ be the monoid $(\{0, 1\}, \Delta\}$, where Δ is defined as $0 \, \Delta \, 0 = 1 \, \Delta \, 1 = 0$ and $0 \, \Delta \, 1 = 1 \, \Delta \, 0 = 1$. The function $h: \mathbf{Z}^+ \to \{0, 1\}$ maps even integers to 0 and odd integers to 1. Show that h is a monoid homomorphism.

2. If G is an abelian group and the group G' is a homomorphic image of G, prove that G' is abelian.

3. Each of the following rules determines a function $\theta: G \to G$, where G is the group of all nonzero real numbers under multiplication. Decide in each case whether or not θ is a homomorphism, and find the kernel for those that are homomorphisms.
   a. $\theta(x) = |x|$ c. $\theta(x) = -x$
   b. $\theta(x) = 1/x$ d. $\theta(x) = x^2$

Exercises 8.3

1. Find all the nontrivial congruence relations on the monoids defined in the two multiplication tables. For each congruence relation R, find the appropriate quotient monoid and the homomorphism guaranteed by the Fundamental Theorem of Homomorphisms.

a.

	p	q	r	s
p	p	q	r	s
q	q	p	s	r
r	r	r	r	r
s	s	s	s	s

b.

	e	a	b	c	d
e	e	a	b	c	d
a	a	b	c	d	d
b	b	c	d	d	d
c	c	d	d	d	d
d	d	d	d	d	d

2. a. Show that properties (2) and (3) in the definition of group homomorphism on page 330 together imply property (1).
   b. Show that property (2) in the definition of group homomorphism on page 330 implies property (3). Hence we can replace the definition by property (2) alone. Why does this not work for monoids?

3. Let $f: \mathbf{Z}_5 \to \mathbf{Z}_5$ be defined by $f([x]) = [2x]$ for $x \in \{0, 1, 2, 3, 4\}$. Is f a group homomorphism? Why or why not?

4. Find homomorphisms of the group of Example 8.15 onto the subgroups $\{e, p, q\}$ and $\{e, r\}$, respectively, or show that none exist. For each homomorphism, find the kernel and carry out the steps to identify the quotient group.

5. Let $(G, \cdot)$ and $(G', \circ)$ be groups, and let H denote the product group $G \times G'$. Find a homomorphism of H onto G and show that the kernel is isomorphic to G'.

6. Let $(G, \cdot)$ be a group and let $x \in G$. Define the function $T_x: G \to G$ by $T_x(y) = x \cdot y \cdot x^{-1}$, for all $y \in G$.

   a. Show that T_x is an isomorphism. Under what circumstances is T_x not the identity function?

   b. Show that $\{T_x: x \in G\}$ is a subgroup of the permutation group of G (see Example 8.8).

7. Let $(G, \cdot)$ be a finite abelian group. Let r be a positive integer, $r \leq |G|$. Define $f: G \to G$ by $f(x) = x^r$. Show that f is a group homomorphism.

8. Let $f: G \to H$ be a surjective group homomorphism. Show that if G is cyclic, H is also cyclic.

Group Codes

An interesting application of groups is encoding messages for transmission. In our context, a message is a string of bits (0 or 1). A message is thus an element of B^m for some m. Errors can be introduced in data transmission through a variety of means such as hardware failure and interference. To offset such errors, certain "redundancy" coding schemes are used that allow for the detection or even correction of the most likely errors.

Instead of sending a message directly, we intervene at both the sending and receiving ends with encoding and decoding mechanisms such as the ones shown in Figure 8.6.

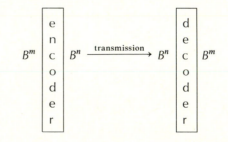

Figure 8.6. Data transmission

Given a binary message $a \in B^m$, the encoder is used to encode it into another binary string $a' \in B^n$. This encoded string is then transmitted, and the decoder is used to reconstruct the message a.

Error-Detecting Codes

Suppose that we are interested in transmitting a single binary digit. That is, we are interested in transmitting either a 1 or a 0. How can we guard against errors? One way to do this is to transmit the message more than once. For example, the message 0 could be encoded as 000 and the encoded message transmitted. Similarly, 1 could be encoded as 111 and then transmitted. These two 3-tuples are the only legitimate words in this code. If no errors occur during transmission, we have the situation shown in Figure 8.7 when we transmit a 0.

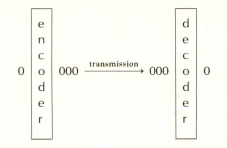

Figure 8.7. Data transmission without error

Suppose that there is a transmission error so that instead of receiving 000 to be decoded, we receive 001. Because this is not a legitimate word, we know that a transmission error has occurred. What if we receive 011, 101, or 110? Here again, the words we receive are not legitimate words, so we will detect the error. In fact, the only time we will be fooled is if we receive 111 when we should have received 000 (or if we receive 000 when we should have received 111). Because this code is able to detect any combination of one or two errors, it is called a *double-error detecting* code.

Error-Correcting Codes

Suppose that we not only want to be able to detect errors but also to correct them. If in the example above we receive 001 to be decoded, we know that a transmission error has occurred, but we do not know if it was a single or a double error. If the actual coded message was 000, then a single error occurred. If the actual coded message was 111, then a double error occurred. If we assume that one error is more likely than two errors, we can assume that the coded message was 000 instead of 111 and, therefore, decode it as 0. This approach minimizes the probability of a mistake in our decoding decision, and is known as the **maximum-likelihood** estimation method.

Using this approach with our encoding scheme, we are able to correct single errors, but we will incorrectly decode double errors. This type of code is called a **single-error correcting** code. Using this approach, the decoder decodes every 3-tuple it receives as the "nearest" legitimate code word. Figure 8.8 illustrates this situation.

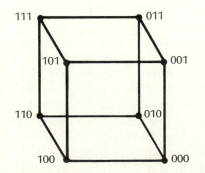

Figure 8.8. Graph of all possible 3-tuples

The vertices of the graph represent all possible 3-tuples. An edge connects two vertices if and only if the corresponding codes differ only in one coordinate. The number of coordinates in which two n-tuples differ is called the **Hamming distance** between the n-tuples. The "nearer" the codes are to each other, the shorter the path between them, and the smaller the Hamming distance between them. For example, the codes connected to 000 by a path of length 1 are 100, 010, and 001. The Hamming distance between these codes and 000 is 1. Since they are "close" to 000, they should be decoded as 000. Also note that the Hamming distance between 000 and 111 is 3. These codes are as far apart as possible.

The problem of designing a good code is essentially one of finding a subset of n-tuples that are far from each other. There are several approaches to solving this problem. One of the most promising ones is based on group theory.

Group Codes

A **group code** is a code where the n-tuple code words form a group with respect to the operation $+$, which is defined as componentwise addition modulo 2. That is, if $x = (x_1, x_2, \ldots, x_n)$ and $y = (y_1, y_2, \ldots, y_n)$, then $x + y = (x_1 +_2 y_1, x_2 +_2 y_2, \ldots, x_n +_2 y_n)$. For example, $101 + 110 = 011$. The code

$$0 \to 000$$
$$1 \to 111$$

is a group code with 000 as the identity element. As another example, consider a code used to encode two bits:

$$00 \to 0000$$
$$01 \to 0101$$
$$10 \to 1010$$
$$11 \to 1111$$

This code is also a group code with identity 0000 and with each element as its own inverse. Note that we constructed this code by using the bits we want to transmit as the initial bits of the code.

Given the operation $+$, as defined above, we can easily compute the Hamming distance between two n-tuples. To compute the distance, we simply add the n-tuples and count the number of 1's. Thus the Hamming distance between 111 and 101 is the number of 1's in $111 + 101 = 010$, or 1.

We can use the Hamming distance to determine how many errors a code can detect and correct. The *minimum distance* of a code whose words are n-tuples is the minimum of the Hamming distances between all possible pairs of n-tuples in that code. It can be shown that a code can detect all combinations of d or fewer errors if and only if its minimum distance is at least $d + 1$. The error-correction capability of a code is also a function of the minimum distance. It can also be shown that a code can correct every combination of d or fewer errors if and only if its minimum distance is at least $2d + 1$.

For the code

$$0 \rightarrow 000$$
$$1 \rightarrow 111$$

the minimum distance is obviously 3. The code can, therefore, detect any combination of two or fewer errors, and can correct one error. Similarly, for the code

$$00 \rightarrow 0000$$
$$01 \rightarrow 0101$$
$$10 \rightarrow 1010$$
$$11 \rightarrow 1111$$

the minimum distance is 2. Thus the code is a single-error detecting code, and it cannot correct any errors.

Decoding

For any code, the problem of decoding reduces to finding the code word nearest to the n-tuple being decoded. If we have a group code that is a single error correcting code, we can use group theory to solve the problem. Consider the case where the true code word is x, but we receive the n-tuple x' to be decoded. If ε is the n-tuple that has 1's in coordinates in which x' has errors, then we can decode x' as $x = x' + \varepsilon$. (Why?) The problem, of course, is finding ε. In order to find ε, we simply enumerate all cosets of the form $C + \varepsilon$, where C is the group code, and each ε is an n-tuple with one or zero 1's.

For example, consider the code

$$0 \rightarrow 000$$
$$1 \rightarrow 111$$

one more time. We can arrange our coset calculations in table form. Each column consists of the coset where $C = \{000, 111\}$, and ε is the first entry at the top of the column. Thus the first column of 3-tuples is $C + 000$, the second column is $C + 001$, and so on. Each row of the table yields all the 3-tuples with Hamming distance 1 from each other. To decode x', a 3-tuple with an error, simply locate it in the table. The last entry in the row in which it is located is the decoded value. Thus 010 decodes as 0, and 110 decodes as 1.

$$0 \rightarrow 000 \quad 001 \quad 010 \quad 100 \rightarrow 0$$
$$1 \rightarrow 111 \quad 110 \quad 101 \quad 011 \rightarrow 1$$

Consider one additional example. The following code is used to encode 2-bit messages:

$$00 \rightarrow 00000$$
$$01 \rightarrow 01110$$
$$10 \rightarrow 10101$$
$$11 \rightarrow 11011$$

It is a 1-correcting code. The following decoding table can be used to decode 5-bit patterns received.

$00 \rightarrow 00000$	10000	01000	00100	00010	$00001 \rightarrow$	00
$01 \rightarrow 01110$	11110	00110	01010	01100	$01111 \rightarrow$	01
$10 \rightarrow 10101$	00101	11101	10001	10111	$10100 \rightarrow$	10
$11 \rightarrow 11011$	01011	10011	11111	11001	$11010 \rightarrow$	11

Thus if we were to receive 11111, we would decode it as 11. If we were to receive 00110, we would decode it as 01, and so on.

Exercises 8.4

1. Show that the coding scheme

$$00 \rightarrow 000$$
$$01 \rightarrow 011$$
$$10 \rightarrow 101$$
$$11 \rightarrow 110$$

used to encode pairs as 3-tuples gives a 1-error detecting code.

2. Show that the coding scheme

$$00 \rightarrow 00000$$
$$01 \rightarrow 01110$$
$$10 \rightarrow 10101$$
$$11 \rightarrow 11011$$

used to encode pairs as 5-tuples gives a single-error correcting code.

3. Show that the coding scheme of exercise 2 is a group code.

4. If $\delta(x, y)$ is the Hamming distance from x to y, show that it is equal to the number of 1's in $x + y$.

5. Prove that a code is d-error detecting if and only if the minimum distance is at least $d + 1$.

SUMMARY

1. A **semigroup** is a pair $(S, \cdot)$, where S is a nonempty set and $\cdot$ is a binary operation on S satisfying the following condition: the operation $\cdot$ is **associative**. That is, for all $x, y, z \in S$,

$$x \cdot (y \cdot z) = (x \cdot y) \cdot z$$

2. A slightly more complex system than a semigroup is a monoid. A **monoid** is a semigroup $(S, \cdot)$ such that S contains a special element e, called the **identity** relative to $\cdot$, satisfying the condition

$$x \cdot e = e \cdot x = x$$

for all $x \in S$.

3. **Theorem 8.1**. Let $(S, \cdot)$ be a monoid with identity element e. Let x be an element of S such that for all $y \in S$, $x \cdot y = y \cdot x = y$. Then $e = x$.

4. Let $(S, \cdot)$ be a monoid with identity e. A subset S' of S is a **submonoid** of S if $e \in S'$, and if $x \cdot y \in S'$ for all $x, y \in S'$. (That is, S' is **closed** under the operation $\cdot$).

5. A monoid always has at least two (*trivial*) submonoids: the singleton set $\{e\}$ and the entire set S.

6. For a monoid $(M, \cdot)$ and an element $x \in M$, we define the **powers** of x as follows:

$$x^0 = e$$

and for all $n \geq 0$,

$$x^{n+1} = x^n \cdot x$$

7. A **group** is a monoid $(G, \cdot)$ with identity e such that for all $x \in G$ there exists an element $y \in G$ such that $x \cdot y = y \cdot x = e$. We call y the **inverse** of x with respect to $\cdot$. It is usually written as x^{-1}.

8. If the group operation is commutative, we call the group a **commutative** group or an **abelian** group.

9. **Theorem 8.2**. Let $(G, \cdot)$ be a group with identity e.
   1. For all $x \in G$, the inverse is unique.
   2. For all $x \in G$, $(x^{-1})^{-1} = x$.
   3. For all $x, y, z \in G$, the **cancellation laws** hold:

$$x \cdot y = x \cdot z \Rightarrow y = z$$

and

$$y \cdot x = z \cdot x \Rightarrow y = z$$

4. If G is finite, then for all x in G there exists a positive integer k such that $0 < k \le |G|$ and $x^k = e$.

10. Let G be a group with respect to the binary operation $\cdot$. A subset H of G is called a **subgroup** of a group $(G, \cdot)$ if H forms a group with respect to the binary operation $\cdot$ that is defined in G.

11. The **order** of a group is the number of elements in the group.

12. For an element x in the group G, the **order of x** is the smallest positive integer k such that $x^k = e$. The **subgroup generated by** an element x of finite order k is the set $\langle x \rangle = \{x, x^2, \ldots, x^k\}$.

13. The set $\langle x \rangle$ is called a **cyclic** subgroup. If $G = \langle x \rangle$, we say that G is a **cyclic group**.

14. Let $A_1, A_2, \ldots, A_n$ be semigroups, and let $A = A_1 \times A_2 \times \cdots \times A_n$ be the product set. We define an operation on elements of A by using each of the component operations. Let $a = (a_1, a_2, \ldots, a_n)$ and $b = (b_1, b_2, \ldots, b_n) \in A$. Define $a \cdot b = (a_1 \cdot b_1, a_2 \cdot b_2, \ldots, a_n \cdot b_n)$. In each coordinate, the operation indicated is taken from the corresponding semigroup. A is a semigroup, called the **product semigroup**.

15. If each of the A_i are monoids and $e_1, e_2, \ldots, e_n$ are the identity elements in $A_1, A_2, \ldots, A_n$, respectively, then the element $e = (e_1, e_2, \ldots, e_n)$ is the identity element in A, and A is a monoid, called the **product monoid**.

16. A relation R on a monoid $(M, \cdot)$ is a **congruence relation** if R is an equivalence relation and $\forall x, y, u, v \in M$ xRu, $yRv \Rightarrow x \cdot yRu \cdot v$.

17. **Theorem 8.3**. Let $(S, \cdot)$ be a monoid with identity element e. Let R be an equivalence relation on S. Then the partition of S induced by R is a monoid under the operation $[x] \cdot [y] = [x \cdot y]$ if and only if R is a congruence relation. In this case, the identity element is $[e]$.

18. **Theorem 8.4**. Let $(G, \cdot)$ be a group with identity element e. Let R be a congruence relation on G.

    a. If xRy, then $x^{-1}Ry^{-1}$.
    b. $[e]$ is a subgroup of G.
    c. For $x, y \in G$, xRy if and only if $x \cdot y^{-1} \in [e]$ (or $x^{-1} \cdot y \in [e]$).
    d. For all x in G, $\{x \cdot y : y \in [e]\} = \{y \cdot x : y \in [e]\} = [x]$.

19. Let H be any subgroup of the group $(G, \cdot)$. We define the **left coset**, xH, of H by $xH = \{x \cdot h : h \in H\}$ and the **right coset**, Hx, of H by $Hx = \{h \cdot x : h \in H\}$.

20. A subgroup H of a group $(G, \cdot)$ is said to be a **normal subgroup** if for all x in G, $xH = Hx$.

21. **Theorem 8.5**. Let $(G, \cdot)$ be a group with identity element e. Let H be a subgroup of G.

    1. The left cosets (or the cosets) of H form a partition of G.
    2. For all $x, y \in G$, $|xH| = |Hy| = |H|$.

22. **Theorem 8.6 (Lagrange's Theorem)**. Let $(G, \cdot)$ be a finite group, and let H be a subgroup of G. Then the order of H divides the order of G.

23. Let $(M, \cdot)$ and $(M', \circ)$ be monoids with identities e and e', respectively. A **homomorphism** between $(M, \cdot)$ and $(M', \circ)$ is a function $f \colon M \to M'$ such that
    a. $f(e) = f(e')$
    b. For all $x, y \in M$, $f(x \cdot y) = f(x) \circ f(y)$.

24. **Theorem 8.7**. Let $(M, \cdot)$ and $(M', \circ)$ be monoids with identity elements e and e', respectively. Let $f \colon M \to M'$ be a monoid homomorphism.
    a. The image of M by f is a submonoid of $(M', \circ)$.
    b. $f^{-1}(e')$ is a submonoid of $(M, \cdot)$.
    c. Let R_f be the relation on M defined by $x R_f y$ if and only if $f(x) = f(y)$. Then R_f is a congruence relation on $(M, \cdot)$.

25. A **monoid isomorphism** is a bijective monoid homomorphism. When monoid $(M, \cdot)$ is isomorphic to monoid $(M', \circ)$, we write $(M, \cdot) \cong (M', \circ)$ (or $M \cong M'$).

26. **Theorem 8.8**. Let f be a surjective homomorphism from the monoid $(M, \cdot)$ onto the monoid $(M', \circ)$. Then R_f is a congruence relation on $(M, \cdot)$ and $M/R_f \cong M'$. Conversely, if R is a congruence on the monoid $(M, \cdot)$, there exists a surjective monoid homomorphism $h \colon M \to M/R$ such that $R = R_h$.

27. Let $f \colon G \to G'$ be a function. We say that f is a **group homomorphism** if the following conditions are met:
    a. $f(e) = e'$
    b. $f(x \cdot y) = f(x) \circ f(y)$ for all $x, y \in G$
    c. $f(x^{-1}) = (f(x))^{-1}$ for all $x \in G$

28. A **group isomorphism** is a bijective group homomorphism.

29. **Theorem 8.9**. Let $(G, \cdot)$ and $(G', \circ)$ be groups with identity elements e and e', respectively. Let $f \colon G \to G'$ be a group omomorphism, and let $K_f = f^{-1}(e')$.
    a. K_f is a normal subgroup of G.
    b. f is an injection if and only if $K_f = \{e\}$.

30. **Theorem 8.10**. Let f be a surjective homomorphism from the group $(M, \cdot)$ to the group $(M', \circ)$. Then K_f is a normal subgroup of $(G, \cdot)$ and $G/K_f \cong G'$. Conversely, if K is a normal subgroup of the group $(G, \cdot)$, there exists a surjective group homomorphism $h \colon G \to G/K$ such that $K = K_h$.

31. A **group code** is a code where the n-tuple code words form a group with respect to the operation $+$, which is defined as componentwise addition modulo 2.

REVIEW PROBLEMS

1. Verify that each of the following is a semigroup.
    a. $(\mathbf{R}^+, +)$, where $+$ represents addition
    b. $(B, \wedge)$, where $\wedge$ represents conjunction

2. a. Show that the pair $(\{a, b, c\}, \cdot)$ is a semigroup, where $\cdot$ is defined by the following table:

$\cdot$	a	b	c
a	a	b	c
b	a	b	c
c	a	b	c

b. Show that the pair $(\{a, b, c\}, *)$ is not a semigroup, where $*$ is defined by the following table:

$*$	a	b	c
a	a	b	c
b	b	a	a
c	c	b	a

3. Consider the set $\{a, b\}$ of two elements called "letters." We are interested in the set of all "words" which can be generated from $\{a, b\}$. However, we will consider all words which are identical in the first three letters as equivalent words. For example, *aabaa* is equivalent to *aabba*. Show that the set of words and the operation of concatenation form a semigroup.

4. Let S be the set $S = \{a, b, c\}$ with the operation $\cdot$ defined by the following table:

$\cdot$	a	b	c
a	a	b	c
b	b	c	a
c	c	a	b

Show that $(S, \cdot)$ is a semigroup.

5. Which of the following are monoids?

   a. even integers with addition
   b. odd integers with addition
   c. even integers with multiplication
   d. odd integers with multiplication

6. Which of the following represent monoids?

a.

·	a	b	c	d
a	b	a	d	c
b	a	b	c	d
c	d	c	b	a
d	c	d	a	b

b.

·	a	b	c	d
a	a	b	c	d
b	b	a	d	c
c	c	c	c	c
d	d	d	d	d

7. Let S be the set $S = \{1, 2, 3, 4, 5\}$, and define the operation θ by the statement: $a \ \theta \ b = $ minimum of a and b.

   a. Construct the operation table for this system.

   a. Show that (S, θ) is a commutative monoid.

8. Assume that $*$ is an associative binary operation on a set S with identity element e. Let $a \in S$. Prove that if b is a right inverse of a, and c is a left inverse of a, then $b = c$.

9. Consider the set $S = \{0, 1, 2, 3, 4\}$ and the operation · as defined by the following table. Is $(S, \cdot)$ a group?

·	0	1	2	3	4
0	0	1	2	3	4
1	1	2	3	4	0
2	2	3	4	0	1
3	3	4	0	1	2
4	4	0	1	2	3

10. Let $(G, \cdot)$ be the monoid with the following multiplication table:

·	e	p	q	r	s	t
e	e	p	q	r	s	t
p	p	e	s	t	q	r
q	q	s	r	e	t	p
r	r	t	e	q	p	s
s	s	q	t	p	r	e
t	t	r	p	s	e	q

   a. Show that $(G, \cdot)$ is a group by finding the inverse of each element of G with respect to ·.

   b. Find the cyclic subgroups of G.

11. Determine whether each of the following is a group, monoid, or semigroup and whether or not it is commutative.

   a. the set of all positive rational numbers under multiplication

   b. the set of all negative integers under addition

c. the set of all nonnegative integers under addition

d. the set S of all functions of a set A into itself under function composition

e. the set of all bijections from A to A

12. Consider the set $\{a, b, c\}$ of three distinct elements. This set has six permutations:

$$P_1 = \{(a, a), (b, b), (c, c)\}$$
$$P_2 = \{(a, b), (b, c), (c, a)\}$$
$$P_3 = \{(a, c), (b, a), (c, b)\}$$
$$P_4 = \{(a, b), (b, a), (c, c)\}$$
$$P_5 = \{(a, c), (b, b), (c, a)\}$$
$$P_6 = \{(a, a), (b, b), (c, b)\}$$

Let the binary operation · denote function composition, and let $P = \{P_1, P_2, P_3, P_4, P_5, P_6\}$.

a. Construct the operation table for $(P, \cdot)$.

b. Show that $(P, \cdot)$ is a noncommutative group.

13. List all the permutations of the set $\{a, b\}$, and use function composition to write the operation table of this group of permutations.

14. Show that a group $(G, \cdot)$ with no more than four elements is abelian. (*Hint*: Compare the two sets $\{e, a, b, a \cdot b\}$ and $\{e, a, b, b \cdot a\}$. The element e is the identity.)

15. By looking at the powers of individual elements, find all the subgroups of the group in problem 12.

16. Let G be the set of all pairs (x, y) of real numbers for which $x \neq 0$. A binary operation $\circ$ is defined in G by $(x, y) \circ (u, v) = (xu, yu + v)$. Verify that G is a group.

17. The **symmetries** of a geometrical figure are bijections of the set of points of the figure onto itself which preserve distance. In geometry, such symmetries are called isometries. Consider the symmetries of the letter H. There are four of them, the identity bijection I, the reflections H and V about the lines H and V, and the $180°$ rotation R about the intersection of lines H and V. For any two symmetry transformations T_1 and T_2, define $T_1 \cdot T_2$ as transformation T_1 followed by transformation T_2.

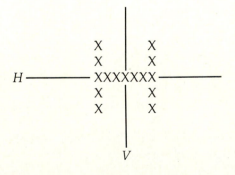

a. Construct the operation table for $(\{I, H, V, R\}, \cdot)$.

b. Show that the symmetry transformations form an abelian group. (This group is known as the *Klein 4-group*.)

18. Let (G, θ) be a group where $G = \{a, b, c\}$, and the incomplete operations table is as follows:

θ	a	b	c
a	a	b	c
b	b	c	
c	c	a	

Complete the operations table for θ.

19. Consider the set $G = \{1, 2, 3, 4, 5, 6\}$. Then $(G, \cdot_7)$, where $\cdot_7$ is multiplication modulo 7, is a group.

a. Construct the multiplication table for $(G, \cdot_7)$.

b. Find 2^2 and 6^2.

c. Find the order of the subgroup generated by 3.

d. Show that $(G, \cdot_7)$ is cyclic.

20. Find the nontrivial subgroups of the group $(G, \cdot)$, and then find their cosets.

$\cdot$	e	a	b	c	d	f
e	e	a	b	c	d	f
a	a	b	e	f	c	d
b	b	e	a	d	f	c
c	c	d	f	e	a	b
d	d	f	c	b	e	a
f	f	c	d	a	b	e

21. Let $(A, *)$ be a monoid such that for every $x \in A$, $x * x = e$, where e is the identity element. Show that $(A, *)$ is an abelian group.

22. Let $(G, \cdot)$ be a group and H a nonempty subset of G. Show that $(H, \cdot)$ is a subgroup if for any $a, b \in H$, $a \cdot b^{-1} \in A$.

23. Let $(G, \cdot)$ be a group. Show that $(a \cdot b)^{-1} = b^{-1} \cdot a^{-1}$.

24. Let $(G, \cdot)$ be the group of all nonzero real numbers under the operation of multiplication, and consider the subgroup $(H, \cdot)$, where $H = \{2^n : n \in \mathbf{Z}\}$. Determine all the left cosets of H in G.

25. Let H be a subgroup of finite group $(G, \cdot)$ with $[G:H] = 2$. Show that $(H, \cdot)$ is a normal subgroup of $(G, \cdot)$.

26. Let $(G, \cdot)$ be an abelian group generated by the elements x and y, where x has order 4 and y has order 3. Find the order of $(G, \cdot)$.

27. Let B^5 be the set of binary strings five bits long. For $(a_1, a_2, \ldots, a_5), (b_1, b_2, \ldots, b_5) \in B^5$, define $(a_1, a_2, \ldots, a_5) + (b_1, b_2, \ldots, b_5) = (a_1 + b_1, a_2 + b_2, \ldots, a_5 + b_5)$ with $+$, the coordinatewise sum given as follows:

+	0	1
0	0	1
1	1	0

   a. Show that $(B^5, +)$ is a group.
   b. Find the subgroups of $(B^5, +)$ of order 2.

28. Describe the quotient group A/H, where A is the group $(\mathbf{Z}_{15}^+, +_{15})$ and H is any subgroup of order 3. Construct the operation table for A/H.

29. Let $(G, \cdot)$ be a group with identity e. Show that the function $h: G \to G$ defined by $h(x) = e$ for all $x \in G$ is a homomorphism.

30. Let f and g be homomorphisms from a group $(G, *)$ to a group $(H, \cdot)$. Show that $(A, *)$ is a subgroup of $(G, *)$, where

$$A = \{x \in G: f(x) = g(x)\}$$

31. Show that the two groups $(\{0, 1, 2, 3\}, \cdot)$ and $(\{1, 2, 3, 4\}, *)$ are isomorphic.

$\cdot$	0	1	2	3
0	0	1	2	3
1	1	2	3	0
2	2	3	0	1
3	3	0	1	2

$*$	1	2	3	4
1	1	2	3	4
2	2	4	1	3
3	3	1	4	2
4	4	3	2	1

32. Let $(G, \cdot)$ and $(H, *)$ be groups. Show that if $f: G \to H$ is an isomorphism, then $f^{-1}: H \to G$ is also an isomorphism.

33. Consider the code

$$0 \to 00000$$
$$1 \to 11111$$

   a. Show that this is a group code.
   b. Set up the coset table to show that this is a double-error correcting code.

34. Consider the following group code, where elements from B^3 are encoded as elements from B^6.

$$000 \rightarrow 000000$$
$$001 \rightarrow 001111$$
$$010 \rightarrow 010011$$
$$011 \rightarrow 011100$$
$$100 \rightarrow 100110$$
$$101 \rightarrow 101001$$
$$110 \rightarrow 110101$$
$$111 \rightarrow 111010$$

Decode each of the following words.
a. 110010
b. 001100
c. 000111

▪ Programming Problems

35. Write a program that inputs an operation table and checks if the operation is associative.

36. Write a program that, given an operation table and any two elements, returns the result of the operation.

37. Write a program that, given an operation table and an element, returns the inverse of the element if it exists.

38. Write a program that, given a group code and a message to be transmitted, will encode the message.

39. Write a program that, given a group code and a code word, will decode the code word.

SELF TEST ANSWERS

Self Test 1

1. a. To show that the operation is associative, it is necessary to check that $x \cdot (y \cdot z) = (x \cdot y) \cdot z$ for all $x, y, z \in \{A, B, C, D\}$. We show a few cases here. (How many cases are there?)

$$A \cdot (B \cdot C) = A \cdot B = D \quad \text{and} \quad (A \cdot B) \cdot C = B \cdot C = D$$
$$C \cdot (B \cdot B) = B \cdot B = C \quad \text{and} \quad (C \cdot B) \cdot B = B \cdot B = C$$

b. From the table, it is clear that $C \cdot x = x \cdot C = x$ for all $x \in \{A, B, C, D\}$, so C is the identity element.

c. $(S, \cdot)$ is not a group since D has no inverse; that is, $D \cdot x = C$ has no solution for $x \in \{A, B, C, D\}$.

2. a. The operation is associative, and has identity element -1. Moreover, for each $x \in \mathbf{Z}$, $x \cdot (-x - 2) = x + (-x - 2) + 1 = -1$. So $(\mathbf{Z}, \cdot)$ is a group.

b. The operation is not associative. For instance, $2 - (3 - 5) = 2 - (-2) = 4$, while $(2 - 3) - 5 = -1 - 5 = -6$.

3. First note that $0 \in \mathbf{Z}^+$. The sum of two even integers is again even, so E is closed with respect to addition. E "inherits" the associativity of addition from $\mathbf{Z}^+$. Hence, $(E, +)$ is a submonoid of $(\mathbf{Z}^+, +)$.

Self Test 2

1. The set $B^2 \times B^3$ consists of pairs (x, y) where $x = (a_1, a_2) \in B^2$ and $y = (b_1, b_2, b_3) \in B^3$. The operation on $B^2 \times B^3$ is the product of the $\vee$ operation in each component. Since for B^2 and B^3 the operation is the coordinatewise $\vee$, we can describe the product as $(B^5, \vee)$.

2. a. R is reflexive, symmetric, and transitive because of the equality in $a^2 - a - 2 = b^2 - b - 2$.

b. R is not a congruence since, in general, aRc and bRd does not guarantee that $(a + b)R(c + d)$. For instance, $1R1$ and $-1R2$. However, $(1 - 1)R(1 + 2)$ means $0^2 - 0 - 2 = 3^2 - 3 - 2$, which is false.

3. If $H = \{[0]\}$, H has index 5 in $(\mathbf{Z}_5, +_5)$. Suppose H contains an element other than $[0]$. By inspecting the table, we can see that every element other than $[0]$ has order 5. Hence $H = \mathbf{Z}_5$, so H has index 1. This means that the only subgroups of $(\mathbf{Z}_5, +_5)$ are the trivial ones.

Self Test 3

1. In a manner similar to that of the third part of Example 8.16, we can examine all the cases in a simple table:

m	n	$n + m$	$h(m) \, \Delta \, h(n)$	$h(m + n)$
even	even	even	$0 \, \Delta \, 0 = 0$	0
even	odd	odd	$0 \, \Delta \, 1 = 1$	1
odd	even	odd	$1 \, \Delta \, 0 = 1$	1
odd	odd	even	$1 \, \Delta \, 1 = 0$	0

Since the last columns of the table agree, h is a monoid homomorphism.

2. There is a surjective monoid homomorphism $h: G \to G'$. Let $x, y \in G'$. There exist $u, v, \in G$ such that $h(u) = x$, and $h(v) = y$. Hence, $x + y = h(u) + h(v) = h(u + v) = h(v + u) = h(v) + h(u) = y + x$.

3. a. $\theta(x) = |x|$ is a monoid homomorphism since $|1| = 1$ and $|x \cdot y| = |x| \cdot |y|$ for all nonzero real numbers x and y. θ is a group homomorphism with kernel $\{1, -1\}$.

    b. $\theta(x) = 1/x$ is a monoid homomorphism since $1/1 = 1$ and $1/(x \cdot y) = (1/x) \cdot (1/y)$ for all nonzero real numbers. θ is a group homomorphism with kernel $[1] = \{1\}$.

    c. $\theta(x) = -x$ is not a monoid homomorphism since $\theta(1) \neq 1$.

    d. $\theta(x) = x^2$ is a monoid homomorphism since $(1)^2 = 1$ and $(x \cdot y)^2 = x^2 \cdot y^2$ for all nonzero real numbers x and y. θ is a group homomorphism with kernel $[1] = \{1, -1\}$.

9 Machines and Computations

Computer scientists are interested in answering two questions: What problems can be solved with a computer, and if a problem is (theoretically) solvable, what is the most effective way to solve it? Answers to the first question arise through development and study of models of computation. In the previous chapter, we studied algebraic structures, which we used to model arithmetic calculations. In this chapter, we deal with the foundations of computation, which primarily addresses the first question, and discuss abstract models of a digital computer. The structure we will study is known as a finite state machine, or finite state automaton. Finite state automata are of great practical interest as well as theoretical interest and are commonly used in compilers to perform lexical analysis. In terms of theoretical modeling capability, however, finite state automata are at the low end of the spectrum. Turing machines, which are conjectured to be the most capable class of automata, are at the other end of the spectrum. If a Turing machine can't be found to model a computation, then we believe there is no effective way to perform the computation. There is, between these two ends of the spectrum, an abundance of other computation models.

The first section of the chapter illustrates the use of finite state automata as a modeling tool. The second section deals with finite state automata with "no outputs." The class of languages "recognized" by finite state automata is of particular interest. The third section deals with finite state automata with outputs, which leads to consideration of automata as "function computers." The fourth section introduces Turing machines, considering them in terms of language recognizers and function computers, and concludes with some remarks on the general theory of computation.

Before getting into the formal material, we begin with an example that shows how automata can serve as models for useful devices.

9.1 Automata as Models

A **finite state automaton**, or **finite state machine**, is a simple mathematical model that has a finite set of discrete inputs and goes through a finite set of states determined by the order in which inputs are received. An automaton may have a finite set of outputs. If so, the automaton will produce a sequence of outputs in response to the sequence of inputs. Coin changers, elevators, vending machines, and garage door openers are examples of machines based on this simple model. Finite state machines are quite useful design models in many areas of computer science, including lexical scanning within compilers, computer network protocol definition, and switching circuits. Example 9.1 shows how an automaton might be used to model a simple device.

EXAMPLE 9.1

Figure 9.1 shows a state model, represented as a labeled digraph, or *state diagram*, for the operation of a garage door opener. Assume that the door control consists of two buttons: one labeled OPEN and one labeled CLOSE. The four states of the door are represented by the circles labeled *shut*, *rising*, *lowering*, and *up*. Describe the operation of the garage door.

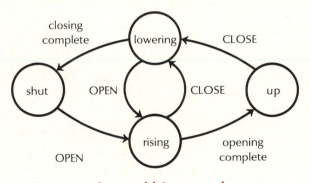

Figure 9.1. State model for garage door opener

Solution

When the door is in the *shut* state, pushing the OPEN button causes the door to go into the *rising* state, during which it gradually opens under motor control. After the door has risen completely, it is in the *up* state. From there, pushing the CLOSE button puts the door in the *lowering* state, during which the door gradually closes. Observe that the state diagram clearly indicates that the door cannot move instantly from the *up* state to the *shut* state, or vice versa. Moreover, it is possible to push the OPEN and CLOSE buttons alternately, causing repeated transitions between the *rising* and *lowering* states so that the door behaves like a yo-yo. Finally, the model shows that pushing the CLOSE button when the door is in the *shut* state or pushing the OPEN button when the door is in the *up* state does not cause any state change, so that nothing happens. Detection of the events "closing complete" and "opening complete" occurs within the motor controlling the door, and is thus separate from the mechanism for the push buttons.

In Example 9.1, we can consider the finite automaton as having outputs consisting of the control signals sent to the motor control when the door is in a particular state. In other cases, there are no identifiable outputs. We are mainly interested in the state that the machine reaches in response to the input sequence. Our study of automata will be divided between automata with outputs and automata without outputs.

9.2   Finite State Automata Without Outputs

Studying finite automata without outputs leads us into the area of language theory, since these automata are frequently used for solving problems concerning membership in a special class of languages. In Example 9.2, we show a machine that appears to be abstract in nature, but which is nevertheless representative of this useful class of recognizers. An automaton without outputs has a designated start state and a designated set of final states. We are interested in the state an automaton reaches when it is activated in response to a particular input.

EXAMPLE 9.2

Describe the activity of the automaton shown in the state diagram of Figure 9.2.

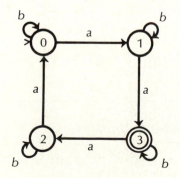

Figure 9.2.

Solution

States are shown as labeled circles. The "start state," state 0, is indicated by the arrowhead attached to its circle. The lone final state, 3, is indicated by a double circle. The state transitions are represented by the edges, which are labeled *a* or *b*. For instance, the edge labeled *a* from state 0 to state 1 means that if the automaton is in state 0 and receives *a* as an input, then the automaton will make a transition to state 1.

An automaton operates according to the following rule: Beginning from the start state, the automaton makes the transition to the state determined by the first input symbol; from that state, it makes a transition to another state based on the next input symbol, and so on, until all input symbols have been used. To understand the action of the automaton in Figure 9.2, consider the transitions for the input sequence *abaa*:

Current state	Current symbol	Next state	Remaining symbols
0	*a*	1	*baa*
1	*b*	1	*aa*
1	*a*	3	*a*
3	*a*	2	
2			

Since our representation of the automaton is a labeled digraph in which the vertices represent the states, it is natural to speak of the *path* traversed in response to a particular input sequence. The path traversed in response to the input sequence *abaa* is 0, 1, 1, 3, 2. Since the state obtained when all symbols have been used is 2, we see that the automaton does not "halt" in state 3 as expected. Thus the input sequence *abaa* is not valid for this automaton. This is analogous to the silent rejection of a vending machine that does not respond when it expects four dimes and is fed only two. But if we simply remove the last input symbol and use the input sequence *aba*, the automaton halts in the terminal state 3. In this case, we say the automaton "accepts" the input sequence *aba*. We could have also extended the input sequence to *abaaaaa* to have the automaton halt in state 3.

In general, the automaton will halt in the designated final state 3 if and only if the input sequence contains *k* *a*'s, where *k* is an integer of the form $4n + 2$, $n \geq 0$. The number of *b*'s in the input is irrelevant.

Definition of a Finite State Automaton

A **finite state automaton**, or **FSA**, is a quintuple $M = (S, A, T, s_0, F)$.

S is a finite set whose elements are called **states**.

A is a finite set, called an **input alphabet**, whose elements are called **input symbols**.

T is a function from $S \times A \to S$, called the **transition** function.

s_0 is an element of S called the **start state**.

F is a subset of S, whose elements are called **final**, or **terminal** states.

The next example should help tie the abstract definition of an automaton to its state diagram.

EXAMPLE 9.3

Let M be the automaton with:

states $S = \{0, 1, 2, 3\}$

alphabet $A = \{a, b\}$

start state $s_0 = 0$

final states $F = \{3\}$

transition function T given by the following table:

		alphabet	
		a	b
s	0	1	0
t	1	3	1
a	2	0	2
t	3	2	3
e			
s			

Give a representation of the automaton as a labeled digraph, and describe the action of the automaton.

Solution

We interpret the table for the transition function as follows: When in state 0, an input of a causes a transition to state 1. Thus, in the graph we need an edge from state 0 to state 1 labeled a. Similarly, we need an edge from state 0 to itself labeled b, and so on. The digraph for the FSA is exactly the one shown in Figure 9.2. In Example 9.2 we showed that the FSA accepted the set of strings of a's and b's that have $4n + 2$ a's, where n is a nonnegative integer.

Finite State Automata as Language Recognizers

Example 9.2 illustrated the use of an FSA for deciding whether or not a string was a member of a given set of strings. Finite state automata play a significant role in compiler design as **lexical analyzers**; that is, they are used to recognize certain patterns of symbols such as identifiers, real numbers, special characters, and so on as being legitimate constructs of a language.

Now let us consider some definitions. You will recall that an alphabet is a set whose elements are called symbols. For example, the set $\{A, B, \ldots, Z, a, b, \ldots, z\}$ is an alphabet, the set $\{0, 1, \ldots, 9\}$ is an alphabet, and the set $\{\alpha, \beta, \gamma\}$ is an alphabet. We define a **word** of length k from an alphabet A by $a_1 a_2 \ldots a_k$, where $a_i \in A$, for $i = 1, 2, \ldots, k$. If x is a word from A, we denote its length by $|x|$. The word of length 0, or the **null** word, is the word that has no elements. The null word is denoted by λ. Words of length 1 are just symbols from the alphabet itself. A word is like a character string in a programming language.

Making a word of length k from an alphabet A amounts to taking a k-sample from A; that is, a word is formed by selecting symbols from A with repetitions allowed, and words are distinguished by the order in which symbols are selected. Hence there are $|A|^k$ words of length k from an alphabet A.

The set of all words from alphabet A is denoted by A^*. A^* is a countably infinite set. If $x = a_1 a_2 \ldots a_k$ and $y = b_1 b_2 \ldots b_m$ are words of length k and m, respectively, from the alphabet A, we define the **concatenation** of x and y to be the word of length $k + m$ given by

$$x \cdot y = a_1 a_2 \ldots a_k b_1 b_2 \ldots b_m$$

Moreover, $x \cdot \lambda = \lambda \cdot x = x$ for all $x \in A^*$.

A **language** over the alphabet A is a subset of A^*. Thus, written English can be considered a language over the set consisting of the uppercase and lowercase alphabetic characters augmented by the "blank," or space, symbol and all necessary punctuation symbols. FORTRAN and Pascal can be considered as languages over the alphabet of ASCII characters. To do so, of course, is to trivialize them. All three of these languages have rules for determining membership. For English, stating the rules is nontrivial, or perhaps just futile. Human communication is usually possible even in the presence of spelling errors, syntax errors, or colloquialisms. For FORTRAN and Pascal, the rules for membership can be stated fairly concisely. Failure to observe the rules leads to frustrating encounters with compilers that do not know what you *mean*, only what you write. It is in the latter domain that finite state automata are used to begin the process of translating source code to machine code. In the remainder of this section, we explore the capabilities of an FSA as a language recognizer; that is, as a tool to answer the question: If L is a language over A and x is a word in A^*, is x a member of L?

The first step is to extend the action of a FSA from (state, symbol) pairs to (state, word) pairs. Let (S, A, T, s_0, F) be an FSA. Define the extension of T to $T': S \times A^* \to S$ recursively by

1. $T'(s, \lambda) = s$ for all $s \in S$
2. $T'(s, ax) = T'(T(s, a), x)$ for all $s \in S$, $x \in A^*$ and $a \in A$

The extension T' is a formal definition of the action of an FSA when it is presented with a word of arbitrary length. In particular, $T'(s, a) = T'(T(s, a), \lambda) = T(s, a)$. Hence T' agrees with T when restricted to words of length 1. Figure 9.3 illustrates the machine's action on a tape whose cells contain a word from A^*. Beginning with the leftmost symbol, the FSA examines the symbol and, based on its current state, determines a new state. Then it moves one cell to the right to the next symbol. We say the FSA **halts** when all symbols have been acted upon. If the FSA halts in one of the designated final states, we say that the word is **accepted**, or **recognized**, by the FSA. The **language** accepted (recognized) by an FSA M is the set of words, $L(M)$, given by

$$L(M) = \{x: x \in A^* \quad \text{and} \quad T'(s_0, x) \in F\}$$

A language which is accepted by some FSA is said to be **regular**. That is, a language L is regular if $L = L(M)$ for some FSA M.

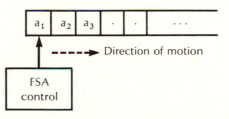

Figure 9.3. A representation of an FSA

EXAMPLE 9.4

Compute $T'(0, abba)$ for the FSA whose transition function is given by the table:

	a	b
0	0	1
1	2	1
2	1	0

Solution

First observe that we are following the convention that states define the rows and symbols define the columns, so that the table implicitly defines $S = \{0, 1, 2\}$ and $A = \{a, b\}$. We need to define only the start state and final states to complete the specification of the FSA.

$$\begin{aligned}
T'(0, abba) &= T'(T(0, a), bba) = T'(0, bba) \\
&= T'(T(0, b), ba) = T'(1, ba) \\
&= T'(T(1, b), a) = T'(1, a) \\
&= T(1, a) \\
&= 2
\end{aligned}$$

EXAMPLE 9.5

	a	b
0	1	0
1	3	2
2	1	3
3	3	3

$s_0 = 0$

$F = \{2\}$

Find the language $L(M)$ accepted by the FSA M given by the table.

Solution

The digraph of the FSA is shown in Figure 9.4. State 0 is the start state, and state 2 is the lone final state. Once the FSA reaches state 3, it will remain there. Beginning in state 0, we may have as many b's as we want, including none. After the first a, the next symbol must be a b, to avoid state 3. Thereafter, reaching state 2 again requires that each a be followed by exactly one b. Thus we can characterize $L(M)$ as the set of words over $A = \{a, b\}$ such that an arbitrary number of b's is followed by a string consisting of repetitions of the string ab.

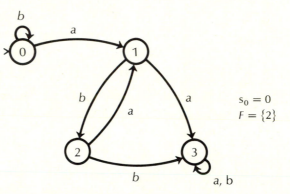

$s_0 = 0$

$F = \{2\}$

Figure 9.4. The digraph for the FSA in Example 9.5

An alternate description is that words in the set of recognized words have the form $b^n(ab)^m$, for $n \geq 0$ and $m \geq 1$. The symbol a^n denotes a word consisting of n a's. Thus, for $n = 3$ and $m = 2$, the word is $bbbabab$.

EXAMPLE 9.6

Find an FSA that accepts all words in $\{0, 1\}^*$ that end in 0 and have length less than or equal to 3.

Solution

Since the alphabet consists of 0's and 1's, we will use state names s, t, u, and so on to avoid confusion. If we use a simple FSA such as the one shown in Figure 9.5, we end up accepting all words in $\{0, 1\}^*$ that end in 0. In states s and t, we can't keep track of the number of symbols received.

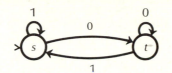

Figure 9.5. An incorrect FSA for Example 9.6

In order to keep track of the length, we'll need a state for each step made. At the same time, we have to be sure that when a 0 is encountered the FSA will move to a final state. Figure 9.6 shows a solution. Observe that there are three final states, one for the word of length 1, one for words of length 2, and one for words of length 3. State y is used to discard any words that have too many symbols.

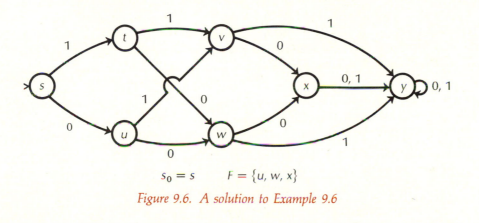

$$s_0 = s \qquad F = \{u, w, x\}$$

Figure 9.6. A solution to Example 9.6

The examples above may seem a bit contrived. The next example brings us a bit closer to the practical use of an FSA as a lexical analyzer.

EXAMPLE 9.7

Find an FSA that recognizes unsigned fixed-point numbers; that is, real numbers consisting of a nonempty string of digits optionally followed by a decimal point and another nonempty string of digits. We require that an integer cannot end with a decimal point, and a number between 0 and 1 must have exactly one 0 before its decimal point.

Solution

One solution is shown in Figure 9.7. As a shortcut, we use d to represent any of the nonzero digits 1, 2, . . . , 9. The digraph shows only the transitions necessary to accept an unsigned number; any unnecessary transitions are omitted to avoid cluttering the

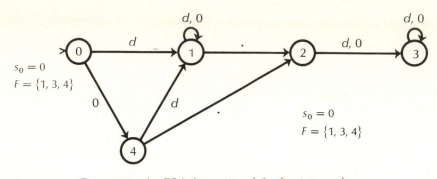

Figure 9.7. An FSA for unsigned fixed-point numbers

digraph. The missing transitions can be regarded as transitions to an error, or "rejecting," state, from which there is no escape. As you might expect, getting two decimal points in a row is illegal, so we do not need to show a transition for a decimal point input symbol from state 2. Note that the FSA in Figure 9.7 accepts fixed-point numbers with an arbitrary number of leading and trailing zero's. For instance, it will accept the number ten in the representations 10.0, 10, 010.000, and so on. Such padding might not be desirable. The FSA in Figure 9.8 eliminates the leading and trailing zeros, and allows at most two representations of a given value.

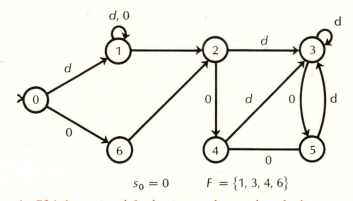

Figure 9.8. An FSA for unsigned fixed-point numbers without leading or trailing zeros

---

EXAMPLE 9.8

In some programming languages an identifier is defined as any sequence of letters and digits that is surrounded by spaces and whose initial symbol is a letter. Let the input alphabet be $A = \{l, d, s, n\}$, where $l = $ letter, $d = $ digit, $s = $ space, and $n = $ none of these. Design an FSA M to recognize identifiers in this language.

Solution

The input alphabet is given. Let $s_0 = 0$. If the input symbol is s, the FSA remains in the initial state 0. If it is l, the FSA transfers to state 1. If it then encounters a blank, it moves to the accepting state 2. On the other hand, an input of d or n to the initial state causes a transition to an error state, 3. We construct the following transition function table:

	l	d	s	n
0	1	3	0	3
1	1	1	2	3
2	3	3	3	3
3	3	3	3	3

$s_0 = 0$

$F = \{2\}$

The graphical representation is shown in Figure 9.9.

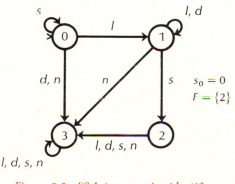

$s_0 = 0$

$F = \{2\}$

Figure 9.9. FSA to recognize identifiers

Having seen some examples of the languages recognized by finite state automata, we can characterize the size of the language accepted by them. We have seen, for instance, that an FSA can recognize a finite language or an infinite language. Theorem 9.1 gives necessary and sufficient conditions for one or the other of these to occur.

Theorem 9.1

Let $M = (S, A, T, s_0, F)$ be an FSA that recognizes language $L(M)$. Let $n = |S|$.

1. $L(M)$ is nonempty if and only if there exists $x \in L(M)$ such that $0 \le |x| < n$.
2. $L(M)$ is infinite if and only if there exists $x \in L(M)$ such that $|x| \ge n$.

Proof

1. If $L(M)$ contains a word of length less than n, then $L(M)$ is not empty. On the other hand, suppose that $L(M)$ contains a word x of length k, where $k \geq n$. Let $s_0, s_1, \ldots, s_k$ be the path traversed through the states when the input word x is recognized. Representing x as $x_1 x_2 \ldots x_k$, we see that $T(s_{i-1}, x_i) = s_i$, for $i = 1, \ldots, k$. Since $k \geq n$, there must be a cycle $s_i, \ldots, s_j$ in the path. Removing the symbols $x_{i+1}, \ldots, x_j$ from x generates a word $x' \in L(M)$ whose length is less than k. If $|x'| < n$, we have found the word sought. If not, the procedure can be repeated a finite number of times to produce a word in $L(M)$ whose length is less than or equal to n.

2. The key to the proof of (1) is that if $L(M)$ contains a word x of length $k \geq n$, then the path taken through the states in recognizing x contains a cycle, as shown in Figure 9.10.

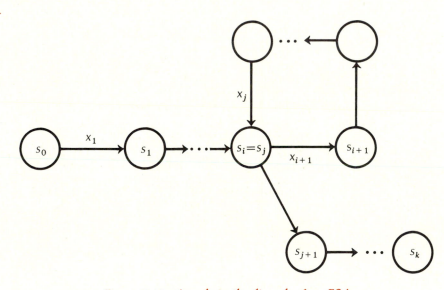

Figure 9.10. A cycle in the digraph of an FSA

We use this intermediate fact in a different way here. Rather than shrinking the word by removing the symbols causing the cycle, we repeat the symbols to find a longer, hence distinct, word that is accepted by M. This process of "pumping up" the word leads to a result called the Pumping Lemma (see exercise 5 on page 373).

Again representing x as $x_1 x_2 \ldots x_k$, we find $i < j$ such that the input symbols $x_{i+1}, \ldots, x_j$ cause a cycle in the state diagram. Let z be the word formed by repeating the inner cycle once:

$$z = x_1 x_2 \ldots x_{i+1} \ldots x_j x_{i+1} \ldots x_j x_{j+1} \ldots x_k$$

Since x_k causes a transition to a final state, $z \in L(M)$. Moreover, $|z| > |x|$. The process can be repeated as many times as we wish to find longer and longer words belonging to $L(M)$. Hence $L(M)$ must be infinite.

The proof of the converse is left as an exercise (see exercise 6 on page 373.)

Limits of Finite State Automata as Language Recognizers

Theorem 9.1 gives us a precise way of determining whether or not a nonempty language recognized by an FSA is infinite or finite. The criterion, in effect, is that the language is infinite if and only if there is a path from the start state to a final state that contains a cycle. Thus an FSA is capable of recognizing arbitrarily long words. In fact, a very simple FSA, as illustrated in Figure 9.11, can recognize *every* word from a given alphabet. The symbol α is used to represent all symbols from the alphabet.

Are there any limits, then, to the recognition power of the class of FSA? The next example shows that finite state automata can't do everything.

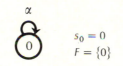

$$s_0 = 0$$
$$F = \{0\}$$

Figure 9.11. An FSA to recognize A*

EXAMPLE 9.9

Let $A = \{a, b\}$. Show that there is no FSA that can recognize the language $L = \{a^n b^n : n \geq 0\}$.

Solution

Clearly L is infinite. We use the idea of "pumping up" the accepted words to show that there is no FSA that can accept L. Suppose, to the contrary, that M is an FSA recognizing L, and that M has p states. Let $x = x_1 x_2 \ldots x_k$ be a word in L of length $k \geq p$. Again we find $i < j$ such that the input symbols $x_{i+1}, \ldots, x_j$ cause a cycle in the state diagram. Let $y = x_{i+1} \ldots x_j$. Let z be the word formed by repeating the inner cycle once:

$$z = x_1 x_2 \ldots x_i \cdot y \cdot y \cdot x_{j+1} \ldots x_k$$

We know z is a member of L, since it ends as x does. We claim that z is not of the form $a^i b^i$ for any i. There are three cases to consider:

1. The word y contains both a and b. Then the subword $y \cdot y$ must have an a following a b, so z cannot have the right form.

2. The word y consists entirely of a's. Then since x has the same number of a's as b's, z must have more a's than b's.

3. The word y consists entirely of b's. Then since x has the same number of a's as b's, z must have more b's than a's.

All three cases lead to a contradiction of the assumption that $L = L(M)$.

SELF TEST 1

1. Let M be the automaton with:

   states $S = \{0, 1, 2\}$
   alphabet $A = \{a, b\}$
   start state $s_0 = 0$
   final states $F = \{0, 1\}$
   transition function T given by the following table:

		alphabet	
		a	b
states	0	0	1
	1	0	2
	2	2	2

   Give a representation of the automaton as a labeled digraph, and describe the action of the automaton.

2. Find the language $L(M)$ accepted by the FSA M given by the following transition function table:

	a	b
0	0	1
1	1	2
2	2	0

   $s_0 = 0$
   $F = \{0\}$

3. Construct a finite automaton M with input symbols a and b that will only accept words in which $aabb$ appears as a substring in the word. For example, it should accept $baaabbb$ and $aaabbb$, but it should not accept $abbaa$ or $aaabaa$.

Exercises 9.1

1. Find an FSA that recognizes each of the following languages.
   a. The words from $\{a, b, c\}$ that contain no c
   b. The words from $\{0, 1\}$ that have an even number of 1's
   c. The words from $\{\alpha, \beta\}$ such that the last two symbols are the same
   d. The words from $\{\alpha, \beta\}$ of the form $\alpha^n \beta^n$, with $0 \le n \le 5$
   e. The words from $\{a, b\}$ of the form $(ab)^n$, $n \ge 0$
   f. All words from $\{a, b\}$ of length less than 5

2. Find $L(M)$ for each of the following finite state automata.
   a. The FSA of Example 9.2, with final states $F = \{0, 2\}$
   b. The FSA given by the following transition table:
   c. The FSA of (b) with final states changed to $\{0, 1, 2, 4\}$

	a	b
0	1	4
1	4	2
2	3	4
3	4	4
4	4	4

$s_0 = 0$

$F = \{3\}$

3. List the words accepted by the FSA in Figure 9.6.

4. Modify the FSA in Figure 9.7 in each of the following ways.
   a. So that the fixed-point numbers recognized can have a sign ($+$ or $-$)
   b. So that only integer values are recognized
   c. So that an exponent of the form $E \pm d$ can be added.

*5. Prove the following version of the **Pumping Lemma**: Let $M = (S, A, T, s_0, F)$. If $L(M)$ contains a word x such that $|x| > |S|$, then there exist strings u, v, and $w \in A^*$ such that
   1. $x = u \cdot v \cdot w$
   2. $|u \cdot v| \le |s|$ and $|v| \ge 1$
   3. $u \cdot v^i \cdot w \in L(M)$ for $i = 0, 1, 2, \ldots$

6. Prove that if $L(M)$ is infinite, then $L(M)$ contains a word x such that $|x| > n$. In fact, x can be chosen so that $n \le |x| < 2n$.

7. Use the technique of Example 9.9 to show that there is no FSA that accepts the language consisting of words of the form a^i, where i is a perfect square ($i = j^2$ for an integer j).

9.3   Finite State Automata with Outputs

The FSA we have considered so far have not had "outputs." Given an input word, they either follow a path to an accepting state or they don't. This is consistent with their use as language recognizers. In this section, we explore the use of automata that generate outputs, chosen from an output alphabet, in response to inputs. Thus, an automaton with outputs can be considered an input-output machine, or function. An automaton with outputs has at its core an FSA as we discussed in the previous section. To distinguish automata with outputs from automata without outputs, we will refer to them as **(finite state) machines**. For finite state machines, we are not concerned with the final state reached, as we were with finite state automata. What we are concerned with is the output word generated in response to the input word. If A is the input alphabet, and O is the output alphabet, a finite state machine M is a model of a function from A^* to O^*.

The only decision is whether to associate the generation of an output with the state the machine is in or with the transition based on an input symbol. There are two separate models determined by the choice: the **Moore machine**, in which outputs are generated according to the state, and the **Mealy machine**, in which outputs are generated according to the transition.

Moore Machines

A **Moore machine** is a quintuple of the form (S, A, T, O, f).

S, A, and T are the same as in the definition of FSA, and S contains a designated start state s_0.

O is the **output alphabet**.

$f: S \rightarrow O$ is the **output function**.

The **output** of a Moore machine in response to input $a_1 \ldots a_k$ is the word $f(s_0) \ldots f(s_k) \in O^*$, where $s_0, \ldots, s_k$ is the path through the state diagram induced by the input. Thus, when each state is entered, a Moore machine determines an output symbol and adds it to the output word. A Moore machine always generates the output $f(s_0)$ in response to the null word λ.

EXAMPLE 9.10

Given the machine described in the following table, determine its output if the input is *abbaaa*.

	Input		Output
	a	b	
0	1	0	x
1	2	1	y
2	2	0	y

$s_0 = 0$

Solution

We can show the output in tabular form as follows:

Input	a	b	b	a	a	a	—
State	0	1	1	1	2	2	2
Output	x	y	y	y	y	y	y

The output is, therefore, *xyyyyyy*. The initial *x* is not the result of any input; it merely reflects the starting state.

The graphical representation of a Moore machine is very similar to that of an FSA. The underlying state diagram is augmented by showing the output of a given state directly above the state.

EXAMPLE 9.11

Let (S, A, T, O, f) be the Moore machine shown in Figure 9.12.

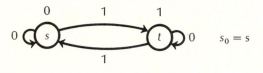

Figure 9.12. *Moore machine*

1. Find the output generated in response to the inputs 001 and 10101.
2. Describe the general input-output behavior of the machine.
3. Represent the machine in tabular form.

Solution

From the digraph in Figure 9.12, it is clear that both the input and output alphabets are $\{0, 1\}$, and that the state set is $\{s, t\}$.

1. In state *s*, the output is 0, and in state *t* the output is 1. The response to input 001 is 0001. The initial 0 is produced from the start state. The response to input 10101 is 011001.
2. Inputs of 0 are effectively ignored by the machine. The first 1 encountered causes a transition to state *t*, where a 1 is produced as output. The next 1, if any, causes a transition back to state *s*, where a 0 is output. This cycle is repeated for every pair of 1's. If the latest symbol output by the machine is a 1, then we know that the word received so far has an odd number of 1's. If a 0 is produced, then we know that the word received so far has an even number of 1's. Hence this machine is a simple *parity* machine.
3. The table for the machine of Figure 9.12 is given below.

state	input 0	input 1	output
s	s	t	0
t	t	s	1

$s_0 = s$

For finite state machines in which both the input alphabet and the output alphabet are the set {0, 1}, the machine is actually a model for the computation of a function from the set of nonnegative integers to the set of nonnegative integers. The integers, of course, are represented in binary form. Example 9.11 fits into this class, although the function computed is a bit unusual. From our study of the limitations of finite state automata, we expect that not all integer functions can be computed by a finite state machine. Moreover, as the next example shows, even simple functions may have very complex descriptions as finite state machines.

EXAMPLE 9.12

Construct an example of a Moore machine that adds 1 to a positive integer. Then trace the execution of the machine when the input is 23.

Solution

First, let's do some empirical work and examine a few cases of binary arithmetic. The three sums below show us, in binary, the addition of 1 to 4, 7, and 11, respectively. In binary, 4 is represented as 100, 7 is represented as 111, and 11 is represented as 1011.

$$
\begin{array}{ccc}
(4 + 1 = 5) & (7 + 1 = 8) & (11 + 1 = 12) \\
100 & 111 & 1011 \\
+\ \ 1 & +\ \ 1 & +\ \ 1 \\
\hline
101 & 1000 & 1100
\end{array}
$$

In discussing each example, we work from right to left. The first example shows us that if the first input digit is 0, the first output digit is 1 and, because there's no carry, all succeeding input digits are merely copied to the output.

In the second example, the first input digit is 1, so the first output digit must be 0. The next input digit is 1, and because of the carry from the first addition, the output is 0. Then we have a carry to the next input, which is the last 1. The last digit is a special case, since we have to generate two binary digits (10) to account for the carry.

In the third example, we begin with two output 0's because of the two input 1's. However, for the third input digit, 0, we output a 1 and the carry process stops, so the remaining input digit is copied to the output.

We now have enough information to construct the finite state machine. Let the input alphabet be {0, 1, Δ), where Δ represents a blank, which we will use to indicate the end of the input. Let the output alphabet be {0, 1, 1Δ, Δ}. The symbol 1Δ is needed to generate the carry of the last digit when all input digits are 1's. The finite state machine is shown in Figure 9.13. When using the machine, remember that the inputs must be fed in from low order to high order (from right to left), and that the outputs are also generated from low order to high order.

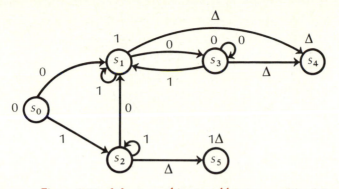

Figure 9.13. Moore machine to add 1 to a positive integer

Since 23 has binary representation 10111, the input is 11101Δ. The answer we expect is 24, which will appear in reverse order as 000011Δ. Just as before, the output will contain an extra initial value, in this case a 0. The sequence of states visited and the output for each is shown below.

input	state	output
	s_0	0
1	s_2	0
1	s_2	0
1	s_2	0
0	s_1	1
1	s_1	1
Δ	s_4	Δ

Moore machines are sometimes called **recognition machines** since they can be made to behave like finite state automata, and thus be used as language recognizers.

We can view an FSA M as a Moore machine with two output symbols, *no* and *yes* (or 0 and 1), where the output *yes* is associated with accepting states and the output *no* is associated with nonaccepting states. We can easily make an FSA into a Moore machine by defining an output function $f: S \to \{no, yes\}$ as follows:

$$f(s_i) = \begin{cases} no, & \text{if} \quad s_i \text{ is a nonaccepting state} \\ yes, & \text{if} \quad s_i \text{ is an accepting state} \end{cases}$$

EXAMPLE 9.13

Let M be the automaton with:
    states $S = \{0, 1, 2, 3\}$
    alphabet $A = \{a, b\}$

start state $s_0 = 0$

final state $F = \{3\}$

transition function T given by the following table:

		alphabet	
		a	b
s t a t e s	0	1	0
	1	3	1
	2	0	2
	3	2	3

Describe M as a Moore machine, and draw the corresponding digraph.

Solution

The corresponding Moore machine is

state	Input a b		Output
0	1	0	*no*
1	3	1	*no*
2	0	2	*no*
3	2	3	*yes*

$s_0 = 0$

The digraph is shown in Figure 9.14.

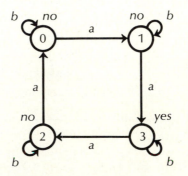

Figure 9.14. FSA and corresponding Moore machine

Mealy Machines

The alternate view of automata with outputs is the Mealy machine, which associates its output with transitions. In graphical form, the output is shown alongside the symbol labeling the edge between states. After we give the formal definition, we will construct Mealy machines that perform the same functions as some of the Moore machines we have already constructed. The two types of machines are capable of solving the same class of problems.

A **Mealy machine** is a 5-tuple (S, A, T, O, f), where

S, A, T are the same as in the definition of FSA, and S contains a designated start state s_0.

O is the **output** alphabet.

$f: S \times A \rightarrow O$ is the **output** function.

Observe that the critical difference between a Moore machine and a Mealy machine is that the former generates its output based only on the current state, whereas the latter requires knowledge of the current state and the current input symbol.

EXAMPLE 9.14

Find Mealy machine representations of each of the following machines.

1. The parity machine of Example 9.11
2. The machine of Example 9.12, which adds 1 to an integer represented in binary form

Solution

1. As before, we want to alternate between two states based on the number of 1's encountered in the input so far. A Mealy machine that does this is shown in Figure 9.15. This particular machine is a direct translation of the Moore version for the parity machine. The labels on the arcs are of the form i/o, where i is the input symbol and o is the output symbol.

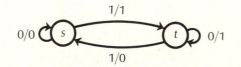

Figure 9.15. Mealy representation of a parity machine

If, for example, the input is 10101, the state path is s, t, t, s, s, t, and the output is 11001. The first 0 output by the corresponding Moore machine is no longer present, because the Mealy machine does not respond to a null input.

2. Again the conversion of the Moore machine to a Mealy machine is straightforward, and the spurious first output symbol is avoided. The machine is shown in Figure 9.16.

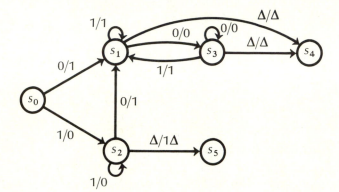

Figure 9.16. A Mealy machine to add 1 to a positive integer

However, if we analyze the problem slightly differently, we can find a Mealy machine that does the same calculation with fewer states. For instance, after the first 0 is encountered in the input, we want to copy all remaining inputs to the output. The Moore model requires two states for this. However, in a Mealy machine, we can use one state for copying, and let the different transitions from that state generate the distinct outputs. With the Mealy machine we also reduce the number of *stop* states by showing the different outputs on the transitions to one *stop* state. The resulting machine is shown in Figure 9.17.

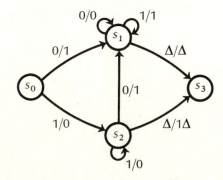

Figure 9.17. A simpler Mealy machine to add 1 to a positive integer

So far, each calculation we have performed with a Moore machine we have been able to do with a Mealy machine, and vice versa. The only difference has been that a Moore machine always generates an output in response to the null word and a Mealy machine never does. Except for this minor difference, every Moore machine has an equivalent Mealy machine, and vice versa. In the examples above, it was quite easy to take a Moore machine and convert it to a Mealy machine. The differences between Figures 9.16 and 9.12 suggest that the conversion in the other direction may not be so easily achieved.

EXAMPLE 9.15

Design a Mealy machine to do binary addition.

Solution

In order to add a pair of binary numbers with the same number of binary digits, we take pairs of binary digits as input. For example, in order to perform the addition

$$11100011$$
$$+\,00111001$$
$$\overline{100011100}$$

we will take as input the pairs of digits to be added, low order digits first:

$$11 \quad 10 \quad 00 \quad 01 \quad 01 \quad 11 \quad 10 \quad 10 \quad \Delta\Delta$$

The symbol Δ denotes a blank space and is used to terminate the input string. The output should be

$$0 \quad 0 \quad 1 \quad 1 \quad 1 \quad 0 \quad 0 \quad 0 \quad 1$$

The input alphabet is $A = \{00, 01, 10, 11, \Delta\Delta\}$. The output alphabet is $O = \{0, 1, \Delta\}$.

We need three states to indicate the status of the calculation at any point in time. The states are $S = \{\text{NoCarry(NC)}, \text{Carry(C)}, \text{Stop(S)}\}$. The initial state is NC.

The transition function T and the output function f are given in the following tables:

T	00	01	10	11	$\Delta\Delta$
NC	NC	NC	NC	C	S
C	NC	C	C	C	S
S	S	S	S	S	S

f	00	01	10	11	$\Delta\Delta$
NC	0	1	1	0	Δ
C	1	0	0	1	1
S	Δ	Δ	Δ	Δ	Δ

The Mealy machine is diagrammed in Figure 9.18.

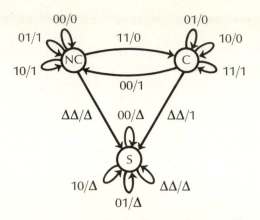

Figure 9.18. A Mealy machine for binary addition

SELF TEST 2

1. For each of the following Moore machines, compute the output string given the corresponding input string. The starting state $s_0 = 0$.

    a. Input string *aaaabababa*

	Input		Output
state	a	b	
0	0	1	a
1	0	2	b
2	2	2	a

    b. Input string 00111100

    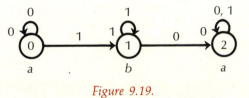

    Figure 9.19.

2. Design a Mealy machine that inputs a binary string and outputs the string multiplied by 2.

Exercises 9.2

1. An electronic door opener works as follows: When the door is closed and some-one steps on the pressure pad, the door opens as long as there is pressure on the pad, and will start closing three seconds after the pressure stops. If the door is closing and someone steps on the pressure pad, the door reopens. Otherwise, it closes completely. Draw a state model similar to the model in Example 9.1 that describes the door's activity. The outputs are the signals from the control to the motor controlling the door. The inputs are the signals from the pressure contact, the door open and door closed contacts, and the 3-second timer signal. You may use either a Moore or a Mealy machine model.

2. a. In Figure 9.13, it appears that states s_0 and s_2 could be combined. Would this work? Why or why not?
   b. Find a Mealy machine that subtracts 1 from a positive integer. Find a Moore machine that performs the same calculation.

*3. Is it possible to have a Moore machine that adds 1 to an integer and accepts the high order inputs first (that is, accepts inputs from left to right)?

4. Devise a Moore machine that does the following task. Then devise a Mealy machine for the task. In both cases, assume that the input and output are the binary representation of the integers, presented from low order to high order, and compute the product of 3 and the positive integer.

9.4   Turing Machines (Optional)

In the previous sections we have covered finite automata, which represent one end of the spectrum of computational capability. At the other end of the spectrum are Turing machines, which, it is conjectured, can model any computation for which an "effective procedure" exists. We will show that every FSA can be represented as a Turing machine, so that every computation that can be modeled by an FSA can be modeled by a Turing machine. Turing machines recognize the class of languages known as recursively enumerable sets, which is a larger class of languages than the class of regular languages recognized by finite state automata. The class of functions modeled by Turing machines is the class of functions known as partial recursive functions.

Effective Procedures

Figure 9.20 contains a general representation of a computation that accepts inputs, performs a calculation, and generates outputs.

Figure 9.20. General input-output computation

As we saw in the previous sections, finite automata are capable of modeling some of these computations. We limit our attention to those computations for which an effective procedure exists. An **effective procedure** for a computation has a finite description consisting of a finite number of instructions, each of which can be performed mechanically. In arithmetic, for instance, there are well-known procedures for performing multiplication, for extracting square roots, for determining if a given number is prime, and so on. In previous chapters of the text, we have presented algorithms for topological sorting, for finding the connectivity number of a graph, and for other computations. All of these algorithms are effective procedures.

For a moment, consider only the class of functions from the integers to the integers. If there is an effective procedure for an integer function, it can be given a finite representation in some alphabet. Hence, there are only a countable number of integer functions with effective procedures. On the other hand, there must be an uncountable number of integer functions. (Exercise 7 on page 393 leads you through the details of the counting arguments needed to establish this fact.) Thus there must exist integer functions for which there is no effective procedure. The fundamental question is this: "What are the functions for which effective procedures exist?" An answer to that question requires a formal model of computation.

Turing's Model of Computation

The mathematician A. M. Turing explored the finite representability of effective procedures and developed a model of computation known as the **Turing machine**. Turing's model is depicted in Figure 9.21.

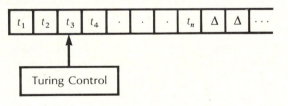

Figure 9.21. The Turing model

Turing's method was to encode each input with a finite number of symbols and place them on an "infinite tape" to be read and acted on by a human computer. Turing assumed that at any point in time the human "computer" has access to a limited number of symbols on the tape. The behavior of the computer is determined by the symbols observed and its "state of mind." The computer performs the calculation by examining symbols on the tape, and, depending on its state of mind, making changes to the tape. The computer can then change its state of mind.

Many apparently different formal definitions have been given of computability. All have been found to be equivalent in modeling power. The particular definition we give is based on the picture in Figure 9.21, which is similar in appearance to the model used for an FSA. The input symbols are placed in cells on a tape, with the first symbol at the left end of the tape. The symbols are scanned one at a time. Given a current state and a current symbol, the Turing machine can rewrite the symbol in the cell, change its own state, and move one cell to the left or the right. The right end of the tape is assumed to be as long as needed, so that a Turing machine could keep moving out to the right forever. It is also assumed that a Turing machine need not have a response to every possible pairing of state and symbol. Thus there are three significant differences between the FSA model and the Turing model. The first is that an FSA may not rewrite the symbols on its tape. The second is that an FSA must always move to the right. Hence, an FSA is incapable of "peeking" at any previously seen symbols. In a card game such as gin rummy, we have the analogous restriction that a player may not look back at any cards in the discard pile to aid in choosing a next move. The third difference is that an FSA has a completely defined transition function, whereas a Turing machine has only a partial transition function.

Formally, a **Turing machine** is a quintuple (S, A_T, N, s_0, F).

S is a finite set whose elements are called **states**.

$$s_0 \in S \text{ and } F \subseteq S$$

A_T is the finite **tape alphabet**, which is partitioned into an input alphabet A_I, an output alphabet A_O, and a special symbol Δ, usually called a *blank*.

$N: S \times A_T \to S \times A_T \times \{L, R\}$ is the **next move** function, which is defined only on a subset of $S \times A_T$. L signifies that the machine is to move to the left, and R signifies that the machine is to move to the right.

At any point in time, the word on the tape will be of the form $x \cdot y$, where x and y belong to A_T^*. The word consisting of the symbols on the tape up to but not including the current scanning position is x. The word consisting of the symbols on the tape from the current scanning position to the last nonblank symbol is y. There are two special cases: x is the null word, λ, if the current scanning position is the left end of the tape; y is the null word if the scanning position is beyond the last nonblank symbol. If s is the current state, we call the triple (x, s, y) an **instantaneous description** **(ID)** of the computation. The initial instantaneous description has the form (λ, s_0, y), as shown in Figure 9.22.

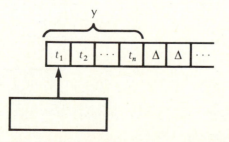

Figure 9.22. The initial instantaneous description

The moves of a Turing machine that we informally described above can now be given a formal description in terms of instantaneous descriptions. Let the symbols on the tape be $t_1, t_2, \ldots, t_n$, where t_n is the last nonblank symbol. Assume that the Turing machine is scanning the symbol t_i. Let x be $t_1 t_2 \ldots t_{i-1}$, let y be $t_i \ldots t_n$, and let s be the state of the machine, so that (x, s, y) is the ID.

Moves to the Left.   If $i = 1$, then $x = \lambda$, and the machine is not allowed to move left and fall off the end of the tape. If $i > 1$ and $N(s, t_i) = (s', t', L)$, the move left produces the ID (x', s', y'), where $x' = t_1, \ldots, t_{i-2}$ and $y' = t_{i-1} t' t_{i+1} \ldots t_n$. We write

$$(t_1 \ldots t_{i-1}, s, t_i \ldots t_n) \rightarrow (t_1 \ldots t_{i-2}, s', t_{i-1} t' t_{i+1} \ldots t_n)$$

to describe the move to the left in terms of the ID's. Figure 9.23 shows a left move.

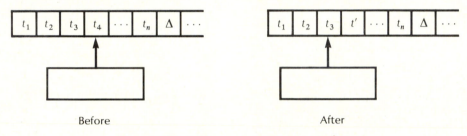

Before   After

Figure 9.23. Left move of a Turing machine

Moves to the Right.   If $i = n$, the machine is at the last nonblank symbol. Let $N(s, t_n) = (s', t', R)$. The new ID is (x', s', y), where $x' = t_1 \ldots t_{n-1} t'$ and $y' = \lambda$. If $i < n$, the new ID is (x', s', y'), where $x' = t_1 \ldots t_{i-1} t'$ and $y' = t_{i+1} \ldots t_n$. We write

$$(t_1 \ldots t_{i-1}, s, t_i \ldots t_n) \rightarrow (t_1 \ldots t_{i-1} t', s', t_{i+1} \ldots t_n)$$

to describe the right move. Figure 9.24 shows a right move.

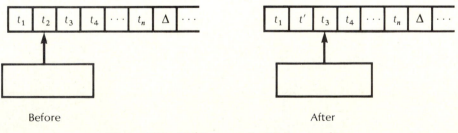

Before   After

Figure 9.24. Right move of a Turing machine

Halting Condition.   A Turing machine **halts** when $N(s, t_i)$ is undefined.

The description of Turing machine operations seems quite complex. To bring it down to earth a bit, we will present four examples. The first, Example 9.16, introduces table representations of the next move function and gives us practice in mechanically tracing the execution of a Turing machine without having to worry about what the machine does in a more abstract sense. Example 9.17 demonstrates that any computation that can be performed by an FSA can also be performed by a Turing machine. In Example 9.18 we construct a Turing machine that recognizes the language of words $a^n b^n$, $n \geq 1$, a language which no FSA can recognize. The fourth example, Example 9.19, shows how Turing machines serve as "function" computers. One lesson to learn from the examples is that Turing machines, while capable of performing any reasonable calculation, are very difficult to program and to understand.

EXAMPLE 9.16

Consider the Turing machine given by the following table. Trace the evolution of ID's when the initial contents of the input tape is *aaba*.

state	a	b	A	B	Δ
0	$(1, A, R)$	—	—	$(3, B, R)$	—
1	$(1, a, R)$	$(2, B, L)$	—	$(1, B, R)$	—
2	$(2, a, L)$	—	$(0, A, R)$	$(2, B, R)$	—
3	—	—	—	$(3, B, R)$	$(4, \Delta, R)$
4	—	—	—	—	—

The input alphabet is $\{a, b\}$, the tape alphabet is $\{a, b, A, B, \Delta\}$, and the start state $s_0 = 0$. When there is no next move defined for a given state and input, the table entry is a dash.

Solution

Evidently, the machine must halt whenever the machine reaches state 4. If the input tape initially contains $aaba\Delta\Delta\Delta \ldots$, then the initial ID is $(\lambda, 0, aaba)$. When in state 0 and scanning an a, the machine writes an A into the cell, changes the state to 1, and moves right to scan the adjacent symbol. Thus the first move is $(\lambda, 0, aaba) \rightarrow (A, 1, aba)$. When in state 1 and scanning an a, the machine writes a on the tape, remains in state 1, and moves right to scan the adjacent symbol. Thus the second move is $(A, 1, aba) \rightarrow (Aa, 1, ba)$. Without further discussion, the sequence of moves, as shown by changes in the ID's, is as follows:

$$(\lambda, 0, aaba) \rightarrow (A, 1, aba) \rightarrow (Aa, 1, ba) \rightarrow$$
$$(A, 2, aBa) \rightarrow (\lambda, 2, AaBa) \rightarrow (A, 0, aBa) \rightarrow$$
$$(AA, 1, Ba) \rightarrow (AAB, 1, a) \rightarrow (AABa, 1, \lambda) \rightarrow \text{HALT}$$

EXAMPLE 9.17

Let $M = (S, A, T, s_0, F)$ be an FSA. Represent M as a Turing machine.

Solution

The workings of an FSA are very simple. The FSA always leaves the tape symbols as it finds them and continually moves to the right, halting when it encounters the first Δ. We need to specify a quintuple (S', A'_T, N', s'_0, F') for an equivalent Turing machine. Let $S' = S, A'_T = A \cup \{\Delta\}, s'_0 = s_0,$ and $F' = F.$ Define $N': S \times A'_T \to S \times A'_T \times \{L, R\}$ by

$$N'(s, a) = (T(s, a), a, R) \text{ for } s \in S \text{ and } a \in A$$

$N'(s, \Delta)$ is not defined, so that when the first Δ on the right of the input word is encountered, the Turing machine halts.

Turing Machines as Language Recognizers

Earlier in this section, we made reference to Turing machines and languages. The **language accepted by a Turing machine** is the set of words $y \in A_I{}^*$ such that the ID (λ, s_0, y) is transformed in a finite number of moves to the ID (x', s', y'), where $s' \in F$ and the machine halts. For words in $A_I{}^*$ but not in L, the machine need not halt. A language accepted by a Turing machine is called a **recursively enumerable set**.

Having shown that Turing machines are at least as powerful as finite state automata, we now show that a Turing machine can perform a task that an FSA cannot. In the process, we demonstrate that there is a recursively enumerable language that is not regular.

EXAMPLE 9.18

Design a Turing machine to recognize the language L consisting of words of the form $a^n b^n, n \geq 1.$ Then trace the evolution of ID's when the input is $a^2 b^2.$

Solution

We'll use the tape alphabet $\{a, b, A, B, \Delta\}.$ The input alphabet is $\{a, b\},$ and the output alphabet is $\{A, B\}.$ Given an initial input on the tape, we move back and forth, matching the outermost a's and b's and replacing them with A's and B's, respectively. At the end, we should have no more input symbols left to consider. Figure 9.25 shows how we want the tape to change over several moves.

Imbalances in the number of a's and b's are detected as follows: If, after moving right to the first B or Δ, the first symbol on the left is an A or a, then there are too few b's. On the other hand, if after moving left to the rightmost A, the next symbol on the right is a b, then there are too many b's. To terminate "normally" after balancing the last a with a b, we expect to move back to the left, find the rightmost A, move right one cell and find a B. The next move partial function of the Turing machine is shown below.

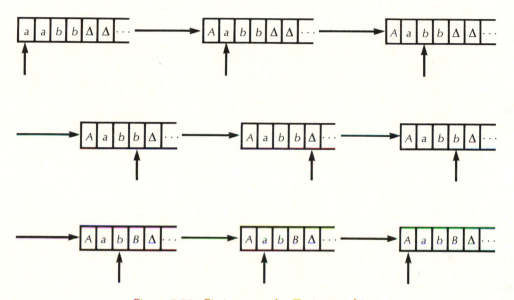

Figure 9.25. First moves of a Turing machine

The start state is 0 and the only final state is 5.

	a	b	A	B	Δ
0	$(1, A, R)$	—	—	—	—
1	$(1, a, R)$	$(1, b, R)$	—	$(2, B, L)$	$(2, \Delta, L)$
2	—	$(3, B, L)$	—	—	—
3	$(3, a, L)$	$(3, b, L)$	$(4, A, R)$	—	—
4	$(1, A, R)$	—	—	$(5, B, R)$	—
5	—	—	—	—	—

The ID's below show how the Turing machine defined above acts on the input a^2b^2. In state 0, the first a is converted to A, and the machine switches to state 1, which is a right-moving state, searching for a Δ, or for the first B. Finding Δ, the machine backs up one cell and checks that it contains a b. The b is converted to a B, and the machine switches to the left-moving state 3, searching for the first A. After going back and forth twice, the machine checks for extra b's in state 4, and finding none, halts in state 5.

$$(\lambda, 0, aabb) \rightarrow (A, 1, abb) \rightarrow (Aa, 1, bb) \rightarrow$$
$$(Aab, 1, b) \rightarrow (Aabb, 1, \lambda) \rightarrow (Aab, 2, b) \rightarrow$$
$$(Aa, 3, bB) \rightarrow (A, 3, abB) \rightarrow (\lambda, 3, AabB) \rightarrow$$
$$(A, 4, abB) \rightarrow (AA, 1, bB) \rightarrow (AAb, 1, B) \rightarrow$$
$$(AA, 2, bB) \rightarrow (A, 3, ABB) \rightarrow (AA, 4, BB) \rightarrow$$
$$(AAB, 5, B) \rightarrow \text{HALT}$$

Turing Machines as Function Computers

The other reason for studying Turing Machines is for modeling functions. We restrict our attention to functions on the set of positive integers where the integers are represented in unary form; that is, each integer is represented as a string of 1's. The string '1' represents the integer 1, '11' represents the integer 2, '111' represents the integer 3, and so on. We will use a 0 to separate one string of 1's from another. Thus the input alphabet is $\{0, 1\}$.

EXAMPLE 9.19

Find a Turing machine that takes two positive integers such that the first integer is preceded by a 0 and the two integers are separated by a 0 as input, computes the minimum, and leaves the minimum at the left end of the tape. Show the evolution of ID's when the initial tape contains $0111011\Delta\Delta\Delta \ldots$.

Solution

We assume that the input tape contains two nonempty strings of 1's separated by a 0. (It should not be hard to devise a Turing machine that inspects the tape to make sure the initial word on the tape is as expected.) The procedure we use is as follows: A 1 at the left end of the first integer is replaced by the symbol A. The machine moves to the right and replaces the last 1 in the second integer with a Δ. Then the machine moves back to the left until it finds the leftmost 1 in the first integer, replaces it with an A, moves right to replace the last 1 in the second integer by a Δ, and so on. There are two special conditions that can occur. If the first integer is larger than the second, then after the machine has changed the last remaining 1 in the second integer, it will move left and encounter the 0 separating the two integers. The machine should then move left and replace all symbols with a Δ until it finds the first A. If the first integer is smaller than the second, then after the machine has moved back to the left, encountered an A, and started back to the right, it will encounter the 0 immediately. At this point, all symbols to the right, before the first Δ, should be replaced by Δ.

After either of the two conditions is encountered, the tape will contain a number of A's, and the current cell position will be somewhere to the right of all the A's. The machine completes the computation by moving to the left end replacing the A's by 1's as it goes. The following table is the next move table.

	0	1	A	Δ
0	$(1, 0, R)$	—	—	—
1	—	$(2, A, R)$		
2	$(2, 0, R)$	$(2, 1, R)$	—	$(3, \Delta, L)$
3	—	$(4, \Delta, L)$		
4	$(6, \Delta, L)$	$(5, 1, L)$	—	—
5	$(5, 0, L)$	$(5, 1, L)$	$(7, A, R)$	—
6	$(9, 0, R)$	$(6, \Delta, L)$	$(6, 1, L)$	$(6, \Delta, L)$
7	$(8, \Delta, R)$	$(2, A, R)$	—	—
8	—	$(8, \Delta, R)$	—	$(6, \Delta, L)$
9	—	—	—	—

When the tape contains $0111011\Delta\Delta\ldots$, the initial ID is $(\lambda, 0, 0111011)$. As the 1's in the second integer are converted to Δ's, the length of the right side of the ID becomes smaller. The moves of the machine are

$$(\lambda, 0, 0111011) \rightarrow (0, 1, 111011) \rightarrow (0A, 2, 11011) \rightarrow$$
$$(0A1, 2, 1011) \rightarrow (0A11, 2, 011) \rightarrow (0A110, 2, 11) \rightarrow$$
$$(0A1101, 2, 1) \rightarrow (0A11011, 2, \lambda) \rightarrow (0A1101, 3, 1) \rightarrow$$
$$(0A110, 4, 1) \rightarrow (0A11, 5, 01) \rightarrow (0A1, 5, 101) \rightarrow$$
$$(0A, 5, 1101) \rightarrow (0, 5, A1101) \rightarrow (0A, 7, 1101) \rightarrow$$
$$(0A5A, 2, 101) \rightarrow \ldots \rightarrow (0AA101, 2, \lambda) \rightarrow$$
$$(0AA10, 3, 1) \rightarrow (0AA1, 4, 0) \rightarrow (0AA, 6, 1) \rightarrow$$
$$(0A, 6, A) \rightarrow (0, 6, A1) \rightarrow (\lambda, 6, 011) \rightarrow$$
$$(0, 9, 11) \rightarrow \text{HALT}$$

SELF TEST 3

1. Trace the evolution of ID's for the Turing machine of Example 9.16 when the initial content of the tape is *aaab*.

2. Represent the FSA given in Figure 9.26 as a Turing machine.

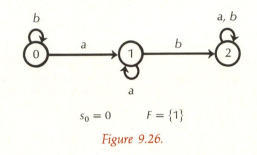

$$s_0 = 0 \qquad F = \{1\}$$

Figure 9.26.

3. Devise a Turing machine that adds 1 to a number, when the number is represented in unary.

Church–Turing Thesis

Earlier in this section, we discussed the intuitive notion of a "computable" function and looked at Turing's model of computation. Turing's work was done in the 1930's, well before the advent of modern computers. Other mathematicians, including Alonzo Church, were also actively investigating formalisms that could identify and explain the limits of computability. All the most powerful models developed then, and since then, have been shown to be equivalent; that is, all have been shown to represent what are called the partial recursive functions. The **Church–Turing Thesis** states that the class of computable functions coincides with the partial recursive functions. We observe that there can be no proof of this thesis, because of the intuitive nature of "computable." The astounding consequence of the thesis is that any computation that can be performed on a powerful modern computer can be modeled by a Turing machine.

Exercises 9.3

1. Trace the evolution of the ID's for the Turing machine of Example 9.16 for each of the following inputs.
   a. *aaabb*
   b. *aabbb*
   c. *aaabbb*

2. Construct a Turing machine that removes the last two input symbols when the input has two or more symbols, and computes forever otherwise.

3. The input tape to a Turing machine consists of two nonempty words from alphabet A, with the special symbol [ ] separating the first word from the second word. Construct Turing machines that concatenate the two words together under each of the following conditions.

   a. by shifting the second word back one cell to the left, so that only the result is left on the tape
   b. if the words and the result are to be left on the tape
   c. if either of the words can be empty

4. a. The Turing machine of Example 9.18 accepts the language consisting of words of the form $a^n b^n$, $n \geq 1$. That machine worked from the outside in. Find a Turing machine that recognizes the same language but works from the inside out; that is, first the innermost a and b are matched, then the a and b outside those are matched, and so on. The pitfall to avoid is allowing an a to follow a b.

   b. Expand the inside-out method above to recognize the language of "well-formed parentheses," which is defined as follows:

      • ( ) is well formed.
      • If f and g are well formed, then so are fg and (f).

      For example, ((( ))), (( )( )), and (( )((( )))) are well formed. You should be able to find a Turing machine that performs the task with only three states.

5. Analyze the behavior of the Turing machine of Example 9.16.

6. Devise a Turing machine to perform each of the following computations.

   a. Compute the sum of two positive integers represented in unary form.
   b. Detect whether a given word in $\{a, b\}^*$ is a palindrome (a word that is the same written from left to right or from right to left).

7. Follow the steps below to verify that the set of integer functions is not enumerable. The method is known as the **Cantor Diagonalization Procedure**.

   a. It is enough to show that the set, F, of functions from N to $\{0, 1\}$ is not enumerable.
   b. Suppose the set F is enumerable. Then it is possible to number the elements of F as $f_1, f_2, f_3, \ldots$.
   c. Define a function $f: N \to \{0, 1\}$ by

$$f(i) = \begin{cases} 0, & \text{if } f_i(i) = 1 \\ 1, & \text{if } f_i(i) = 0 \end{cases}$$

   Show that f is not equal to any of the f_i's, contrary to the assumption that F is enumerable.

Problem Solving

We have seen several applications of finite state machines and their graphs in this chapter, including recognition of valid identifiers in a programming language and design of binary adders. Finite state machines can also be used to model more general problems. As an illustration of such a use of finite state machines, let us consider the well-known problem of the cannibals and the missionaries.

In some remote place, there is a tribe of cannibals who are visited by a group of missionaries. On a certain day, two missionaries and two cannibals end up on the same side of a river that they all wish to cross. The boat that is available can transport only two people: one rower and one passenger. The missionaries are safe as long as there are as many of them in any place as there are cannibals. Should there be more cannibals than missionaries, disaster would strike the missionaries. The problem is to devise a sequence of moves that enables the cannibals and the missionaries to cross the river without mishap. Let us assume that the two cannibals and the two missionaries are initially on the right bank of the river and that they wish to get to the left bank.

Finite State Automata as Problem-Solving Models

The idea is to construct a finite state automaton that will take the number of cannibals and missionaries in the boat as input and thereby allow us to determine which sequences of passengers and rowers will lead to the solution of the problem. The various states of the problem will define the states of the FSA. The set of states in the problem contains the number of cannibals and missionaries on the right side of the river at any point in time and the location of the boat. The starting state is (2 Cannibals, 2 Missionaries, Right). The only accepting state is (0 Cannibals, 0 Missionaries, Left). This is the case in which everyone has crossed the river. We can now design an FSA to model our problem. The input alphabet is the set of all possible pairs of passengers (m, c), where m represents the number of missionaries and c represents the number of cannibals.

$$A = \{(0, 1), (0, 2), (1, 0), (1, 1), (2, 0)\}$$

Since there must always be at least one person in the boat, and since there can never be more cannibals than missionaries unless there are no missionaries, these are the only permissible combinations of missionaries and cannibals in the boat.

We can designate the finite number of states of the FSA as ordered triples (m, c, l), where m represents the number of missionaries on the right bank (0, 1, or 2), c represents the number of cannibals on the right bank (0, 1, or 2), and l represents the location of the boat (Left bank or Right bank).

We now may construct the FSA given in the following table.

		Input alphabet				
		(0, 1)	(0, 2)	(1, 0)	(1, 1)	(2, 0)
	(2, 2, R)	(2, 1, L)	(2, 0, L)	(End)	(1, 1, L)	(0, 2, L)
	(2, 1, R)	(2, 0, L)	(End)	(1, 1, L)	(1, 0, L)	(0, 1, L)
	(2, 0, R)	(End)	(End)	(1, 0, L)	(End)	(0, 0, L)
	(1, 1, R)	(1, 0, L)	(End)	(0, 1, L)	(0, 0, L)	(End)
s	(0, 2, R)	(0, 1, L)	(0, 0, L)	(End)	(End)	(End)
t	(0, 1, R)	(0, 0, L)	(End)	(End)	(End)	(End)
a	(2, 1, L)	(2, 2, R)	(End)	(End)	(End)	(End)
t	(2, 0, L)	(2, 1, R)	(2, 2, R)	(End)	(End)	(End)
e	(1, 1, L)	(End)	(End)	(2, 1, R)	(2, 2, R)	(End)
s	(0, 2, L)	(End)	(2, 2, R)	(End)	(End)	(2, 2, R)
	(0, 1, L)	(0, 2, R)	(End)	(1, 1, R)	(End)	(2, 1, R)
	(0, 0, L)	(0, 1, R)	(0, 2, R)	(1, 0, R)	(1, 1, R)	(2, 0, R)
	(End)	(End)	(End)	(End)	(End)	(End)

The state (End) is used as a dead state for all nonacceptable actions. All we need to do now is find a path from the starting state (2, 2, R) to the accepting state (0, 0, L). A solution path is shown in Figure 9.27.

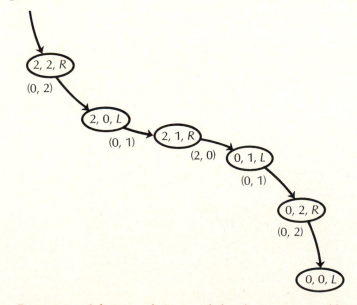

Figure 9.27. Solution path for cannibal and missionary problem

Exercises 9.4

1. Find all paths that lead to the solution of the cannibal and missionaries problem.

2. A puzzle similar to the cannibals and missionaries problems involves a farmer, a wolf, a goat, and a cabbage. The farmer must get the whole group across the river but can carry only one additional passenger. If he leaves the wolf with the goat, the wolf will eat the goat. If he leaves the goat with the cabbage, the goat will eat the cabbage. Devise an FSA to solve the problem and find the solution.

SUMMARY

1. A **finite state automaton**, or **FSA**, is a quintuple $M = (S, A, T, s_0, F)$, where

   S is a finite set whose elements are called **states**.
   A is a finite set, called an **input alphabet**, whose elements are called **input symbols**.
   T is a function from $S \times A \to S$, called the **transition** function.
   s_0 is an element of S called the **start state**.
   F is a subset of S, whose elements are called **final**, or **terminal** states.

2. The transition function T for an FSA M is extended to a function $T': S \times A^* \to S$ defined recursively by

   $$T'(q, \varepsilon) = q$$
   $$T'(q, ax) = T'(T(q, a), x)$$

3. Finite state automata play a significant role in compiler design as **lexical analyzers**; that is, they are used to recognize certain patterns of symbols such as identifiers, real numbers, special characters, and so on as being legitimate constructs of a language.

4. A **word** of length k from an alphabet A is $a_1 a_2 \ldots a_k$, where $a_i \in A$, for $i = 1, 2, \ldots, k$.

5. A **language** over the alphabet A is a subset of A^*.

6. The **language accepted** (recognized) **by** an FSA M is the set of words, $L(M)$, given by

   $$L(M) = \{x: x \in A^* \text{ and } T'(s_0, x) \in F\}$$

7. A language which is accepted by some FSA is said to be **regular**. That is, a language L is regular if $L = L(M)$ for some FSA M.

8. **Theorem 9.1**. Let $M = (S, A, T, s_0, F)$ be an FSA that recognizes language $L(M)$. Let $n = |S|$.

   1. $L(M)$ is nonempty if and only if there exists $x \in L(M)$ such that $0 \le |x| < n$.
   2. $L(M)$ is infinite if and only if there exists $x \in L(M)$ such that $|x| \ge n$.

9. A **Moore machine** is a quintuple of the form (S, A, T, O, f).

   S, A, and T are the same as in the definition of FSA, and S contains a designated start state s_0.
   O is the **output alphabet**.
   $f: S \to O$ is the **output function**.

10. Moore machines are also sometimes called **recognition machines** since they can be made to behave like finite state automata and thus be used as language recognizers.

11. We can view an FSA M as a Moore machine with two output symbols, *no* and *yes* (or 0 and 1), where the output *yes* is associated with accepting states and the output *no* is associated with nonaccepting states. We can easily make an FSA into

a Moore machine by defining an output function $f: S \rightarrow \{no,\ yes\}$ as follows:

$$f(s_i) = \begin{cases} no, & \text{if} \quad s_i \text{ is a nonaccepting state} \\ yes, & \text{if} \quad s_i \text{ is an accepting state} \end{cases}$$

12. A **Mealy machine** is a 5-tuple (S, A, T, O, f).

S, A, T are the same as in the definition for an FSA, and S contains a designated start state s_0.

O is the **output alphabet**.

$f: S \times A \rightarrow O$ is the **output function**.

13. The difference between a Moore machine and a Mealy machine is that the former generates its output based only on the current state, whereas the latter requires knowledge of the current state and the current input symbol.

14. A **Turing machine** is a quintuple (S, A_T, N, s_0, F).

S is a finite set whose elements are called **states**.

$$s_0 \in S \text{ and } F \subseteq S.$$

A_T is the finite **tape alphabet**, which is partitioned into an input alphabet A_I, an output alphabet A_O, and a special symbol Δ, usually called a *blank*.

$N: S \times A_T \rightarrow S \times A_T \times \{L, R\}$ is the **next move** function, which is defined only on a subset of $S \times A_T$. L signifies that the machine is to move to the left, and R signifies that the machine is to move to the right.

15. The **language accepted by a Turing machine** is the set of words $y \in A_I^*$ such that the ID (λ, s_0, y) is transformed in a finite number of moves to the ID (x', s', y'), where $s' \in F$ and the machine halts. For words in A_I^* but not in L, the machine need not halt.

16. Turing machines are also used to model functions.

17. The **Church–Turing Thesis** states that the class of computable functions coincides with the class of partial recursive functions.

REVIEW PROBLEMS

1. Describe the activity of the automaton given in the state diagram of Figure 9.28.

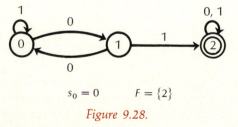

$$s_0 = 0 \qquad F = \{2\}$$

Figure 9.28.

2. Let M be the automaton with:

states $S = \{0, 1, 2, 3\}$
alphabet $A = \{a, b\}$
start state $s_0 = 0$
final states $F = \{2, 3\}$
transition function T given by the following table:

		alphabet	
		a	b
s	0	0	2
t	1	3	1
a	2	1	3
t	3	3	2
e			
s			

Give a representation of the automaton as a labeled digraph. Then describe the action of the automaton.

3. Show that the FSA of Figure 9.29 will accept each of the following words. (*Note:* Any transitions not shown are to be considered as moving to a dead state.)

a. $-.12$ c. $+.0001$
b. 12.34 d. 00005

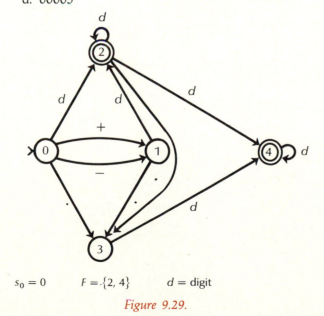

$s_0 = 0$ $F = \{2, 4\}$ $d = $ digit

Figure 9.29.

4. Show that the FSA of Figure 9.29 will not accept each of the following words.
   a. $+12-3$ c. $-3+$
   b. $.00.2$ d. $1..23$

5. Find the language $L(M)$ accepted by the FSA M given by the table on the left:

6. Describe informally the language accepted by the FSA of Figure 9.30.

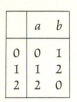

	a	b
0	0	1
1	1	2
2	2	0

$s_0 = 0$

$F = \{1\}$

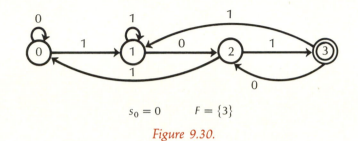

$s_0 = 0$ $F = \{3\}$

Figure 9.30.

7. Describe in English the words in Letter $\{$Letter, Digit$\}$*, where Letter $=$ $\{A, B, \ldots, Z, a, b, \ldots, z\}$ and Digit $= \{0, 1, \ldots, 9\}$.

8. Find an FSA that accepts all words in $\{0, 1\}$* that have both '01' and '10' as substrings.

9. Find an FSA that accepts all words in $\{a, b\}$*, where the number of a's is divisible by 3.

10. In the programming language FORTRAN, an identifier is defined to be a letter followed by up to 5 letters or digits. Here Letter $= \{\$, A, B, \ldots, Z\}$ and Digit $= \{0, 1, \ldots, 9\}$. Design an FSM M to recognize identifiers in FORTRAN.

11. Let $A = \{a, b\}$. Show that there is no FSA that can recognize the language $L = \{a^n: n$ is prime$\}$.

12. Given the machine described in the following table, determine its output if the input is $abbaaa$.

Current state	Next state		
	Input		Output
	a	b	
0	0	2	x
1	1	0	y
2	2	0	x

$s_0 = 0$

13. Let (S, A, T, O, f) be the Moore machine shown in Figure 9.31.

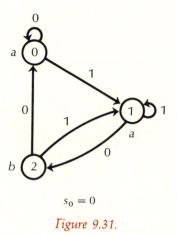

$$s_0 = 0$$

Figure 9.31.

    a. Find the output generated in response to the inputs 1001 and 010101.
    b. Describe the general input-output behavior of the machine.
    c. Represent the machine in tabular form.

14. a. Construct a Moore machine that computes the remainder upon division by 3. The input is in binary form.
    b. Trace the execution of the machine when the input is ten.

15. Let M be the FSA of Figure 9.32. Describe it as a Moore machine, and draw the corresponding digraph.

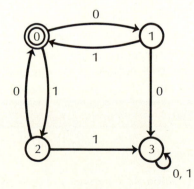

Figure 9.32.

16. Find a Mealy machine representation of the Moore machine of problem 15.

17. Design a Mealy machine for the following process: for input from $\{0, 1\}^*$, if input is 01, output is a; if input is 10, output is b; if input is 11, output is c; otherwise output is d.

18. Consider the Turing machine given by the following table. Trace the evolution of ID's when the initial contents of the input tape is *aabb*.

state	a	b	A	B	Δ
0	$(1, A, R)$	—	—	$(3, B, R)$	—
1	$(1, a, R)$	$(2, B, L)$	—	$(1, B, R)$	—
2	$(2, a, L)$	—	$(0, A, R)$	$(2, B, R)$	—
3	—	—	—	$(3, B, R)$	$(4, \Delta, R)$
4	—	—	—	—	—

The input alphabet is $\{a, b\}$, the tape alphabet is $\{a, b, A, B, \Delta\}$, and the start state $s_0 = 0$. When there is no next move defined for a given state and input, the table entry is a dash.

19. a. Design a Turing machine to recognize the language L consisting of words of the form $a^n b c^n$, $n \geq 1$.
    b. Trace the evolution of ID's when the input is $a^2 b c^2$.

20. Find a Turing machine that takes as input two positive integers such that the first integer is preceded by a 0 and the two integers are separated by a 0, computes the maximum, and leaves the maximum at the left end of the tape. Show the evolution of ID's when the initial tape contains $0111011\Delta\Delta\Delta \ldots$.

Programming Problems

21. Write a program to simulate an FSA.

22. Write a program to simulate a Mealy machine.

23. Program the cannibals and missionaries problem for three cannibals and three missionaries, assuming the boat holds only two people.

SELF TEST ANSWERS

Self Test 1

1. The representation of the automaton is given in Figure 9.33. The FSA rejects any word in $\{a, b\}^*$ with two or more consecutive b's.

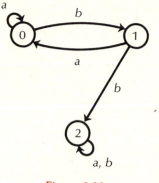

Figure 9.33.

2. The only way to begin and end in state 0 is to have *3k b's*, *k* = 0, 1, 2, 3, ...
3. The automaton is shown in Figure 9.34.

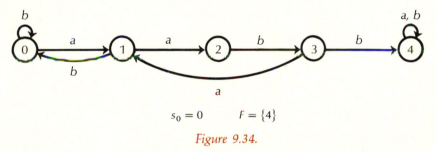

$$s_0 = 0 \qquad F = \{4\}$$

Figure 9.34.

Self Test 2

1. a. The sequence of states reached in response to the input *aaaabababa* is

$$0, 0, 0, 0, 1, 0, 1, 0, 1, 0$$

    The output is *aaaaababab a*. The initial *a* is the output from state 0.
   b. The output is 000111100.
2. The Mealy machine is shown in Figure 9.35. The symbol Δ is a blank at end of the string.

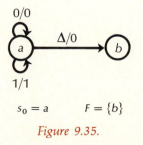

$$s_0 = a \qquad F = \{b\}$$

Figure 9.35.

Self Test 3

1. $(\lambda, 0, aaab) \rightarrow (A, 1, aab) \quad \rightarrow (Aa, 1, ab) \rightarrow (Aaa, 1, b)$
$\rightarrow (Aa, 2, aB) \quad \rightarrow (A, 2, aaB) \rightarrow (\lambda, 2, AaaB)$
$\rightarrow (A, 0, aaB) \quad \rightarrow (AA, 1, aB) \rightarrow (AAa, 1, B)$
$\rightarrow (AAaB, 1, \Delta) \rightarrow \text{HALT}$

2.

	a	b	Δ
0	$(1, a, R)$	$(0, b, R)$	—
1	$(1, a, R)$	$(2, b, R)$	—
2	$(2, a, R)$	$(2, b, R)$	—

3. We simply move right and insert a 1 on the tape in place of the first blank.

	1	Δ
0	$(0, 1, R)$	$(1, 1, R)$
1	—	—

10 Probability

Probability is a branch of mathematics that is often associated with games of chance. Laws of probability also affect weather prediction, insurance rates, and public opinion polls. It is curious that society regards gambling as a vice but insurance as a virtue since both are based on some expectation or likelihood of an event as measured by the event's frequency of occurrence.

As a formal discipline, probability has its origins in the early analyses of games of chance by mathematicians Pascal and Fermat. Probability has its "laws," which give it a certain rigor, but it is applied to areas about which there is uncertainty. In contrast with logic, which says that a proposition is *true* or *false*, probability yields only a "maybe."

The abuses of probability in applications to statistical phenomena are legion. One often hears that you can prove anything with statistics. Mark Twain observed, "There are lies, damn lies, and statistics." The primary reason for the frequent misapplication of the laws of probability is that a legitimate experiment requires very careful observance that all conditions of a hypothesis be met before valid conclusions can be drawn.

Elementary Properties of Probability

The notion of assigning a probability to an event is generally tied to the intuitive idea of **frequency of occurrence** of the event. Let us consider an example.

EXAMPLE 10.1

What is the probability that when three coins are tossed there will be exactly two heads and one tail?

Solution

We must assume that the tosses of the coins are independent; that is, the result of the first toss has no effect on the result of the second, and so on. We also want to look at the outcomes of tossing the coins—whether the first coin is heads or tails, whether the second coin is heads or tails, and whether the third coin is heads or tails. Hence there are eight possible results from tossing three coins. We denote heads by H and tails by T:

H, H, H   H, H, T   H, T, H   H, T, T

T, H, H   T, H, T   T, T, H   T, T, T

Of the eight possibilities, three have two heads and one tail. Thus we say that the probability of getting two heads and one tail is 3/8. What we imply is that if we conduct an experiment that consists of tossing the three coins eight times, then three times we "should" get two heads and one tail. It would be instructive for you to carry out such an experiment and compare your results with the predicted results.

It is not too difficult to grasp the interpretation of probability as frequency of occurrence of an event. But what would it mean to say that the probability of your getting an A in a certain course is 3/8? Using the frequency of occurrence model of probability, we would say that, if you took the course eight times under exactly the same set of circumstances, then you would "probably" get an A three times. Undoubtedly, you are not willing or able to conduct this experiment. However, it is the theoretical repeatability that is the key to the interpretation here.

Probability, then, is concerned with the occurrence of **random**—that is, unpredictable or nondeterministic—events. The events are assumed to have a certain regularity, in that they occur with a certain relative frequency when an experiment producing the events is performed. The collection of all possible events connected with an experiment is called a **sample space**. Each event has associated with it a real number, the **probability** of the event, which is a measure of the expected relative frequency of the event from repetitions of the experiment. Thus we say that the probability of getting heads as a result of a coin toss is 1/2 because, if we toss a coin a large number of times, we expect that approximately half the tosses will result in heads. In reality, tossing a coin 1000 times can also have the much less likely result of 999 heads and 1 tail.

A **finite probability space** consists of a finite set, S, called a **sample space**, and a **probability function** p from S to the real numbers, such that for all $s \in S$, $0 \leq p(s) \leq 1$, and

$$\sum_{s \in S} p(s) = 1$$

We refer to the elements of the sample space as **elementary events** and to subsets of the sample space as **compound events**; that is, compound events are sets of elementary events. We extend the definition of probability to compound events $A \in$

$\mathscr{P}(S)$ by defining

$$p(A) = \sum \{p(s): s \in A\}$$

It is easy to check that the probabilities of compound events satisfy the following three conditions:

1. For every event $A \in \mathscr{P}(S)$, $0 \le p(A) \le 1$.
2. $p(\varnothing) = 0$ and $p(S) = 1$.
3. For all $A, B \in \mathscr{P}(S)$ such that $A \cap B = \varnothing$,

$$p(A \cup B) = p(A) + p(B)$$

We will refer to condition (3) as the **sum rule** for a probability space.

Uniform Probability Space

The simplest example we can give of a probability space is known as the **uniform probability space**. Let S be a finite, nonempty set, with $|S| = n$. The uniform probability function on S is defined by

$$p(s) = 1/n, \text{ for all } s \in S$$

Extending this to compound events, we see that

$$p(A) = |A|/n, \text{ for all } A \le S$$

If we interpret the uniform probability in terms of selecting elements from S at random, each element is equally likely to be chosen. This is the case, for instance, in tossing a six-sided die, where $S = \{1, 2, 3, 4, 5, 6\}$ and the probability of getting any one of the values is uniformly $1/6$.

EXAMPLE 10.2

Let S be a finite, nonempty set. Show that (S, p), where p is the uniform probability function on S, is a probability space, and verify conditions (1) through (3) directly.

Solution

Since $0 \le 1/n \le 1$ for all $A \subseteq S$, and

$$\sum_{s \in S} 1/n = n \cdot (1/n) = 1$$

the conditions for a probability space are verified easily. Moreover, $p(A) = |A|/n$, so $0 \leq p(A) \leq 1$, and $p(\emptyset) = 0$ and $p(S) = 1$, which establishes conditions (1) and (2). To establish condition (3), the sum rule, let A and B be subsets of S. By the Principle of Inclusion and Exclusion,

$$|A \cup B| = |A| + |B| - |A \cap B|$$

When $A \cap B = \emptyset$, we obtain condition (3) exactly.

Theorem 10.1 contains some properties of finite probability spaces that are derived from the sum rule. Part (3) extends the sum rule to an arbitrary finite collection of mutually disjoint sets, and part (4) extends the sum rule to the union of two nondisjoint sets.

Theorem 10.1

Let (S, p) be a finite probability space. Let $A, B \subseteq S$.

1. If $A \subseteq B$, then $p(A) \leq p(B)$.
2. $p(A') = 1 - p(A)$
3. Let $A_1, \ldots, A_k$ be a collection of pairwise disjoint subsets of S. Then

$$p(A_1 \cup \ldots \cup (A_k) = p(A_1) + \cdots + p(A_k)$$

4. $p(A \cup B) = p(A) + p(B) - p(A \cap B)$

Proof

1. If $A \subseteq B$, then $B = A \cup (B - A)$. Thus we have

$$p(B) = p(A) + p(B - A)$$

  Since $p(B - A) \geq 0$, it follows that $p(B) \geq p(A)$.
2. The proof of (2) is left as an exercise.
3. Not surprisingly, the proof will use induction on the number of sets, k. When $k = 2$, (3) reduces to

$$p(A_1 \cup A_2) = p(A_1) + p(A_2)$$

which is just a restatement of the sum rule for a probability space. Next assume that (3) holds for some $k \geq 2$. Let $A_1, \ldots, A_{k+1}$ be a collection of pairwise disjoint subsets of S. Applying the sum rule, we get

$$p((A_1 \cup \ldots \cup A_k) \cup A_{k+1}) = p(A_1 \cup \ldots \cup A_k) + p(A_{k+1})$$
$$= p(A_1) + \cdots + p(A_k) + p(A_{k+1})$$

which is what we're trying to show.

4. The proof of (4) is also left as an exercise.

EXAMPLE 10.3

A card is drawn from a standard deck of 52 cards. What is the probability that it is either a face card or a heart?

Solution

We'll work this two ways. The first is an example of the use of the formal results of the theorem. Let A be the event that the card is a face card. Let B be the event that the card is a heart. We seek $p(A \cup B)$, which is given by the formula

$$p(A \cup B) = p(A) + p(B) - p(A \cap B)$$

Now $p(A) = 12/52$, $p(B) = 13/52$, and $p(A \cap B) = 3/52$. Hence,

$$p(A \cup B) = 12/52 + 13/52 - 3/52$$
$$= (25 - 3)/52 = 22/52$$

The second solution uses a more direct analysis. There are 13 hearts and 9 face cards that are not hearts. Using the uniform probability, the probability of obtaining a heart or a face card is just $(13 + 9)/52$, or $22/52$.

EXAMPLE 10.4

Let A be an event from a probability space (S, p). The **odds** that A occurs is defined as the ratio $p(A)/p(A')$. Find the odds of getting a 7 as the sum when two six-sided dice are tossed.

Solution

The sample space is the set of 36 pairs (i, j), where i and j take on the values 1 through 6. Since all pairs are equally likely, p is just the uniform probability function on S, so each pair has probability $1/36$. A 7 can be obtained as the sum of the pairs $(1, 6)$, $(2, 5)$, $(3, 4)$, $(4, 3)$, $(5, 2)$, and $(6, 1)$. Hence, the probability of getting a 7 is $6/36$, or $1/6$, and the probability of getting something besides a 7 is $5/6$. Thus the odds of getting a 7 are

$$\frac{1/6}{5/6} = 1/5$$

which we read as "1 to 5."

EXAMPLE 10.5

A wheel is divided into five wedges as shown in Figure 10.1. When the wheel is spun,

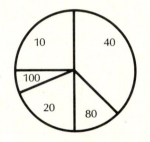

Figure 10.1. A wheel of fortune

the score is determined by the position obtained. Because of the relative widths of the segments, the probabilities are not uniform but are given by the following table:

value	probability
10	1/4
20	3/16
40	3/8
80	1/8
100	1/16

What is the probability of getting less than 80 points?

Solution

By part (3) of Theorem 10.1, the probability is the sum of the probabilities of getting 10, 20, or 40 points. Thus the result is

$$\frac{1}{4} + \frac{3}{16} + \frac{3}{8} = \frac{4 + 3 + 6}{16} = \frac{13}{16}$$

SELF TEST 1

1. A coin is to be tossed three times. Find each of the following probabilities.
   a. The probability that there is exactly one head in the three tosses
   b. The probability that the result of the first toss is a head
   c. The probability that there are two heads in a row
2. Verify part (2) of Theorem 10.1.
3. Using the uniform probability, find the probability of obtaining a sum of 8 when two six-sided dice are tossed.

Exercises 10.1

1. What is the probability that when three coins are tossed there will be exactly three heads? One head and two tails?
2. Prove part (4) of Theorem 10.1.
3. Let (S, p) be a probability space. Let A, B, C be subsets of S. Show that

$$p(A \cup B \cup C) = p(A) + p(B) + p(C) - p(A \cap B)$$
$$- p(B \cap C) - p(A \cap C) + p(A \cap B \cap C)$$

4. Let S be the set $\{a, b, c, d\}$. It is known that $p(a) = 1/3$, $p(b) = 1/4$, and $p(c) = 1/5$. What value must be assigned to $p(d)$ to make p a valid probability function on S?
5. Find the odds of drawing a Jack from each of the following.
   a. a standard 52-card deck
   b. a deck consisting of only face cards
   c. a deck consisting of only hearts
6. a. The odds that an event A occurs are known to be 5 to 2. What is the probability that A occurs?
   b. What is the probability that A occurs if the odds that A occurs are 1 to 1?
7. In Dungeons and Dragons™, tosses of multisided dice are used to determine results of actions in a random way. In some situations players roll a ten-sided die twice to obtain a value between 1 and 100, which is then used to determine the outcome. The first toss determines the first digit (0–9), and the second toss determines the second digit. The result 00 is considered to be 100. Verify that this method of selecting a number from 1 to 100 is just the uniform probability function; that is, all numbers between 1 and 100 are equally likely results.

10.2   Conditional Probability

So far, the probabilities we have computed have been straightforward. In some circumstances we are not interested in looking at the entire sample space because we have some prior knowledge that the event of interest is conditional on some other event. For instance, in a card game, you tend to reduce your estimate of the probability that your opponent has an ace if you already have one ace in your hand. We begin our discussion of conditional probability with an example.

EXAMPLE 10.6

A coin is to be tossed three times. Find the probability that the result includes exactly two heads under each of the following conditions.

1. The first toss is a head.
2. The first toss is a tail.

Solution

As we discovered in Example 10.1, there are only eight possible results from tossing a coin three times. We repeat the list of results:

$$H, H, H \quad H, H, T \quad H, T, H \quad H, T, T$$
$$T, H, H \quad T, H, T \quad T, T, H \quad T, T, T$$

Of these eight, three have two heads, so without taking other conditions into account, the probability of getting two heads is 3/8. But we get different answers if we assume we know the results of the first toss.

1. In the context of our problem here, the sample space is reduced to the results (H, H, H), (H, H, T), (H, T, H), (H, T, T). Of these four possibilities, two have exactly two heads. Hence, given that the result of the first toss is H, the probability of getting exactly two heads in three tosses is 2/4, or 1/2.
2. If the first toss yields tails, the sample space is reduced to (T, H, H), (T, H, T), (T, T, H), (T, T, T). Of these four possibilities, only one has exactly two heads, so the probability we seek is 1/4.

The cases considered in Example 10.6 are typical of conditional probabilities. For events A and B, we want to know the probability that A has occurred under the assumption that we know B has occurred. The Venn diagram in Figure 10.2 illustrates the situation. The outer rectangle represents the sample space S, and the two circles represent A and B. If B has occurred, then we are interested only in the relative proportion of A that is a subset of B, namely $A \cap B$. For this reason, we define the **conditional probability of A given B** by

$$p(A \mid B) = \frac{p(A \cap B)}{p(B)}$$

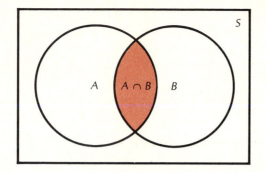

Figure 10.2. Venn diagram of conditional probability

In the first part of Example 10.6, we wanted the conditional probability of A given B, where A is the event of getting two heads, and B is the event that the first toss is heads. Using the formula in the definition, we get

$$p(A|B) = \frac{p(A \cap B)}{p(B)} = \frac{2/8}{4/8} = \frac{2}{4} = \frac{1}{2}$$

which agrees with our analysis in the example.

From the definition of conditional probability, we can see that

$$p(A \cap B) = p(A|B)p(B) = p(B|A)p(A)$$

This can be generalized, by induction, to the following result.

Theorem 10.2

Let (S, p) be a probability space. Let $A_1, \ldots, A_k$ be a collection of compound events from S. Then

$$p(A_1 \cap \ldots \cap A_k) = p(A_1)p(A_2|A_1)p(A_3|A_1 \cap A_2) \ldots p(A_k|A_1 \cap \ldots \cap A_{k-1})$$

EXAMPLE 10.7

Three cards are to be drawn from a deck. What is the probability that not one of them is a club?

Solution

Let A_1 be the event that the first card is not a club, A_2 the event that the second card is not a club, and A_3 the event that the third card is not a club. We seek $p(A_1 \cap A_2 \cap A_3)$. We can apply the equation of Theorem 10.2.

Since there are 13 clubs in the original deck, $p(A_1) = 39/52$. If one card is drawn and is not a club, then there are 51 cards left, of which 13 are still clubs. Thus $p(A_2 | A_1) = 38/51$. Similarly, if two cards have been drawn and neither is a club, there are 13 clubs left among the 50 remaining cards. Hence $p(A_3 | A_1 \cap A_2) = 37/50$. The final result is

$$p(A_1 \cap A_2 \cap A_3) = \frac{39 \cdot 38 \cdot 37}{52 \cdot 51 \cdot 50}$$

$$= \frac{19 \cdot 37}{4 \cdot 17 \cdot 25} = \frac{703}{1700}$$

EXAMPLE 10.8

It is known that 50 percent of the computer science majors at VLSI Tech take at least two math courses, but only 25 percent of the rest of the students take at least two math courses. Ten percent of the students are computer science majors. A student is picked at random and is found to have two math courses. What is the probability that the student is a computer science major?

Solution

Let S be the set of students, A the set of students who have taken at least two math courses, and B the set of computer science majors. We know that $p(A | B) = 0.5$, $p(A | B') = 0.25$, and $p(B) = 0.1$. What we want is $p(B | A)$. By definition,

$$p(B | A) = \frac{p(A \cap B)}{p(A)} = \frac{p(A | B)p(B)}{p(A)}$$

We do not know $p(A)$ directly. However, $A = (A \cap B) \cup (A \cap B')$, so

$$p(A) = p(A \cap B) + p(A \cap B')$$
$$= p(A | B)p(B) + p(A | B')p(B')$$

Hence,

$$p(B | A) = \frac{(0.5)(0.1)}{(0.5)(0.1) + (0.25)(0.9)}$$

$$= \frac{0.05}{0.05 + 0.225}$$

$$= \frac{0.05}{0.275}$$

$$= \frac{2}{11}$$

Bayes' Theorem

The last example illustrates the principle known as **Bayes' Theorem**, which is useful for determining certain kinds of conditional probabilities.

Theorem 10.3

Let (S, p) be a probability space. Let $B_1, \ldots, B_n$ be a partition of S. Let A be an event in S. Then

$$p(B_j|A) = \frac{p(A|B_j)p(B_j)}{p(A|B_1)p(B_1) + p(A|B_2)p(B_2) + \cdots + p(A|B_n)p(B_n)}$$

for all $j = 1, 2, \ldots, n$.

Proof

By definition of conditional probability, we see that

$$p(B_j|A) = \frac{p(A|B_j)p(B_j)}{p(A)}$$

By part (3) of Theorem 10.1,

$$p(A) = p(A \cap B_1) + p(A \cap B_2) + \cdots + p(A \cap B_n)$$
$$= p(A|B_1)p(B_1) + p(A|B_2)p(B_2) + \cdots + p(A|B_n)$$

so the result follows.

EXAMPLE 10.9

Medical researchers have been working on a test for detecting a disease known to afflict 5 percent of the population. Applying the test to people known to have the disease (by other diagnoses) yields positive results 95 percent of the time. Applying the test to people known not to have the disease yields positive results 5 percent of the time. A person chosen at random is given the test. The results are positive. What is the probability that this person has the disease?

   Solution

Let A stand for the event that the patient has the disease, B the event that the test is positive. We know the following probabilities:

$$p(A) = 0.05 \qquad p(B|A) = 0.95 \qquad p(B|A') = 0.05$$

We are trying to determine $p(A|B)$.

By Bayes' theorem,

$$p(A\,|\,B) = \frac{p(B\,|\,A)p(A)}{p(B\,|\,A)p(A) + p(B\,|\,A')p(A')}$$

$$= \frac{(0.95)(0.05)}{(0.95)(0.05) + (0.05)(0.95)}$$

$$= \frac{1}{2}$$

Independent Events

An idea closely related to the idea of conditional probability is the idea of independence of events. We have already used an intuitive idea of independence when we considered multiple coin tosses, and observed that the first toss had no effect on the outcome of the second, and so on. We compute, for example, the probability of two heads in a row by multiplying the probability of getting heads on each toss. We generalize this by saying that the events A and B are **statistically independent** if

$$p(A \cap B) = p(A)p(B)$$

A trivial consequence is that if A and B are independent events, then

$$p(A\,|\,B) = p(A)$$

The last equality can be interpreted as saying that prior knowledge about the occurrence of B has no impact on the computation of the probability of A. This is quite different from saying that A and B are mutually exclusive, since in that case $A \cap B = \varnothing$, and $p(A \cap B) = 0$.

EXAMPLE 10.10

Let S be the sample space consisting of the results of tossing a coin three times. Let

$$B_1 = \{(H, H, H), (H, H, T), (H, T, H), (H, T, T)\}$$

and

$$B_2 = \{(H, H, H), (H, H, T), (T, T, H), (T, H, H)\}.$$

Verify that B_1 and B_2 are independent.

Solution

$B_1 \cap B_2 = \{(H, H, H), (H, H, T)\}$, so $p(B_1 \cap B_2) = 1/4$. Moreover, $p(B_1) = 1/2$ and $p(B_2) = 1/2$. Hence,

$$p(B_1 \cap B_2) = p(B_1)p(B_2)$$

so the two events are independent.

The next example uses probabilistic independence to analyze the likelihood of errors in transmitting a message over a communication channel.

EXAMPLE 10.11

When bits of information (strings of 0's and 1's) are transmitted over a communications channel, there is always a small, but nonzero, probability that a bit will be in error when it arrives; that is, a 0 will be converted to a 1, or vice versa, perhaps because of noise or interference on the channel. Characters are usually encoded as 7-bit strings. If any bit in the representation of the character is corrupted during transmission, the character will not be interpreted properly when it arrives at its destination. Let p be the probability that a single bit will be corrupted. Assuming that the errors are introduced randomly with probability p and that the corruption of one bit has no impact on the corruption of any other bits, show that the probability that a character will be transmitted incorrectly is approximately $7p$ when p is very small.

Solution

The probability that a bit will arrive without error is $(1 - p)$. Assuming independence, the probability that the seven consecutive bits in a character are received correctly is then $(1 - p)^7$. Thus the probability that the character is transmitted incorrectly is $1 - (1 - p)^7$. Expanding the seventh power of $(1 - p)$ we get

$$\begin{aligned}
1 - (1 - p)^7 &= 1 - (1 - 7p + 21p^2 - 35p^3 + 35p^4 - 21p^5 + 7p^6 - p^7) \\
&= 7p - 21p^2 + 35p^3 - 35p^4 + 21p^5 - 7p^6 + p^7
\end{aligned}$$

Since p is very small, we can regard the higher powers of p as negligible and ignore them. This reduces the expression on the right to $7p$, as desired. Observe that $7p$ is only an approximation of the correct solution, not the exact answer.

EXAMPLE 10.12

Three distinct values are chosen from a set of numbers so that each is equally likely. Find the probability that the third is larger than the first two, and then find the probability that the three elements are chosen in increasing order.

Solution

In general, if we have three numbers, there are 3!, or 6, ways in which the three numbers can be chosen. Of these, 2! have the largest number last. Thus the probability of having the third number larger than the other two is 2!/3!, or 1/3. By similar reasoning, only one of the permutations has the numbers arranged in increasing order. Hence the probability that the three elements are chosen in increasing order is 1/6.

We could find the solution in another manner. Let the numbers be denoted by a, b, and c, where a is the first, b the second, and c the third. Then the first part of the problem requires $p(c > a, b)$ and the second part requires $p(c > b > a)$.

Now

$$p(c > b > a) = p((c > a, b) \text{ and } (a > b))$$
$$= p(c > a, b) \cdot p(a > b) \quad \text{[by independence of the choices]}$$
$$= (1/3) \cdot (1/2)$$
$$= 1/6$$

SELF TEST 2

1. Two dice are tossed. Find the probability that the result is a double (both dice have the same value) given that the sum of the face values is 8.

2. A test consists of four multiple choice questions, each having three choices. A student who has not studied decides to guess randomly. What is the probability that the student answers all the questions correctly?

3. Two dice are tossed. Let A be the event that the first die has a face value of 3, and let B be the event that the sum of the two face values is 8. Are A and B independent events?

Exercises 10.2

1. An urn contains three red balls, two white balls, and four blue balls. A ball is drawn at random from the urn. It is not white. What is the probability that it is red? What is the probability that it is blue?

2. Three people are in a room. What is the probability that at least two of them have the same birthday? (*Hint*: Use Theorem 10.2 to find the probability that they have distinct birthdays.)

3. There are three fraternities at VLSI Tech: fraternity X, which has 40 members, fraternity Y, which has 30 members, and fraternity Z, which also has 30 members. Five percent of the members of fraternity X are on probation, while 10 percent of fraternity Y and 20 percent of fraternity Z are on probation. A student picked at random from the three fraternities is on probation.

   a. What is the probability that he is from fraternity X?
   b. What is the probability that he is from fraternity Y?
   c. What is the probability that he is from fraternity Z?
   d. What is the sum of the three results? Why?

4. Two dice are tossed. The sum of the face values is 7. What is the probability that one of the dice has face value of 2?

5. A software house is so flooded with applications for programming positions that only 60 percent of the applicants are qualified. A programming aptitude test is given to help screen applicants. Of the qualified applicants, 80 percent pass the test, but 20 percent of unqualified applicants also pass the test. If an applicant passes (or fails) the exam, what is the probability that he or she is qualified?

6. Six members of a special military team are to draw straws to determine who will go on a sure suicide mission. The person drawing the shortest of six straws must go. Where should a member stand in order to maximize his chances of avoiding the short straw? ("By the door" is not an adequate answer.)

7. Let S be the sample space consisting of the results of tossing a coin four times. Let A be the event that the first two tosses yield heads. Find an event B that is independent of A.

8. Generalize the result of Example 10.12 to show that if k distinct values $x_1, \ldots, x_k$ are chosen at random from a uniform distribution,

$$p(x_k > x_{k-1}, \ldots, x_1) = 1/k$$

and

$$p(x_k > x_{k-1} > \cdots > x_1) = 1/k!$$

10.3   Repeated Trials and Expected Values

Often we have a situation where an experiment or event is to take place many times under the same circumstances, and each repetition is independent of all the others. We can regard tossing a given number of coins as providing the same results as an equal number of repetitions of tossing a single coin. Each occurrence of the experiment or event is called a **trial**. Each trial is assumed to have only two outcomes; one is called a *success*, the other a *failure*. We are interested in the number of successes in, say, n trials and in the probability that the number of successes takes on a specific value. We begin with an example.

EXAMPLE 10.12

The dial on the weighted wheel shown in Figure 10.3 is to be spun four times. Let
S be the sample space consisting of all 16 possible results (4-tuples), considering only
whether or not the dial lands in the shaded area. Let $X: S \to \mathbf{R}$ be the number of
times the dial ends in the shaded area, for each element of S. For $k = 0, 1, 2, 3, 4$,
find the probability that $X = k$.

Figure 10.3. Weighted wheel

TABLE 10.1

Outcomes of Spinning
the Wheel Four Times

Outcome	Probability	X
(n, n, n, n)	81/256	0
(n, n, n, s)	27/256	1
(n, n, s, n)	27/256	1
(n, n, s, s)	9/256	2
(n, s, n, n)	27/256	1
(n, s, n, s)	9/256	2
(n, s, s, n)	9/256	2
(n, s, s, s)	3/256	3
(s, n, n, n)	27/256	1
(s, n, n, s)	9/256	2
(s, n, s, n)	9/256	2
(s, n, s, s)	3/256	3
(s, s, n, n)	9/256	2
(s, s, n, s)	3/256	3
(s, s, s, n)	3/256	3
(s, s, s, s)	1/256	4

Solution

Because of the relative areas in the wheel, the probability of the dial ending in the shaded area is $1/4$ for each spin. We can find the probabilities for the multiple spins in two ways. In either way, we must realize that the outcome of each spin of the dial is independent of all the others. The first way to make the computation is just to list all 16 elements of S, and then to count the number of times $X = 0$, $X = 1$, and so on. The results are listed in Table 10.1. Each outcome is shown as a 4-tuple whose elements are either s, to indicate landing in the shaded area, or n, to indicate landing in the nonshaded area. The other two columns contain the probability of obtaining the outcome and the corresponding number of s's in the outcome.

Now that we know all outcomes and their probabilities explicitly, it is a simple matter to add up each of the probabilities associated with the various values of X to obtain the desired results. The final answers are shown in the following table:

k	$p(X = k)$
0	81/256
1	108/256
2	54/256
3	12/256
4	1/256

The second way to find the probabilities is to make use of our knowledge about probabilities of independent events. We represent the results of the four spins as entries in the four slots shown.

—— —— —— ——

Each slot can hold one of the results n or s. For $k = 0$ to 4, then, it is just a question of how many ways k of the slots can be filled with s. From our work with combinatorics, we know that k of the four slots can be chosen in $C(4, k)$ ways. The probability of each of the choices is

$$(1/4)^k \cdot (3/4)^{4-k}$$

Hence the probability of having k spins in the shaded area is

$$C(n, k) \cdot (1/4)^k \cdot (3/4)^{4-k}$$

The values obtained from this formula are given in Table 10.2. These results agree with the probabilities found by the first method.

TABLE 10.2

k	$C(4, k) \cdot (1/4)^k \cdot (3/4)^{4-k}$
0	$\dfrac{4!}{0!4!} \cdot \dfrac{1^0 \cdot 3^4}{4^4} = \dfrac{81}{256}$
1	$\dfrac{4!}{1!3!} \cdot \dfrac{1^1 \cdot 3^3}{4^4} = \dfrac{108}{256}$
2	$\dfrac{4!}{2!2!} \cdot \dfrac{1^2 \cdot 3^2}{4^4} = \dfrac{54}{256}$
3	$\dfrac{4!}{3!1!} \cdot \dfrac{1^3 \cdot 3^1}{4^4} = \dfrac{12}{256}$
4	$\dfrac{4!}{4!0!} \cdot \dfrac{1^4 \cdot 3^0}{4^4} = \dfrac{1}{256}$

The Binomial Distribution

In Example 10.12, we computed the probabilities of landing in the shaded area k times in four spins, for $k = 0, 1, 2, 3, 4$, where the probability of landing in the shaded area is $1/4$ every time we spin. Moreover, we assumed that each spin is independent of the others. We can generalize the result as follows: Assume that an experiment is *binomial*, in that there are only two complementary outcomes to be considered. One outcome is called a **success** and the other is called a **failure**. Denote the probability of success by p and the probability of failure by q (so $q = 1 - p$). Suppose the experiment is repeated n times. We want to compute the probability of obtaining k successes in the n trials, for $k = 0, 1, 2, \ldots, n$. We represent the outcome of the n trials as n slots that are to be filled with either p or q, depending on whether the corresponding trial produced a success or a failure.

$$\underline{\quad}\quad\underline{\quad}\quad\underline{\quad}\quad\underline{\quad}\quad \cdots \quad\underline{\quad}$$

For a fixed value of k, we can choose the k positions for successes in $C(n, k)$ ways. What is the probability for an individual choice? Since there are k successes, each occurring with probability p, and there are $(n - k)$ failures, each occurring with probability q, the probability of a particular occurrence is

$$p^k \cdot q^{n-k}$$

Finally, the probability of k successes is given by the formula

$$B(n, k; p) = C(n, k) \cdot p^k \cdot q^{n-k}$$

for $k = 0, 1, \ldots, n$. $B(n, k; p)$ is called the **binomial distribution function**.

EXAMPLE 10.13

A coin is tossed ten times.

1. Find the probability that there are two heads.
2. Find the probability that there are two or more heads.

Solution

1. There are ten independent trials, where success on a given trial is obtaining heads, with uniform probability 1/2. Hence the probability of obtaining exactly two heads is $B(10, 2; 1/2)$, or

$$\frac{10!}{2!8!} \cdot (1/2)^2 \cdot (1/2)^8 = \frac{10 \cdot 9}{2} (1/2)^{10}$$

$$= \frac{45}{1024}$$

2. In order to calculate the probability of two or more heads, we could calculate the sum

$$B(10, 2; 1/2) + B(10, 3; 1/2) + \cdots + B(10, 10; 1/2)$$

For the sake of completeness, we show all the values of $B(10, 2; 1/2)$ in Table 10.3. However, it is much easier to find the probability of obtaining 0 heads or 1 head and subtract this number from 1. This approach yields

$$1 - \frac{1}{1024} - \frac{10}{1024} = \frac{1013}{1024}$$

TABLE 10.3
Values of Binomial Distribution

x	$B(10, x; 1/2)$	x	$B(10, x; 1/2)$	x	$B(10, x; 1/2)$
0	$\dfrac{1}{1024}$	4	$\dfrac{210}{1024}$	8	$\dfrac{45}{1024}$
1	$\dfrac{10}{1024}$	5	$\dfrac{252}{1024}$	9	$\dfrac{10}{1024}$
2	$\dfrac{45}{1024}$	6	$\dfrac{210}{1024}$	10	$\dfrac{1}{1024}$
3	$\dfrac{120}{1024}$	7	$\dfrac{120}{1024}$		

In Example 10.11, we considered the case of a communications channel in which the probability that a single bit arrives in error is p, and showed that the probability that a 7-bit character arrives with an error is $7p$. In the next example, we use the binomial distribution to show that the addition of a *parity* bit will decrease the error rate in transmitting characters.

EXAMPLE 10.14

For the same transmission problem as in Example 10.11, assume that every 7-bit character has an eighth bit added to it that gives the character *even parity*; that is, the number of 1-bits among the 8 bits is even. If the character arrives with an odd number of 1's, the receiver will be able to detect that an error has occurred. Find the probability that a character will be transmitted without detection of the error, assuming that the bit error rate p on the channel is 10^{-5}.

Solution

The only time an error can go undetected in an even parity character is if two, four, six, or eight of the bits have been altered, since in these cases the received character will have even parity. Again assuming independence of the bits, the probability of two bits being in error is just $B(8, 2; p)$, since we can regard the arrival of the bits in a character as eight independent trials of the arrival of a single bit. Similarly, the probability of four bits being changed is $B(8, 4; p)$. Continuing in this fashion, the probability we seek is given by the sum

$$B(8, 2; p) + B(8, 4; p) + B(8, 6; p) + B(8, 8; p)$$

which is equal to

$$28p^2(1 - p)^2 + 70p^4(1 - p)^4 + 28p^6(1 - p)^2 + p^8$$

Using the value $p = 10^{-5}$ and ignoring terms smaller than 10^{-16}, we see that the probability of an even number of bits in error is approximately

$$28(10^{-10})$$

This value is considerably smaller than $7 \cdot 10^{-5}$, which was our estimate of the probability that the character would be transmitted incorrectly without the parity bit.

Random Variables and Distributions

When we count the number of successes in n repeated trials of an experiment, we are looking at a compound event determined by a function from the sample space to the real numbers. In Example 10.12, we referred to the function $X: S \rightarrow \mathbf{R}$, where for each 4-tuple t, $X(t)$ is the number of successes in t. For each value k, there is a set of events $t \in S$ such that $X(t) = k$. The set can be empty. A similar situation arises when we find the sum of the face values of two six-sided dice. Here the sample space $\mathscr{U}$ consists of sets of ordered pairs (i, j), where each of i and j range from 1 to 6. We can define the function $Y: \mathscr{U} \rightarrow \mathbf{R}$ by $Y(i, j) = i + j$. The functions X and Y are examples of *random variables*. A **random variable** on a sample space S is a function from S to the real numbers.

As in the examples above, we are interested in the probability that a random variable takes on particular values. If $X: S \rightarrow \mathbf{R}$ is a random variable, then for each real number k, we denote the probability that X takes on the value k by $p(X = k)$. More precisely,

$$p(X = k) = p(\{s \in S : X(s) = k\})$$

Observe that if $X(t) \neq k$ for all $t \in S$, then $p(X = k) = 0$. Using this notation, we can rephrase the key question in Exercise 10.12 as "Find the probabilities $p(X = k)$ for $k = 0, 1, 2, 3, 4$." In probability theory, the function that assigns to each k the probability that $X = k$ is called a *distribution* function. We give a precise definition here, and use distribution functions to define the expected value of a random variable. Let $[0, 1]$ denote the interval of real numbers between 0 and 1. Let X be a discrete random variable on a sample space S. The **probability distribution function** for X is the function $f_X : \mathbf{R} \rightarrow [0, 1]$ given by

$$f_X(x) = p(X = x) = p(\{s : X(s) = x\})$$

The distribution function tells us the relative frequency with which a random variable takes on a particular value. The **cumulative distribution function** for a random variable X is the function $F_X : \mathbf{R} \rightarrow [0, 1]$ given by

$$F_X(x) = p(X \leq x)$$

It is easy to see that

$$F_X(x) = \sum \{f_X(t) : f_X(t) \neq 0 \text{ and } t \leq x\}$$

Before going any further, we illustrate the computation of a distribution function.

EXAMPLE 10.15

Let (S, p) be the probability space determined by tossing a fair, six-sided die. Let X be the face value of the die. Find f_X and F_X.

Solution

X takes on the values $1, \ldots, 6$, with equal probability for each. Hence,

$$f_X(x) = \begin{cases} 1/6 & \text{if} \quad x \in \{1, \ldots, 6\} \\ 0, & \text{otherwise} \end{cases}$$

The cumulative distribution function F_X is the function given in the following table:

x	$F_X(x)$
$x < 1$	0
$1 \leq x < 2$	1/6
$2 \leq x < 3$	1/3
$3 \leq x < 4$	1/2
$4 \leq x < 5$	2/3
$5 \leq x < 6$	5/6
$6 \leq x$	1

The functions f_X and F_X are shown in Figure 10.4. Because of its shape, F_X is called a *step* function.

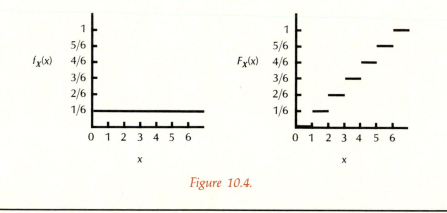

Figure 10.4.

Expected Values

Examine the values of X in Table 10.3 on page 425. These are just the values of another random variable; namely the random variable X whose value is the number of heads in ten tosses of a fair coin. The most likely event is the occurrence of five heads in the ten coin tosses, and we would be surprised if there were ten heads. In general, we often have preconceived ideas about what results to *expect* as outcomes of an experiment. Our objective in this section is to make the ideas of *expected* value of a random variable more precise. We begin with two examples.

EXAMPLE 10.16

A fair, six-sided die is tossed. What is the average face value expected?

Solution

The probability of getting any one of the six face values is $1/6$. In n tosses, we would expect to get $n/6$ 1's, $n/6$ 2's, and so on. Thus we expect the sum of all values for the n tosses to be

$$(n/6) \cdot 1 + (n6) \cdot 2 + (n/6) \cdot 3 + (n/6) \cdot 4 + (n/6) \cdot 5 + (n/6) \cdot 6$$
$$= (n/6) \cdot (1 + 2 + 3 + 4 + 5 + 6)$$
$$= (n/6) \cdot 21$$

Thus, we expect an average face value of $21/6$, or 3.5. This in spite of the fact that 3.5 does not appear on any face of the die!

In effect, the average of the face values is what we might expect to get if we roll the die a large number of times. Each of the values were weighted equally because each value is equally likely. In the next example, we look at the computation of an average where the individual values are not equally likely.

EXAMPLE 10.17

A study was taken of the undergraduate student body to determine the average number of courses taken per semester. The results, as obtained per class, are shown in the following table:

Class	Number of courses	Number of students
Freshmen	5	1000
Sophomores	5.25	800
Juniors	5.5	750
Seniors	5.5	650

Find the average number of courses taken by the entire student body.

Solution

Here the numbers of students per class are not distributed evenly, so that we would be in error if we simply added the numbers in the middle column and divided by 4. What we have to do is to multiply the average number of courses for each class by the fraction of that class in the student body. This yields the weighted average

$$5.0 \cdot \frac{1000}{3200} + 5.25 \cdot \frac{800}{3200} + 5.5 \cdot \frac{750}{3200} + 5.5 \cdot \frac{650}{3200}$$

Multiplying and collecting the terms, we get

$$\frac{16,900}{3200}$$

which is approximately 5.28. We would now expect that if we selected a large number of students at random, the average number of courses per semester would be close to 5.28.

We hope that the two examples above will help you grasp the idea of the **expected value** of a random variable, which we now define formally. Let S be a sample space. Let X be a random variable on S, with values $x_1, x_2, \ldots, x_k$, and probability distribution function f_X. Formally, we define the expected value, or **mean**, of X by

$$\mu = E[X] = \sum_{i=1}^{k} x_i \cdot f_X(x_i)$$

The mean is also called the **average** of the random variable X. Formally, the **variance** of the random variable X with mean μ is

$$\sigma^2 = \sum_{i=1}^{k} (x_i - \mu)^2 f_X(x_i)$$

The square root, σ, of the variance is called the **standard deviation** of X. The standard deviation measures the deviation of a random variable from its mean.

In Example 10.28, we compute the mean and variance of a single binomial experiment. This will be useful to us in later calculations.

EXAMPLE 10.18

A single trial of an experiment is to be carried out. The probability of a success is p, and the probability of a failure is $q = 1 - p$. Define the random variable Z to be 1 if the trial yields a success and define it to be 0 if the trial yields a failure. Find the mean and variance of Z.

Solution

Using the definitions given above, we see that

$$E[Z] = 1 \cdot p + 0 \cdot q = p$$

and

$$
\begin{aligned}
\sigma^2 &= (1 - p)^2 \cdot p + (0 - p)^2 \cdot q \\
&= q^2 p + p^2 q \\
&= pq(q + p) \\
&= pq
\end{aligned}
$$

Theorem 10.4 shows some simple properties of distribution functions and of the mean and variance of a random variable.

Theorem 10.4

Let X be a random variable on the sample space S, with values $x_1, x_2, \ldots, x_n$. Let r be a real number. Then

1. $\sum f_X(x_i) = 1$
2. $E[rX] = rE[X]$
3. $\sigma^2 = E[X^2] - (\mu)^2$

Proof

1. X is defined on S, and for each x_i there exists a set of $s \in S$ such that $X(s) = x_i$. Hence,

$$\sum f_X(x_i) = \sum \{p(s) : s \in S\} = 1$$

2. By the definition of expected value,

$$E[rX] = \sum rx_i f_X(x_i) = r \cdot \sum x_i f_X(x_i) = rE[x]$$

3. $\sigma_X{}^2 = \sum (x_i - \mu)^2 \cdot f_X(x_i)$
   $$= \sum x_i{}^2 \cdot f_X(x_i) - 2\mu \sum x_i \cdot f_X(x_i) + \mu^2 \sum f_X(x_i)$$

The first summand is, by definition of $E[\ ]$, $E[X^2]$. The second summand yields $-2\mu^2$, and the third summand yields μ^2. Hence $\sigma^2 = E[X^2] - \mu^2$, as desired.

EXAMPLE 10.19

Let X be the random variable whose value is the number of successes in n independent trials of an experiment with sample space S, where the probability of success on each experiment is p. Find the mean and variance of X.

Solution

X has probability distribution function $B(n, x; p)$. Thus the mean is given by

$$\mu = \sum_{x=0}^{n} x \cdot B(n, x; p)$$

and by Theorem 10.4 the variance is given by

$$\sigma^2 = \sum_{x=0}^{n} x^2 \cdot B(n, x; p) - \mu^2$$

Intuitively, one might say that if the probability of success on a single trial is p, then one expects np successes in n trials. We will prove this directly from the definition of the expected value. The case $n = 1$ was already handled in Example 10.17.

For $n + 1$ trials, the mean is

$$\sum_{x=0}^{n+1} xB(n + 1, x; p) = \sum_{x=1}^{n+1} x \frac{(n + 1)!}{x!(n + 1 - x)!} p^x(1 - p)^{(n + 1 - x)}$$

$$= \sum_{x=1}^{n+1} \frac{(n + 1)!}{(x - 1)!(n - (x - 1))!} p^x(1 - p)^{(n + 1 - x)}$$

Now we want to do a little sleight of hand. Factoring $(n + 1)$ out of the numerator and one p out of p^x reduces the last sum to

$$(n + 1)p \sum_{x=1}^{n+1} \frac{n!}{(x - 1)!(n - (x - 1))!} p^{x-1}(1 - p)^{(n - (x - 1))}$$

Replacing $x - 1$ by k yields

$$(n + 1)p \sum_{k=0}^{n} \frac{n!}{k!(n - k)!} p^k(1 - p)^{(n - k)} = (n + 1)p \sum_{k=0}^{n} B(n, k; p)$$

Now the last sum is exactly 1, so

$$\mu = (n + 1)p$$

We leave the computation of σ^2 as an exercise.

EXAMPLE 10.20

1. Let X be the value selected from the set S_n with uniform probability. Find the mean and variance of X.

2. Let Z be the sum of the face values obtained from k tosses of a fair, six-sided die. Use the result in (1) to compute the expected value and the variance of Z.

Solution

1. Each of the values $1, 2, \ldots, n$ can occur with probability $1/n$. Hence

$$E[X] = 1 \cdot \frac{1}{n} + 2 \cdot \frac{1}{n} + \cdots + n \cdot \frac{1}{n}$$

$$= \frac{1}{n} \cdot \sum_{i=1}^{n} i$$

$$= \frac{n(n+1)}{n \cdot 2}$$

$$= \frac{(n+1)}{2}$$

A similar argument shows that $E[X^2] = (n+1)(2n+1)/6$. Now we use part (3) of Theorem 10.4 to compute the variance.

$$\sigma^2 = E[X^2] - \mu^2$$

$$= \frac{(n+1)(2n+1)}{6} - \frac{(n+1)^2}{4}$$

$$= \frac{(n+1)(4n+2-3n-3)}{12}$$

$$= \frac{(n+1)(n-1)}{12} = \frac{n^2-1}{12}$$

2. For the case $k = 1$, we have a single die tossed, so the results of (1) apply with $n = 6$. Therefore, the mean is $7/2$, and the variance is $35/12$.

   For the general case, we use the fact that the tosses are mutually independent, so that we can simply multiply the results of (1) by k to get the mean and variance. That is, for k repetitions,

$$\mu = \frac{7k}{2}$$

and

$$\sigma^2 = \frac{35k}{12}$$

SELF TEST 3

1. Compute the values of $B(4, k; p)$ for $k = 0, 1, 2, 3, 4$, with $p = 0.1$. Use these values to compute the mean and variance from their definitions. Compare your results to those given by the formula in Example 10.19.

2. A fair coin is tossed ten times. Find the probability of each of the following.
   a. All the tosses result in heads.
   b. There is at least one head.

3. Two six-sided dice are tossed. Find the distribution of the sum of the two values.

4. A fair coin is to be tossed. When it shows heads, you will get $5.00, and when it shows tails, you will lose $4.00. What is your expected gain?

Exercises 10.3

1. Consider the following simple game. A fair, six-sided die is to be tossed. If the face value is odd, you win $1.00 for each dot showing. If the face value is even, you lose $1.00 for each dot showing. Find your expected winnings under each of the following conditions.
   a. The die is tossed only once.
   b. The die is tossed five times.

2. Let X and Y be random variables. Verify that

$$\sum_i p(X = x_i, Y = y_j) = p(Y = y_j), \text{ for all } j$$

3. Finish the solution of Example 10.19 by finding the variance of the number of successes in n trials, where on each trial, the probability of a success is p.

4. a. A coin is to be tossed an indefinite number of times. Let X be the number of tosses needed before a heads appears. Find the distribution function for X.

b. Generalize the result of (a) to any binomial experiment, where p represents the probability of a success and X is the number of times the experiment must be repeated before a success occurs. The distribution function of X is called the **geometric distribution**.

5. A random variable has probability distribution function given by the following table. Compute its mean and standard deviation.

x	$f_X(x)$
-5	0.3
-3	0.2
0	0.1
2	0.2
4	0.1
8	0.1

6. A friend offers to play either of the following games with you. In the first a fair coin will be tossed seven times. If the number of heads, X, is odd, you pay X dollars. If the value of X is even, you receive X dollars. In the second game, you are to spin the dial on the wheel shown in Figure 10.3 on page 422. If the dial ends in the shaded area, you pay $6.00. If the dial lands in the other area, you receive $2.00. Which game would be better for you to play?

An Average Case Analysis

In Chapter 4 on combinatorics, we discussed analysis of algorithms from a *worst case* perspective; that is, we counted the largest number of operations that would be necessary for a given problem size. In this application, we examine the *average*, or expected number, of operations that are required in sorting lists of items. We will examine two sorting methods: *selection sorting* and *insertion sorting*. For each method, we will find the average behavior of the algorithm under the assumption that the list to be sorted is randomly selected from the set of all possible permutations of the list. We assume that the items in the list contain a *key* value used for sorting (for instance, last names could be used in a list of employees). The list will be denoted by an array $L[1], \ldots, L[n]$, where n is the size of the list. The estimates of performance will then be functions of n. We will be interested in finding the average number of comparisons and the average number of data movements required by each algorithm.

Selection Sort

Our version of selection sorting works by selecting the smallest item in the list, moving it to the front of the list, and then finding the smallest of the remaining items and moving it to the second position in the list, and so on. When two items in the list, say $L[k]$ and $L[m]$, have to be interchanged, we write

$$switch(L[k], L[m])$$

Selection Sort

{The variables *i* and *j* are loop indices.   }
{The variable *min* is used to keep the position}
{of the smallest element between $L[i]$ and $L[n]$. }

```
begin
     for i = 1 to n − 1 do
        begin
           min ← i
           for j = i + 1 to n do
              if L[min] > L[j] then
                 min ← j
           switch(L[min], L[i])
        end
end.
```

The first observation we have is that there is no difference between the worst case and the average case, because the algorithm makes no use of the way the items in the list are distributed. The outer loop will always be executed $(n - 1)$ times and, for each $i = 1$ to n, the inner loop will always be executed $n - i$ times. Thus the number of data movements, as determined by the *switch* procedure, is $n - 1$. The number of comparisons occurring in the **if** statement within the inner loop is

$$(n - 1) + (n - 2) + (n - 3) + \cdots + 1 = \frac{1}{2} n(n - 1) \in O(n^2)$$

Insertion Sort

Sorting by insertion works by keeping the first part of the list in sorted order. If an element in the latter part of the list is encountered out of order, the algorithm searches backwards to find the proper place in which to insert the element. A space is opened for the element, and all the following items in the list are shuffled forward one position. Thus this method takes advantage of any ordering that may already be in the list. We will show that its average case behavior is better than its worst case behavior.

Insertion Sort

```
{The variables i and j are loop counters.      }
{temp is used to store the value of L[i] when   }
{L[i] is out of order and must be moved back.}
{The variable done is a logical variable that is}
{true when the position at which temp is to     }
{be inserted has been found.                    }

begin
      for i = 2 to n do
          if L[i] < L[i − 1] then
             begin
                 j ← i
                 temp ← L[i]
                 repeat
                     j = j − 1
                     L[j + 1] ← L[j]
                     if j = 1 then
                         done ← true
                     else
                         done ← (L[j − 1] ≤ temp)
                 until done
                 L[j] ← temp
             end
end.
```

We analyze three kinds of lists as input to the algorithm. First, assume that the list is already sorted. The expression

$$L[i] < L[i - 1]$$

is always *false*, and therefore no other statements are executed. This is the best possible case, requiring $n - 1$ comparisons to make sure the list is sorted.

Second, suppose the list is sorted in reverse order; that is, $L[n] < L[n - 1] < \ldots L[1]$. The expression

$$L[i] < L[i - 1]$$

in the **if** statement is always *true*. Hence the **repeat** loop, which finds the appropriate place to insert the out of place value, $L[i]$, will be executed $i - 1$ times. Thus the number of assignment statements required to place the ith item is $i + 1$, and the number of comparisons is $i - 1$. Hence, the total number of assignment statements is

$$3 + 4 + 5 + \cdots + (n + 1) = \frac{1}{2}(n + 1)(n + 2) - 3$$

and the total number of comparisons is

$$1 + 2 + 3 + \cdots + (n - 1) = \frac{1}{2}n(n - 1)$$

For the third case, we assume that the items are distinct and are randomly selected from a uniform distribution. We compute the expected number of operations performed. In order to calculate the expected number of comparisons, observe that we always need one comparison for each value of $i = 2, \ldots, n$. Other comparisons can be made only if the element is out of order, and then the number of comparisons is based on the number of times the **repeat** loop is executed. To be precise, we have

$$E[\text{comparisons}] = \sum_{i=2}^{n} (1 + p_i \cdot (n_i - 1))$$

where p_i is the probability that the ith element is out of order, and n_i is the average number of times the **repeat** loop is executed when $L(i)$ is out of order.

Similarly, observe that when the ith element is out of order, there are two assignment statements performed outside the **repeat** loop and one assignment statement performed inside the **repeat** loop. Hence, the expected number of assignment statements executed is

$$E[\text{assignments}] = \sum_{i=2}^{n} p_i \cdot (n_i + 2)$$

where p_i and n_i are as defined above.

Consider the case $i = 2$. The probability that the second element is out of order is $1/2$, since it is as likely that $L[1] > L[2]$ as $L[2] > L[1]$. When $i = 2$, the **repeat** loop is executed only once. Thus $n_2 = 1$. In general, the probability that the ith element is in order is the probability that $L[i]$ is greater than the first $i - 1$ elements, which is $1/i$. Hence $p_i = 1 - 1/i = (i - 1)/i$.

To compute n_i, we make use of the fact that if the ith element is out of order, then it is equally likely to belong before each of $L[1], \ldots, L[i - 1]$. Hence

$$n_i = \frac{1}{i - 1} + \frac{2}{i - 1} + \cdots + \frac{i - 1}{i - 1}$$

$$= \frac{1}{i - 1} \sum_{k=1}^{i} k$$

$$= \frac{1}{i - 1} \cdot \frac{i(i - 1)}{2} = \frac{i}{2}$$

Putting the values of n_i and p_i together in the formulas above, we get

$$E[\text{comparisons}] = \sum_{i=2}^{n} \left(1 + \frac{i - 1}{i} \cdot \frac{i}{2} \right)$$

$$= n - 1 + \sum_{i=2}^{n} \frac{i - 1}{2}$$

$$= n - 1 + \frac{1}{2} \sum_{i=2}^{n} \frac{i}{2} - \sum_{i=2}^{n} \frac{1}{2}$$

$$= \frac{n - 1}{2} + \frac{1}{2} \left(\frac{n(n + 1)}{2} - 1 \right)$$

$$= \frac{2n - 2 + n^2 + n - 2}{4}$$

$$= \frac{n^2 + 3n - 4}{4} = \frac{(n + 4)(n - 1)}{4}$$

Similarly,

$$E[\text{assignments}] = \sum_{i=2}^{n} \frac{i - 1}{i} \cdot \left(\frac{i}{2} + 2 \right)$$

$$= \sum_{i=2}^{n} \left(1 - \frac{1}{i} \right) \cdot \left(\frac{i}{2} + 2 \right)$$

$$= \sum_{i=2}^{n} \left(\frac{i}{2} + \frac{3}{2} - \frac{2}{i} \right)$$

$$= \frac{(n + 8)(n - 1)}{4} - 2 \left(1 + \frac{1}{2} + \frac{1}{3} + \cdots + \frac{1}{n} \right)$$

While the latter sum, which is called the *harmonic series*, can in theory be calculated directly for any finite value of n, it is known to be approximated by the logarithmic term

$$\ln(n + 1)$$

Hence the expected number of assignments is approximately

$$\frac{(n + 8)(n - 1)}{4} - 2 \cdot \ln(n + 1)$$

SUMMARY

1. A **finite probability space** consists of a finite set, S, called a **sample space**, and a **probability function** p from S to the real numbers such that for all $s \in S$, $0 \le p(s) \le 1$, and

$$\sum_{s \in S} p(s) = 1$$

2. For a compound event $A \in \mathscr{P}(S)$

$$p(A) = \sum \{p(s) : s \in A\}$$

3. The probabilities of compound events satisfy the following three conditions:
   1. For every event $A \in \mathscr{P}(S)$, $0 \le p(A) \le 1$.
   2. $p(\varnothing) = 0$ and $p(S) = 1$.
   3. For all $A, B \in \mathscr{P}(S)$ such that $A \cap B = \varnothing$,

   $$p(A \cup B) = p(A) + p(B)$$

4. Let S be a finite, nonempty set, with $|S| = n$. The **uniform probability function** on S is defined by

$$p(s) = 1/n \text{ for all } s \in S.$$

5. **Theorem 10.1.** Let (S, p) be a finite probability space. Let $A, B \subseteq S$.
   1. If $A \subseteq B$, then $p(A) \le p(B)$.
   2. $p(A') = 1 - p(A)$
   3. Let $A_1, \cdots, A_k$ be a collection of pairwise disjoint subsets of S. Then

   $$p(A_1 \cup \ldots \cup A_k) = p(A_1) + \cdots + p(A_k)$$

   4. $p(A \cup B) = p(A) + p(B) - p(A \cap B)$.
6. The **conditional probability** of A **given** B is

$$p(A|B) = \frac{p(A \cap B)}{p(B)}$$

7. **Theorem 10.2.** Let (S, p) be a probability space. Let $A_1, \ldots, A_k$ be a collection of compound events from S. Then

$$p(A_1 \cap \ldots \cap A_k) = p(A_1)p(A_2|A_1)p(A_3|A_1 \cap A_2) \ldots p(A_k|A_1 \cap \ldots \cap A_{k-1})$$

8. **Theorem 10.3. Bayes' Theorem** Let (S, p) be a probability space. Let $B_1, \ldots, B_n$ be a partition of S. Let A be an event in S. Then

$$p(B_j|A) = \frac{p(A|B_j)p(B_j)}{p(A|B_1)p(B_1) + p(A|B_2)p(B_2) + \cdots + p(A|B_n)p(B_n)}$$

for all $j = 1, 2, \ldots, n$.

9. The events A and B are **statistically independent** if

$$p(A \cap B) = p(A)p(B)$$

10. A **random variable** on a sample space S is a function from S to the real numbers.

11. Let $[0, 1]$ denote the interval of real numbers between 0 and 1. Let X be a discrete random variable on a sample space S. **The probability distribution function** for X is the function $f_X: \mathbf{R} \to [0, 1]$ given by

$$f_X(x) = p(X = x) = p(\{s: X(s) = x\})$$

12. The **cumulative distribution function** for a random variable X is the function $F_X: \mathbf{R} \to [0, 1]$ given by

$$F_X(x) = p(X \leq x)$$

13. Let S be a sample space. Let X be a random variable on S, with values x_1, $x_2, \ldots, x_k$, and probability distribution function f_X. The expected value, or **mean**, of X is

$$\mu = E[X] = \sum_{i=1}^{k} x_i \cdot f_X(x_i)$$

14. The **variance** of the random variable X with mean μ is

$$\sigma^2 = \sum_{i=1}^{k} (x_i - \mu)^2 f_X(x_i)$$

15. The square root, σ, of the variance is called the **standard deviation** of X. The standard deviation measures the deviation of a random variable from its mean.

16. **Theorem 10.4.** Let X be a random variable on the sample space S, with values $x_1, x_2, \ldots, x_n$. Let r be a real number. Then
   1. $\sum f_X(x_i) = 1$
   2. $E[rX] = rE[X]$
   3. $\sigma^2 = E[X^2] - (\mu)^2$

REVIEW PROBLEMS

1. Find the probability of drawing two jacks in five cards from a standard 52-card deck.

2. When a fair die is tossed once, the sample space is the set $S = \{1, 2, 3, 4, 5, 6\}$. Let $A = \{2, 3, 4\}$, $B = \{1, 2\}$, $C = \{5, 6\}$. Find each of the following probabilities.

   a. $p(A \cap B)$ c. $p(A \mid B)$
   b. $p(A \cap C)$ d. $p(B \mid A)$

3. A pair of fair dice is tossed. Find the probability for each of the following results.

   a. The sum is 4.
   b. The sum is greater than 5.

4. Find the probability that the sum of two dice is greater than 7 given that one of the dice shows 2.

5. Find the probability that one of two dice shows a 2 given that the sum of the two dice is greater than 7.

6. Samples of computer chips coming off the assembly line show that 10 percent are defective. In a lot of twenty, what is the probability that two are defective?

7. Four urns contain a mixture of red, green, and blue balls. The first contains three red, two green, and two blue balls. The second contains two red, four green, and no blue balls. The third contains four red, four green, and four blue balls. The fourth contains five red, one green, and two blue balls. An urn is picked at random, and a ball is chosen from the urn. Find the probability for each of the following results.

   a. The ball is red.
   b. The ball is not blue.
   c. The ball came from the fourth urn, given that it is blue.

8. A 5-card poker hand is dealt at random from a 52-card deck. Find the probability for each of the following results.

   a. The hand contains no hearts.
   b. All the cards in the hand come from the same suit.
   c. The hand contains no face cards.

9. Let $S = \{a, b, c, d, e, f\}$. Suppose that each of a, b, and c are assigned probability $1/4$, and that d and e are assigned probability $1/12$. What value must f be assigned so that a valid probability space results?

10. A coin is tested extensively and found to be unfair, in that 60 percent of the tosses yield heads. What is the probability of three heads in four tosses?

11. A class contains ten males and fifteen females. If three students are selected at random, what is the probability they all are males?

12. Show that if A and B are independent events, then their complements A' and B' are also independent events.

13. A and B are events with probability 1/2 and 1/4, respectively. If $p(A \cup B) = 5/16$, find $p(A \cap B)$.

14. Compute $B(10, 5; 0.3)$ and $B(6, 1; 0.2)$.

15. Let S be the sample space resulting from tossing four coins. Let A be the event "the second and third tosses are heads." Find an event independent of A.

16. A random variable takes on the values 1, 2, 3, 4, 5, 6, 7, 8, with probability distribution given by the following table. Compute the mean and the variance.

s	$p(s)$
1	0.15
2	0.10
3	0.05
4	0.25
5	0.10
6	0.10
7	0.15
8	0.10

17. Two dice are tossed. Let X be the larger of the two. Find the probability distribution of X and the mean and the variance of X.

18. A fair coin is tossed ten times, and six heads result. Find the probability that the first toss was a head.

19. A fair coin is tossed ten times. If the first toss is a head, what is the probability that six of the ten tosses will be heads? Tails?

20. Many lotteries require players to pick a set of numbers from a given set; for instance, in one state players pick six numbers in the range from 1 to 36. The order of selection of the numbers does not matter. What is the probability of picking the right set of numbers? What is the probability if the order in which the numbers are drawn does matter?

*21. A fair coin is tossed 100 times. Estimate the probability that there will be between 40 and 60 heads.

*22. When we determined the probability that k items could be drawn in sorted order from a uniformly distributed set of items, we assumed that the items were distinct. Compute the probability of drawing k items in sorted order if repetitions are allowed.

▓ Programming Problems

23. Implement the *Selection Sort* algorithm.

24. Implement the *Insertion Sort* algorithm.

SELF TEST ANSWERS

Self Test 1

1. We refer to the list of outcomes in Example 10.1.
   a. p(exactly one head) $= 3/8$
   b. p(head on first toss) $= 1/2$
   c. p(two heads in a row) $= 3/8$

   (*Note*: (H, H, H) is counted here since it includes two heads in a row, although not exactly two heads.)

2. Let (S, p) be a finite probability space.
   Since $1 = p(S) = \sum_{x \in A} p(x) + \sum_{x \in A'} p(x) = p(A) + p(A')$, we see that

   $$p(A') = 1 - p(A)$$

3. The sum of two six-sided dice is 8 when the following ordered pairs of values show on the dice:

   $$(2, 6), (6, 2), (3, 5), (5, 3), (4, 4)$$

   Hence, p(sum is 8) $= 5/36$.

Self Test 2

1. Of the five pairs whose sum is 8, only one is a double. Hence the conditional probability is $1/5$.

2. The probability of a correct guess on a given question is $1/3$. Assuming the guesses are independent, the probability of guessing all four correctly is

   $$\frac{1}{3} \cdot \frac{1}{3} \cdot \frac{1}{3} \cdot \frac{1}{3} = \frac{1}{81}$$

3. $p(A) = 1/6$ and $p(B) = 5/36$; but $p(A \cap B) = 1/5$, so the events are not independent.

Self Test 3

1. $B(4, k; 0.1) = \dfrac{4!(0.1)^k(0.9)^{4-k}}{k!(4-k)!}$, for $k = 0, 1, 2, 3, 4$

   Thus,

   $$B(4, 0; 0.1) = 0.6561$$
   $$B(4, 1; 0.1) = 0.2916$$
   $$B(4, 2; 0.1) = 0.0486$$
   $$B(4, 3; 0.1) = 0.0036$$
   $$B(4, 4; 0.1) = 0.0001$$

Observe that the sum of the values is 1.

$$\mu = (0)(0.6561) + (1)(0.2916) + (2)(0.0486) + (3)(0.0036) + (4)(0.0001)$$
$$= 0.4$$

$$\sigma^2 = (0 - 0.4)^2(0.6561) + (1 - 0.4)^2(0.2916) + (2 - 0.4)^2(0.0486)$$
$$+ (3 - 0.4)^2(0.0036) + (4 - 0.4)^2(0.0001)$$
$$= 0.36$$

2. a. p(all tosses yield heads) $= B(10, 10; 0.5) = 1/2^{10}$
   b. p(at least one head) $= 1 - p$(no toss yields heads) $= 1 - 1/2^{10}$

3. The sum of two six-sided dice can be any of the integers 2 through 12. The distribution is shown in the following table:

Sum	p
2	1/36
3	2/36
4	3/36
5	4/36
6	5/36
7	6/36
8	5/36
9	4/36
10	3/36
11	2/36
12	1/36

4. Expected gain $= (1/2)(\$5.00) + (1/2)(\$4.00)$
   $$= \$4.50$$

Glossary

Terms in italics are defined elsewhere in the Glossary.

abelian group   a *group* whose *binary operation* is commutative

acyclic digraph   a *directed graph* that contains no *cycles*

algorithm   a precise and unambiguous sequence of instructions that lead to a problem's solution in a finite amount of time

alphabet   a finite *set* whose elements are called symbols

average   the *expected value* of a *random variable*

bijection   a function that is both an *injection* and a *surjection*

binary operation   a *function* from $A \times A$ to A

binary relation   between sets A and B, a subset of $A \times B$

binary tree   a rooted *tree* in which each vertex is the root of at most two subtrees

Boolean expression   defined recursively as follows:

(1) 0, 1, and Boolean variables are Boolean expressions when f and g are Boolean expressions.

(2) $f \lor g$, $f \land g$, and (f) are Boolean expressions.

cardinality   the number of elements in a *set*

cartesian product   the *set* $A \times B$ of ordered pairs from A and B

combination   a list of elements from a *set* chosen without repetition, equating lists with the same elements, independently of the order in which the elements are selected

complete graph   a *graph* with an edge between every pair of vertices

complete set of operators   a *set* of Boolean operators sufficient for representing every *Boolean expression*

composition   Let R be a *relation* between A and B, S a *relation* between B and C. The composition of R and S is the relation:

$$R \circ S = \{(x, z): x \in A, z \in C \text{ and there exists } y \in B \text{ such that } xRy \text{ and } ySz\}$$

conditional probability   for events A and B, the conditional probability of A given B is $p(A \mid B) = p(A \cap B)/p(B)$

congruence relation   a *binary relation* R on a *monoid* $(M, *)$ such that for all x, y, z, w in M, xRy and $zRw \Rightarrow x * zRy * w$

connected graph   a *graph* such that every pair of vertices is connected by a *path*

connectivity number   the number of connected components in a *graph*

consistent labeling   a labeling $v_1, v_2, \ldots, v_n$ of the vertices in a *digraph* such that if (v_i, v_j) is an edge, then $i < j$

cycle   in a *graph*, a *path* with the same first and last vertices such that no edge in the path is repeated; in a *digraph*, a path with the same first and last vertex

cyclic group   a *group* G with an element x such that $G = \{e, x, x^2, \ldots, x^n\}$

DeMorgan's Laws

for *propositions*:	(P **and** Q)$'$ = P$'$ **or** Q$'$
	(P **or** Q)$'$ = P$'$ **and** Q$'$
for *sets*:	$\overline{(A \cap B)} = \tilde{A} \cup \tilde{B}$
	$\overline{(A \cup B)} = \tilde{A} \cap \tilde{B}$
for *Boolean expressions*:	$(p \wedge q)' = p' \vee q'$
	$(p \vee q)' = p' \wedge q'$

digraph   a pair (V, E), where V is a finite *set* whose elements are called vertices, and E is a *binary relation* on V whose elements are called edges

distribution function   for a *random variable* x with values $x_1, x_2, \ldots, x_n$,

$$f_X(x_i) = p(X = x_i)$$

duality principle   a rule for obtaining new identities about sets from existing ones by replacing each instance of the universal set with the empty set and vice versa and by replacing each union by intersection and vice versa

equivalence class   the set of elements related by an *equivalence relation*

equivalence relation   a *relation* that is reflexive, symmetric, and transitive

Eulerian path   a *path* in a *graph* that includes all the edges of the graph exactly once

expected value   Let $f(x)$ be the *distribution function* of a *random variable x*. The expected value of x is

$$\mu = \sum x_i f_X(x_i)$$

finite probability space   a pair (S, p), where S is a finite set and $p: S \to \mathbf{R}$ satisfies the conditions $0 \leq p(s) \leq 1$ for all $s \in S$

finite state automaton   a quintuple (S, A, T, s_0, F) where:

S is a finite *set* whose elements are called states.
A is a finite *set* called an input alphabet.
$T: S \times A \to S$ is the transition function.
s_0 is an element of S called the start state.
F is a subset of A, whose elements are called final states.

function   a *relation* f between *sets* A and B such that for all $x \in A$, there exists $y \in B$ such that $(x, y) \in f$ and such that for all $x \in A$ and $y, z \in B$, $(x, y) \in f$ and $(x, z) \in f$ implies $y = z$

graph   a pair (V, E), where V is a finite *set*, and E is a symmetric, irreflexive *relation* on V

group   a *monoid* M with identity e such that for all $x \in M$, there exists an element $y \in M$ such that $x \cdot y = y \cdot x = e$

Hamiltonian circuit   a *cycle* in a *graph* that uses every vertex, except the first and last ones, exactly once

homomorphism   a *function* $h: M \to M'$ from a *monoid* $(M, \cdot)$ to monoid $(M', \circ)$ such that $h(e) = e'$ and for all $x, y \in M$, $f(x \cdot y) = f(x) \circ f(y)$

indegree   the number of edges arriving at a vertex of a *digraph*

independent events   subsets A and B of a *probability space* (S, p) such that $p(A \cap B) = p(A)p(B)$

induction   see *principle of mathematical induction*.

injection   a *function* $f: A \to B$ such that if $f(x) = f(y)$, then $x = y$

inverse relation   for a *relation* R, the relation consisting of all pairs (x, y) such that yRx

isomorphism for monoids   a *homomorphism* that is a *bijection*

isomorphism for graphs (V, E) and (V', E'), a *bijection* $f: V \to V'$ such that for all $x, y \in V$, xEy if and only if $f(x)Ef(x)$

Karnaugh maps   rectangular arrays of boxes labeled with *minterms* that are used to help minimize *Boolean expressions*

language   a subset of A^* for an *alphabet* A

Mealy machine   a quintuple (S, A, T, O, f) where S, A, T are as in the definition of *finite state automaton*, O is a finite *set* called the output *alphabet*, and $f: S \times A \to O$ is the output *function*

mean   the *expected value* of a *random variable*

minterm   a *Boolean expression* that is 1 for exactly one of its inputs

monoid   a *semigroup* $(M, \circ)$ with an element e such that for all $x \in M$, $e \circ x = x \circ e = x$

Moore machine   a quintuple (S, A, T, O, f) where S, A, T are as in the definition of *finite state automaton*, O is the output *alphabet*, and $f: S \to O$ is the output *function*

order of a group   the number of elements in a *group*

outdegree   the number of edges leaving a vertex in a *digraph*

partition   for a *set* S, a collection of nonempty, mutually disjoint subsets of S whose union is S

path   a finite sequence of vertices $v_1, v_2, \ldots, v_k$ in a *graph* or a *digraph*

permutation   a bijection from a *set* to itself; a list of elements from a *set*, chosen without repetition, equating lists with the same elements chosen in the same order

poset   a *set* with a reflexive, antisymmetric, and transitive *relation*

principle of mathematical induction   Let $P(n)$ be a *proposition* that is valid for $n \geq k$, n, k integers. Assume that $P(k)$ is *true* and that $P(n) \Rightarrow P(n + 1)$ for all $n \geq k$. Then $P(n)$ is *true* for all $n \geq k$

probability space   see *finite probability space*

proposition   a statement that has truth value (*true* or *false*)

random variable   a function from a sample space to the real numbers

reachability matrix   the logical matrix whose ijth entry is T when there is a path from vertex v_i to vertex v_j

recurrence relation   an equation that defines the ith value of a sequence in terms of the preceding values in the sequence

regular language   a *language* that is accepted by a *finite state automaton*

relation   see *binary relation*

sample   a list of elements from a *set*, chosen with repetition allowed, equating lists with the same elements chosen in the same order

selection   a list of elements from a *set*, chosen with repetition allowed, equating lists with the same elements, independently of the order in which the elements are selected

semigroup   a pair $(M, \circ)$ where M is a *set* and $\circ$ is an associative *binary operation* on M

set   a collection of objects

shortest path   a *path* between two vertices in a *weighted digraph* with the least weight among all paths between the vertices

sink   a vertex in a *digraph* with *outdegree* equal to 0

source   a vertex in a *digraph* with *indegree* equal to 0

standard deviation   the square root of the *variance* of a *random variable*

statistical independence   Two events A and B are statistically independent if $p(A \cap B) = p(A)p(B)$

surjection   a *function* $f: A \rightarrow B$ such that for all $y \in B$, there exists $x \in A$ such that $f(x) = y$

topological sort   the process of finding a *consistent labeling* for a *digraph*

transitive closure   For a *relation* R on a *set* A, the transitive closure is the relation R^* on A given by xR^*y if and only if there exists a sequence $x_1, x_2, \ldots, x_n$ in A such that $x = x_1$ and $y = x_n$, and $x_{i-1}Rx_i$ for $i = 2, 3, \ldots, n$.

tree   a connected, *acyclic graph*

Turing machine   a quintuple (S, A_T, N, s_0, F), where:

S is a finite *set* whose elements are called states.

$s_0 \in S$ and F is a subset of S.

A_T is the finite tape *alphabet* consisting of an input alphabet A_I, an output alphabet A_O and a special symbol delta.

$N: S \times A_T \rightarrow S \times A_T \times \{L, R\}$ is the next move *function*, where L stands for left and R stands for right.

variance   For a *random variable* x with *mean* μ and *distribution function* f_x, the variance is

$$s^2 = \sum (x_i - \mu)^2 f_x(x_i)$$

weighted digraph   a *digraph* in which each edge has been assigned a nonnegative real number called the weight of the edge

word   a string of symbols from an *alphabet*

Solutions

CHAPTER 1

Exercises 1.1

1. The network is unique.

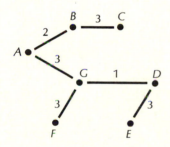

Exercises 1.2

1. **begin**
     $sum \leftarrow 0$
     $squares \leftarrow 0$
     $i \leftarrow 1$
     Input n
     while $i \leq 2n$ **do**
         begin
             $sum \leftarrow sum + i$
             $squares \leftarrow squares + i^2$
             $i \leftarrow i + 2$
         end
   output $sum, squares$
   end.

3. a. The **while** loop resets the value of *sum* to 0 ten times.
   b. The algorithm generates the numbers 1, 1, 2, 3, 5, 8, 13.

Review Problems

1. 6

3. The variable x has no assigned value.

5. If the value of n is 0, a division error results. There should be a test to ensure that $n > 0$.

7. Fill the 3-liter container and pour its contents into the 5-liter container. Refill the 3-liter container and pour 2 liters into the 5-liter container, so that there will be only 1 liter left in the 3-liter container. Next, empty the 5-liter container, empty the 3-liter container into the 5-liter container, and refill the 3-liter container. Pour the 3 liters into the 5-liter container, which will now contain 4 liters.

9. {Algorithm to determine how many of the *n* values}
   {of x are positive, negative, or zero.   }
   begin
       negatives ← 0
       positives ← 0
       zeros ← 0
       Obtain a value for *n*
       for i ← 1 **to** *n* **do**
           begin
               Input (*x*)
               if $x > 0$ **then**
                   positives ← *positives* + 1
               else
                   if $x < 0$ **then**
                   negatives ← *negatives* + 1
                   else
                     zeros ← *zeros* + 1
           end
       Output (*positives, negatives, zeros*)
   end.

CHAPTER 2

Exercises 2.1

1. a. P **and not** Q   d. P $\Rightarrow$ Q
   b. R **and** P   e. **not** P $\Rightarrow$ **not** Q
   c. R $\Rightarrow$ P

3.

P	Q	not (P or Q)	(not P) and (not Q)
T	T	F	F
T	F	F	F
F	T	F	F
F	F	T	T

5. (c) and (d) are logically equivalent.

Exercises 2.2

1. a. $P(x)$: x is an integer and $x = x/2$. There exists x such that $P(x)$.
   b. For all c, $P(c)$, where $P(c)$ is 'c drinks milk and c is a cat.'
   c. For all x, $P(x)$, where $P(x)$ is 'x is an integer and $x < x + 1$.'
3. a. Some dogs do not bark.
   b. All birds don't fly.
   c. Some cats like to swim.
   d. Some computer science students do not know mathematics.

Exercises 2.3

1. a. *Direct*: Assume n^2 is even. Then $nn = 2k$ for some k, so n must be divisible by 2.
      Contrapositive: Assume n is not even, so $n = 2m + 1$, for some integer m. Then
      $n^2 = (2m + 1)(2m + 1) = 4m^2 + 4m + 1$, which is not even.
   b. *Direct*: Suppose that $n = 2k$ and $m = 2j$. Then $n + m = 2(k + j)$, which is even.
      Contrapositive: Suppose that $n + m$ is odd, say $n + m = 2r + 1$. We need to
      show that either n is odd or m is odd.
      If n is even, then $n = 2s$, so $m = 2(r - s) + 1$, and m is odd.
      If m is even, then $m = 2p$, so $n = 2(r - p) + 1$, and n is odd.
   c. *Direct*: If $n = 2k$ and $m = 2p$, then $n - m = 2(k - p)$, which is even.
      Contrapositive: Redo (b), replacing $n + m$ by $n - m$.
3. Let $p = 3$. Then $x = 9 + 1 = 10$, which is not a prime number.

Exercises 2.4

1. a. $P(n)$: $\displaystyle\sum_{i=1}^{n} i^2 = \frac{n(n + 1)(2n + 1)}{6}$, $n > 0$

   $P(1)$ is *true*, since

   $$1^2 = \frac{1(2)(3)}{6}$$

If P(n) is true, then

$$\sum_{i=1}^{n+1} i^2 = \sum_{i=1}^{n} i + (n+1)^2 = \frac{n(n+1)(2n+1)}{6} + \frac{6(n+1)^2}{6}$$

$$= \frac{(n+1)(2n^2+7n+6)}{6} = \frac{(n+1)(n+2)(2n+3)}{6}$$

Hence P(n) $\Rightarrow$ P($n+1$).

b. P(n): $n^3 - n$ is divisible by 3, $n \geq 0$.

P(0) is clearly true. Show that P(n) $\Rightarrow$ P($n+1$) as follows. When P(n) is true, then $n^3 - n = 3k$, for some integer k. Then

$$(n+1)^3 - (n+1) = n^3 + 3n^2 + 3n + 1 - n - 1 = 3(k + n^2 + n),$$

which is divisible by 3.

c. P(n): $\displaystyle\sum_{i=0}^{n} a^i = \frac{a^{n+1} - 1}{a - 1}$, $n \geq 0$

P(0) is true, since

$$1 = \frac{a-1}{a-1}$$

Now assume that P(n) is true.

$$\sum_{i=0}^{n+1} a^i = \sum_{i=0}^{n} a^i + a^{n+1} = \frac{a^{(n+1)} - 1}{a - 1} + a^{n+1}$$

$$= \frac{a^{n+1} - 1 + a^{n+2} - a^{n+1}}{a - 1} = \frac{a^{n+2} - 1}{a - 1}$$

Hence P(n) $\Rightarrow$ P($n+1$).

3. 3, 2, 1, 3, 0, -1

5. Fix $n \geq 0$. Let P(m) be

$$a^{n+m} = a^n \cdot a^m, \, m \geq 1$$

For P(1) is *true*, since

$$a^{n+1} = a^n \cdot a$$

by definition of exponentiation.

Now assume that

$$a^{n+m} = a^n \cdot a^m \text{ for some } m \geq 1.$$

Then

$$a^{n+(m+1)} = a^{(n+m)+1}$$
$$= a^{n+m} \cdot a$$
$$= a^n \cdot a^m \cdot a$$
$$= a^n \cdot a^{(m+1)}$$

Hence $P(m) \Rightarrow P(m+1)$.

Exercises 2.5

The intermediate assertions needed for verification of the algorithms are shown below. The correctness of each algorithm follows easily from the expanded form.

1. Algorithm *Quadratic*

```
{x is a real number}
begin
    y ← ax
    {y = a · x}
    y ← (y + b) x
    {y = ax² + bx}
    y ← y + c
end.
{y = ax² + bx + c}
```

3. Algorithm *Factorial*

```
{n = n₁ and n₁ ≥ 0}
begin
    i ← 0
    Fact ← 1
    {i ≤ n₁ and Fact = i!}
    while i < n do
        begin
            i ← i + 1
            {i ≤ n}
            Fact ← Fact · i
            {Fact = i!}
        end
    {i = n₁ and Fact = i!}
end.
```

Exercises 2.6

1. a. $A = \{10, 11, 12, 13, 14, 15, 16, 17\}$
   b. $A = \{\text{positive integers } x: x = 2 + 3n, n = 0, 1, 2, 3, \ldots\}$
3. a. $\{1, 2, 3, 4, 5, 6, 7, 8, 9, 10, 11, 12\}$
   b. $\{\text{USA, China, Russia}\}$
   c. $\{\varnothing, \{\text{USA}\}, \{\text{China}\}, \{\text{Russia}\}, \{\text{USA, China}\}, \{\text{USA, Russia}\},$
      $\{\text{China, Russia}\}, \{\text{USA, China, Russia}\}\}$
5. a. $A \cap B$ b. $\sim A \cup (B \cup C)$

Exercises 2.7

1. $\{\varnothing, \{1\}, \{2\}, \{3\}, \{4\}, \{1, 2\}, \{1, 3\}, \{1, 4\}, \{2, 3\}, \{2, 4\}, \{3, 4\},$
   $\{1, 2, 3\}, \{1, 2, 4\}, \{2, 3, 4\}, \{1, 2, 3, 4\}\}$
3. a. $|A \cup B \cup C| = |A| + |B| + |C| - |A \cap B| - |A \cap C| - |B \cap C| + |A \cap B \cap C|$
   b. 36, 4
5. $\{\{\alpha\}, \{\beta, \delta\}\}$ $\{\{\alpha, \delta\}, \{\beta\}\}$ $\{\{\alpha, \beta\}, \{\delta\}\}$ $\{\{\alpha, \beta, \delta\}\}$
   $\{\alpha\}, \{\beta\}, \{\delta\}$

Review Problems

1. a. It is not dark outside.
   b. It is dark outside and the stars are out.
   c. It is dark outside or the stars are out.
   d. It is not dark outside and the stars are not out.
   e. The stars are out.
   f. The stars are out or it is not dark outside.

3.

P	Q	R	(P **or** Q) **and** R
T	T	T	T
T	T	F	F
T	F	T	T
T	F	F	F
F	T	T	T
F	T	F	F
F	F	T	F
F	F	F	F

5.

P	Q	(not P) or Q $\Rightarrow$ P
T	T	T
T	F	T
F	T	F
F	F	F

7.

P	Q	R	(P and Q) $\Rightarrow$ R	P $\Rightarrow$ (Q $\Rightarrow$ R)
T	T	T	T	T
T	T	F	F	F
T	F	T	T	T
T	F	F	T	T
F	T	T	T	T
F	T	F	T	T
F	F	T	T	T
F	F	F	T	T

9. All truth tables contain only F's in the results column.

11. P or Q = not (not P and not Q)

13. a.

P	Q	P xor Q
T	T	F
T	F	T
F	T	T
F	F	F

   b. P xor Q is logically equivalent to (P and not Q) or (not P and Q).

15. a. $\forall x$, P(x)
   b. $\forall x$, such that not P(x)
   c. not ($\exists x$ such that not P(x)), or, equivalently, $\forall x$, P(x)
   d. $\forall x$, P(x)

17. a. Let x be divisible by 15, say $x = 15 \cdot n$. Then $x = 3 \cdot (5 \cdot n)$, so x is divisible by 3.
   b. Let $x = 2n + 1$ and $y = 2m + 1$. Then $x \cdot y = 4nm + 2n + 2m + 1$, which is odd.
   c. Let $x > y$. Then $x = y + p$, where $p > 0$, so $x - y = p > 0$.

19. Use the fact that if $y = 0$, then $yx = y$.

21. a. $1 + \dfrac{1}{2} + \dfrac{1}{3} + \dfrac{1}{4} \cdots$

    b. $(x_1 - \mu)^2 - (x_2 - \mu)^2 + (x_3 - \mu)^2 + (x_4 - \mu)^2 + \cdots$

23. Define $a^0 = 1$, and for all $n > 0$, define

$$a^{n+1} = a^n \cdot a$$

25. Inserting the appropriate intermediate assertions yields:

begin
    $y \leftarrow ax$
    $\{y = ax_1\}$
    $y \leftarrow y + b$
end.
$\{y = ax_1 + b\}$

The verification is now straightforward.

27. The first, third, and fourth sets are equal.

29. a. $\{2, 4, 6\}$ e. $\{1, 2, 3, 4, 5, 6, 10, 17, 26\}$
    b. $\{3, 5\}$ f. $\{1, 2, 3, 4, 5, 6, 10, 17, 26\}$
    c. $\varnothing$ g. $\{1, 2, 3, 4, 5, 6\}$
    d. $\{1, 2, 3, 4, 5, 6\}$

31. If $x \in B$, then x is odd and greater than 1, so $x \notin A$.

33. $(A \cup B) \cap (A \cap \sim B) = A \cap \mathcal{U}$

35. $\begin{aligned}
|A \cup B \cup C| &= |(A \cup B) \cup C| \\
&= |A \cup B| + |C| - |(A \cup B) \cap C| \\
&= |A| + |B| - |A \cap B| + |C| - |(A \cap C) \cup (B \cap C)| \\
&= |A| + |B| + |C| - |A \cap B| - (|A \cap C| + |B \cap C| \\
&\quad - |A \cap B \cap C|) \\
&= |A| + |B| + |C| - |A \cap B| - |A \cap C| - |B \cap C| \\
&\quad + |A \cap B \cap C|
\end{aligned}$

37. The sets are $\{1, 5\}$, $\{1, 6\}$, $\{1, 7\}$, $\{2, 5\}$, $\{2, 6\}$, $\{2, 7\}$, $\{3, 5\}$, $\{3, 6\}$, $\{3, 7\}$, $\{4, 5\}$, $\{4, 6\}$, $\{4, 7\}$.
    The elements of the product set are $(1, 5)$, $(1, 6)$, $(1, 7)$, $(2, 5)$, $(2, 6)$, $(2, 7)$, $(3, 5)$, $(3, 6)$, $(3, 7)$, $(4, 5)$, $(4, 6)$, $(4, 7)$.

39. Let $A = \{1, 2, 3\} = B$. $|A| + |B| < |A \cup B|$ is impossible by the Principle of Inclusion and Exclusion.

41. 32 (including the pizza with none of the toppings)

43. Algorithm *FindPowerSet(n)*

```
begin
    if n < 0 then
        Output ('Invalid value of n')
    else
        begin
            if n = 0 then
                P ← {Ø, {0}}
            else
                begin
                    P ← FindPowerSet(n − 1)
                    Q ← Ø
                    for p ∈ P do
                        Q ← Q ∪ {p ∪ {n}}
                    P ← P ∪ Q
                end
            Output (P)
        end
end.
```

CHAPTER 3

Exercises 3.1

1. a.   b.   c.

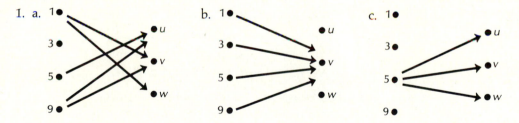

3. $\{(x, B): x$ is an element of A, B is an element of $P(A)$, and $x \in B\}$
5. a. $\{(1, 1), (1, 4)\}$

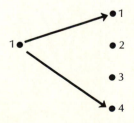

b. $\{(1, 2), (2, 2)\}$

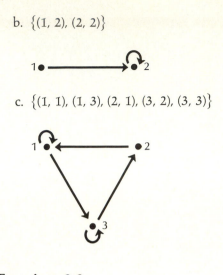

c. $\{(1, 1), (1, 3), (2, 1), (3, 2), (3, 3)\}$

Exercises 3.2

1. Let $S = \{1, 2, 3\}$.

$\{(1, 1), (2, 2), (3, 3)\}$ is reflexive, transitive, and symmetric.

$\{(1, 1), (2, 2), (3, 3), (1, 2), (2, 1), (2, 3), (3, 2)\}$ is reflexive, symmetric, but not transitive.

$\{(1, 1), (2, 2), (3, 3), (1, 2), (2, 3), (1, 3)\}$ is reflexive, not symmetric, and transitive.

$\{(1, 1), (2, 2), (3, 3), (1, 2), (2, 3)\}$ is reflexive, not symmetric, and not transitive.

$\{(1, 2), (2, 1), (1, 1), (2, 2)\}$ is not reflexive, symmetric, and transitive.

$\{(1, 2), (2, 1)\}$ is not reflexive, symmetric, and not transitive.

$\{(1, 2), (2, 3), (1, 3)\}$ is not reflexive, not symmetric, and transitive.

$\{(1, 2), (2, 3)\}$ is not reflexive, not symmetric, and not transitive.

3. a. not reflexive, symmetric, transitive, irreflexive, not antisymmetric
   b. reflexive, symmetric, transitive, not irreflexive, not antisymmetric
   c. reflexive, not symmetric, not transitive, not irreflexive, not antisymmetric
   d. reflexive, not symmetric, transitive, not irreflexive, antisymmetric

5. R is irreflexive if and only if for all x in A, $x\cancel{R}x$.

7. The relation in (d) is a partial ordering. None is a linear order.

9. a. $\{(A, B), (A, C), (B, C), (C, E), (D, C), (E, B), (A, E), (B, E), (B, B), (C, B), (C, C), (D, E), (D, B), (E, C), (E, E)\}$
   b. B, C, and E are recursive.

Exercises 3.3

1. a. $\{(2, 1), (2, 3), (3, 2), (3, 4), (4, 1), (4, 3)\}$
   b. $\{(1, 3), (1, 4), (3, 1), (4, 1)\}$

c. $\{(1, 1), (1, 2), (1, 3), (1, 4), (3, 2)\}$
d. $\{(3, 3), (3, 4), (3, 1)\}$

3. The inverse of R is $\{(1, 1), (2, 1), (2, 2), (3, 1), (3, 3), (4, 1), (6, 1), (4, 2), (6, 3),$ $(6, 2), (6, 6)\}$.

5. The claim follows from the fact that R is reflexive if and only if $(x, x) \in R$ for all $x \in A$.

Exercises 3.4

1. a. not a function
   b. A function with domain $= \{a, b, c, d, e\}$ and codomain and range $= \{\alpha, \delta, \eta\}$.
   c. The relation in Figure 3.28(a) is a bijection with domain $= \{a, b, c\}$ and codomain $= \{1, 2, 3\}$.
      The relation in Figure 3.28(b) is a function with domain $= \{a, b, c\}$, codomain $= \{x, y, z, w\}$, and range $= \{z, w\}$. The function is not an injection and is not a surjection.

3. If $g \textbf{ o } f$ is a bijection, then f and g must be injective and surjective.

5. $f(x) = \begin{cases} x/2 & \text{if} \quad x \text{ is even} \\ (x - 1)/2 & \text{if} \quad x \text{ is odd} \end{cases}$

7. $f^{-1}(j) = \begin{cases} j + 1, & \text{for } j = 1, 2, \ldots, n - 1 \\ 1, & \text{for } j = n \end{cases}$

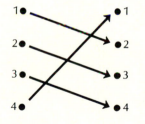

Review Problems

1. a. $A \times B = \{(a, 1), (a, 2), (a, 3), (a, 4),$
   $(b, 1), (b, 2), (b, 3), (b, 4),$
   $(c, 1), (c, 2), (c, 3), (c, 4)\}$
   b. $B \times A = \{(1, a), (1, b), (1, c),$
   $(2, a), (2, b), (2, c),$
   $(3, a), (3, b), (3, c),$
   $(4, a), (4, b), (4, c)\}$

3. a.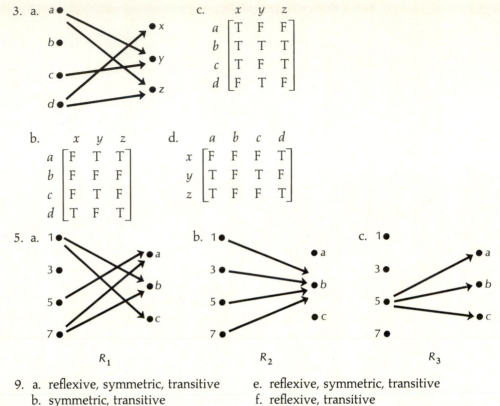

c.

	x	y	z
a	T	F	F
b	T	T	T
c	T	F	T
d	F	T	F

b.

	x	y	z
a	F	T	T
b	F	F	F
c	F	T	F
d	T	F	T

d.

	a	b	c	d
x	F	F	F	T
y	T	F	T	F
z	T	F	F	T

5. a.   b.   c.

R_1 R_2 R_3

9. a. reflexive, symmetric, transitive
   b. symmetric, transitive
   c. symmetric
   d. symmetric

   e. reflexive, symmetric, transitive
   f. reflexive, transitive
   g. reflexive, transitive
   h. symmetric

11. a. Apply the definition directly.
    b. 5

13. Yes

15. $\{(1, 1), (1, 2), (2, 2), (3, 1), (3, 2)\}$

17. a. $\{(1, 1), (1, 3), (3, 2), (4, 2), (4, 3)\}$
    b. $\{(1, 1), (1, 2), (1, 3), (2, 1), (2, 4)\}$
    c. $\{(1, 1), (1, 2), (1, 3), (2, 1), (2, 2), (2, 3)\}$

19. $[1] = \{1, 5\} = [5]$ $[2] = \{2, 6\} = [6]$
    $[3] = \{3\}$ $[4] = \{4\}$

21. a. $\forall m, n \in \mathbf{Z}, m \geq n$ and $n \geq m \Rightarrow m = n$
    b. None of (a, a), (b, b), (c, c) belong to the relation.

23. The relations shown in (a), (b), and (f)

25. a. injective   b. surjective   c. bijective

27. This follows from the facts that composition of injections (respectively, surjections) is an injection (respectively, a surjection).

29. Only the function in (b) is invertible.

31. a.

A	B	C	D	A
1	2	3	4	1
1	1	1	1	2
2	1	2	1	2
1	1	1	1	3
2	1	2	1	3

b.

A	B	C	D
4	3	2	1
2	1	2	1
3	4	2	1
3	4	1	2

c.

A	C	D
1	3	4
1	1	1
4	2	1
2	2	1
3	2	1
3	1	2

CHAPTER 4

Exercises 4.1

1. 2-samples: 1, 1   1, 2   1, 3   1, 4   2, 1   2, 2   2, 3   2, 4
   3, 1   3, 2   3, 3   3, 4   4, 1   4, 2   4, 3   4, 4

   2-selections: 1, 1   1, 2   1, 3   1, 4
   2, 2   2, 3   2, 4
   3, 3   3, 4
   4, 4

   2-permutations: 1, 2   1, 3   1, 4   2, 1   2, 3   2, 4
   3, 1   3, 2   3, 4   4, 1   4, 2   4, 3
   2-combinations: 1, 2   1, 3   1, 4   2, 3   2, 4   3, 4

3. There are more 4-selections.

5. a.

i	0	1	2	3	4	5	6
subsets with i elements	1	6	15	20	15	6	1

   b. When $n = 3$, Equation (4. 3) becomes

$$(a + b)^3 = a^3 + 3a^2b + 3ab^2 + b^3 = \sum_{i=0}^{3} C(3, i)a^i b^{3-i}$$

   c. Use Equation (4.3) with $a = b = 1$.

7. 6048

9. 1 6 15 20 15 6 1

Exercises 4.2

1. a. 2520   b. 30   c. 6720   d. 5040   e. 120
3. 4200
5. a. 90   b. 360

Review Problems

1. a. 240   b. 1/504   c. 156
3. 2-samples: α, α α, β α, γ α, δ β, α β, β β, γ β, δ
γ, α γ, β γ, γ γ, δ δ, α δ, β δ, γ δ, δ
   2-selections: α, α α, β α, γ α, δ β, β β, γ β, δ γ, γ γ, δ δ, δ
   2-permutations: α, β β, α α, γ γ, α α, δ δ, α β, γ γ, β β, δ δ, β
γ, δ δ, γ
   2-combinations: α, β α, γ α, δ β, γ β, δ γ, δ
5. 5
7. a. 64   b. 32   c. 32   d. 48
9. a. 936   b. 34,658
11. 10!
13. 8!
15. 3
17. 100
19.   1   8   28   56   70   56   28   8   1
    1   9   36   84   126   126   84   36   9   1
21. a. 1287   b. 9024   c. 1872
23. a. missing the element g
    b. e is in two subsets
    c. a partition
25. 251
27. 15
29. a. 2520   b. 840
31. $2^{2 \log n} = (2^{\log n})^2 = n^2$
33. None

CHAPTER 5

Exercises 5.1

1. a. There is an edge between every pair of vertices.
    b. Each vertex is connected by an edge to the other $n - 1$ vertices.
    c. $2|E| = \sum_{v \in V} \delta(v) = n(n - 1)$, so $|E| = n(n - 1)/2$
    d. Let $v_1, v_2, \ldots, v_n$ be the vertices. Then $v_1, v_2, \ldots, v_n$ is a Hamiltonian circuit.
3. The sums of the degrees are, respectively, 14, 6, 0.
5. a, b, c, a a, c, e, a a, e, d, a a, b, c, e, a
   a, c, e, d, a a, b, c, a, e, d, a a, b, c, e, d, a
7. K_n has an Eulerian path for $n = 2$, and an Eulerian circuit for odd $n > 2$.

Exercises 5.2

1. By inspection
3. *Hint*: Use induction to show that if G is connected, then $|E| \geq |V| - 2$.
7. Seven guesses.

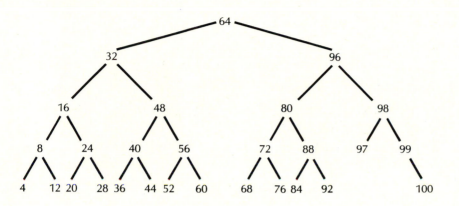

9. a. *search* (*root, searchkey*) returns a logical value.

    begin
        if (*searchkey* = *currentkey*) **then**
            return *true*
        else
            if (*leftsubtree* ≠ *null*) **then**
              search (*leftsubtree, searchkey*)
            else
              if (*rightsubtree* ≠ *null*) **then**
                search (*rightsubtree, searchkey*)
              else
                return *false*
    end.

    b. *in-order traversal* (*root*)

    begin
        if (*root* ≠ *null*) **then**
          begin
              traversal in-order (*leftsubtree*)
              print the current element
              traversal in-order (*rightsubtree*)
          end
    end.

Review Problems

1. Given $G = (V, E)$,

$$\sum_{v \in V} \delta(v) = 2|E|.$$

Let $V_1 = \{v: \delta(v) \text{ is odd}\}$ and $V_2 = \{v: \delta(v) \text{ is even}\}$. Then

$$\sum_{v \in V_1} \delta(v) = 2|E| - \sum_{v \in V_2} \delta(v)$$

The right side is even, and the sum of a set of odd numbers is even only when the number of odd numbers is even.

3. a. The sum of the degrees is 14, and the number of edges is 7.

   b.

2	T				
3	F	T			
4	F	T	T		
5	F	F	T	F	
6	F	F	F	T	T
	1	2	3	4	5

   c. 2, 3, 4, 2;   2, 4, 6, 5, 3, 2;   3, 4, 6, 5

5. Let $\delta(v_0) \geq k$, and let $v_0 E v_i$, $i = 1, \ldots, k$. Then $v_1, v_0, v_2, v_0, \ldots, v_0, v_k$ has length $2k$.

7. There are 19 spanning trees, but only three are distinct.

9. The connectivity components are $\{1, 2\}$ and $\{3, 4, 5, 6\}$.

11. G_1 has a degree of vertex 4, but G_2 does not.

13. *Hint*: Let $f: G \to G'$ be a graph isomorphism. Show that if C is a component in G, then the image of C by f is a component in G'.

15. 13 (respectively, $(3^{(n+1)} - 1)/2$)

17. *Hint*: Let $|V| = n$. Let P(k) be the statement:

$$C(G) = n - k \text{ when } G \text{ is acyclic with } k \text{ edges, } k \leq n.$$

Use an induction argument on k.

19. a. Not bipartite. Vertices 3 and 4 must be put into opposite partitions because they are connected by an edge. But they cannot be in opposite partitions because both are connected to 2.

    b. Let $G = (V, E)$ be a tree. Let v_0 be any vertex. Let $V_1 = \{v: \text{length of the path from } v_0 \text{ to } v \text{ is even}\}$, and $V_2 = \{v: \text{length of the path from } v_0 \text{ to } v \text{ is odd}\}$.

21. Suppose G is not connected. Let v be a vertex with $\delta(v) = 0$. The remaining vertices can be connected to at most $(n - 1)(n - 2)/2$ vertices.

CHAPTER 6

Exercises 6.1

1. a. 1, 4, 6, 2   b. all of them
3. The order of 1, 2, and 3 can't be switched. Another consistent labeling is 1 2 3 6 10 9 8 5 11 12 4 7.
5. Suppose $G = (V, E)$ has no sink. Let $v_1 \in V$. Since G has no sink, there exists $v_2 \in V$ such that $(v_1, v_2) \in E$. By induction, we can construct a sequence of vertices $v_1, v_2, \ldots, v_n$ such that for $i = 1, 2, \ldots, n - 1$, $(v_i, v_{i-1}) \in E$. When $n > |V|$, the Pigeonhole Principle says that there exist $i, j, i < j$, such that $v_i = v_j$. The path $v_i, \ldots, v_j$ is a cycle.
7. Let $V = \{1, \ldots, n\}$, and let $E(i, j)$ be the logical matrix for the digraph. Let $A(i)$ be a subset of V, for $i = 1, \ldots, n$.

```
for i ← 1 to n do
    begin
        A(i) = ∅
        for j = 1 to n do
            if E(i, j) then
                A(i) = A(i) ∪ {j}
end.
```

9. a. 46 days
   b. 1, 2, 3, 6, 10, 11, 12. If any of the weights along this path increases, the completion time increases.
   c. From 4 to 5 days, there is no change. From 4 to 8 days, the longest path has length 48.

Exercises 6.2

1. a. $\begin{vmatrix} T & T & T & T \\ T & T & T & T \\ T & T & T & T \\ T & T & T & T \end{vmatrix}$ b. $\begin{vmatrix} T & T & T & T & T \\ T & T & T & T & T \\ T & T & T & T & T \\ T & T & T & T & T \\ F & F & F & F & F \end{vmatrix}$

3. For $n = 10$, the conventional method requires 10,000 **and**'s, Warshall's only 1000. For $n = 100$, the numbers required are 100,000,000 and 1,000,000, respectively.
5. A vertex v_i is in a cycle if and only if $E^*(i, i) = T$.

7. $N^1 = \begin{vmatrix} 0 & 1 & 0 & 0 \\ 0 & 0 & 1 & 0 \\ 0 & 0 & 0 & 1 \\ 1 & 0 & 0 & 0 \end{vmatrix}$ $N^2 = \begin{vmatrix} 0 & 0 & 1 & 0 \\ 0 & 0 & 0 & 1 \\ 1 & 0 & 0 & 0 \\ 0 & 1 & 0 & 0 \end{vmatrix}$

$N^3 = \begin{vmatrix} 0 & 0 & 0 & 1 \\ 1 & 0 & 0 & 0 \\ 0 & 1 & 0 & 0 \\ 0 & 0 & 1 & 0 \end{vmatrix}$ $N^4 = \begin{vmatrix} 1 & 0 & 0 & 0 \\ 0 & 1 & 0 & 0 \\ 0 & 0 & 1 & 0 \\ 0 & 0 & 0 & 1 \end{vmatrix}$

Review Problems

1. a.

	p	q	r	s	t	u
p	F	T	F	F	F	F
q	F	T	T	T	F	F
r	F	F	F	T	F	F
s	T	F	F	F	T	T
t	F	F	F	F	F	F
u	F	F	T	F	F	F

b.

vertex	p	q	r	s	t	u
outdegree	1	3	1	3	0	1
indegree	1	2	2	2	1	1

c. t is a sink. There are no sources. One cycle is p, q, s, p.

3. There is no consistent labeling.

5. $\begin{vmatrix} T & T & T & T & T \\ T & T & T & T & T \\ T & T & T & T & T \\ T & T & T & T & T \\ T & T & T & T & T \end{vmatrix}$

7.

	A	B	C	D	E	F
A	0	3	∞	∞	8	∞
B	∞	0	12	9	∞	∞
C	∞	∞	0	16	∞	∞
D	∞	∞	∞	0	7	4
E	∞	∞	∞	∞	0	∞
F	∞	∞	∞	∞	∞	0

9.

vertex	A	B	C	D	E	F
distance	∞	∞	9	0	7	4

11. a. Yes.   b. Hint: Use induction on the number of vertices.

13. a.

	e_1	e_2	e_3	e_4	e_5	e_6	e_7	e_8	e_9
A	1	0	0	-1	1	0	0	0	0
B	-1	1	1	0	0	0	0	0	0
C	0	-1	-1	0	0	0	0	1	-1
D	0	0	0	1	0	1	0	-1	0
E	0	0	0	0	-1	-1	1	0	0
F	0	0	0	0	0	0	-1	0	1

b. Each column contains exactly one 1 and one -1, so each column sums to 0. For each i, the ith row has a 1 for each edge leaving v_i, and a -1 for each edge arriving at v_i. Thus the sum of the elements in the ith row is the number of edges leaving v_i — the number of edges arriving at v_i.
c. A vertex is a sink (respectively, a source) if its row in the incidence matrix contains no 1's (respectively, no -1's).

15. 1 2 3 4 5 6 7 8
    b c d a e h f g
    a b c d e h f g
    a b c d e f g h

17.   From a:

vertex	a	b	c	d	e	f	g	h
distance	0	∞	3	3	2	1	7	4

From b:

vertex	a	b	c	d	e	f	g	h
distance	3	∞	3	0	5	4	9	7

CHAPTER 7

Exercises 7.1

1. a.

p	1	$p \vee 1$
0	1	1
1	1	1

b.

p	p'	$p \wedge p'$
0	1	0
1	0	0

c.

p	0	$p \wedge 0$
0	0	0
1	0	0

3. Theorem 7.1, identity 4b: Theorem 7.1, identity 4a:

p	$p \vee p$
0	0
1	1

p	$p \wedge p'$
0	0
1	0

Theorem 7.1, identity 5a: Theorem 7.1, identity 5b:

p	q	$p \vee q$	$p \wedge (p \vee q)$
0	0	0	0
0	1	1	0
1	0	1	1
1	1	1	1

p	q	$p \wedge q$	$p \vee (p \wedge q)$
0	0	0	0
0	1	0	0
1	0	0	1
1	1	1	1

5. a. $(p \vee q')' \wedge r' = p' \wedge q \wedge r'$ by DeMorgan's laws

b. $(z \wedge w) \vee (z' \wedge w) \vee (z \wedge w') = (1 \wedge w) \vee (z \wedge w')$

$$= ((z \vee z') \wedge w \vee (z \wedge w')$$
$$= (w \vee z) \wedge (w \wedge w')$$
$$= z \vee w$$

c. $((p \vee q') \vee (r \wedge (p \vee q')))' = (p \vee q')' \wedge (r \wedge (p \vee q')')$

$$= p' \wedge q \wedge (r' \vee (p \vee q')')$$
$$= p' \wedge q \wedge (r' \vee (p' \wedge q))$$
$$= p' \wedge q \text{ (by absorption property)}$$

d. $x \wedge y \vee x' \wedge y \vee x' \wedge y' \vee x \wedge y' = ((x \vee x') \wedge y) \vee ((x' \vee x) \wedge y')$

$$= y \vee y'$$
$$= 1$$

Exercises 7.2

1. $f_1 \wedge f_3 = f_1$
$f_2 \vee (f_3 \wedge f_2)' = f_2 \vee f'_2 = f_0$

3. 2^n

5. $g(x, y, z, w) = x'y'zw' \vee x'y'zw \vee x'yzw \vee xy'zw \vee xyz'w'$

7. a. $\{ \vee, ' \}$ is a complete set of operators since $x \wedge y = x' \vee y'$.
   b. $x \wedge y \wedge z = (x' \vee y' \vee z')$

Exercises 7.3

1. The expression for Figure 7.14 is $p'qr \vee pq' \vee q'r'$.
   The expression for Figure 7.15 is $z'w' \vee xyw \vee xz'$.

3. a. The truth table for $p \wedge (q' \vee r) \vee (p' \wedge q \vee r')$ is

p	q	r	f
0	0	0	1
0	0	1	0
0	1	0	1
0	1	1	1
1	0	0	1
1	0	1	1
1	1	0	1
1	1	1	1

The Karnaugh map is

The simplified expression is $p \vee q \vee r'$.

b. The truth table for expression $rs \vee (q'r)' \vee (p' \vee r)'$ is

p	q	r	s	f
0	0	0	0	1
0	0	0	1	1
0	0	1	0	0
0	0	1	1	1
0	1	0	0	1
0	1	0	1	1
0	1	1	0	1
0	1	1	1	1
1	0	0	0	1
1	0	0	1	1
1	0	1	0	0
0	0	1	1	1
0	1	0	0	1
0	1	0	1	1
0	1	1	0	1
0	1	1	1	1

The Karnaugh map is

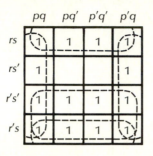

The simplified expression is $q \vee s \vee r'$.

Exercises 7.4

1. a.

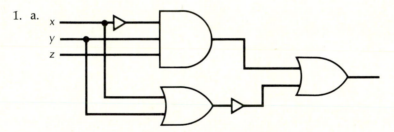

b. Hint: Use the following relationships to convert the gates.

$$x' = x\textbf{NAND}x$$
$$x \wedge y = (x\textbf{NAND}y)\textbf{NAND}(x\textbf{NAND}y)$$
$$x \vee y = (x\textbf{NAND}x)\textbf{NAND}(y\textbf{NAND}y)$$

c.

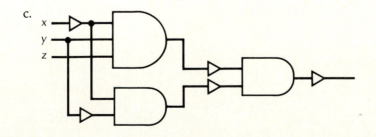

Review Problems

1. Theorem 7.1, identity 1a.

p	q	$p \wedge q$	$q \wedge p$
0	0	0	0
0	1	0	0
1	0	0	0
1	1	1	1

Theorem 7.1, identity 1b.

p	q	$p \vee q$	$q \vee p$
0	0	0	0
0	1	1	1
1	0	1	1
1	1	1	1

Theorem 7.1, identity 2a.

p	q	r	$q \wedge r$	$p \wedge q$	$p \wedge (q \wedge r)$	$(p \wedge q) \wedge r$
0	0	0	0	0	0	0
0	0	1	0	0	0	0
0	1	0	0	0	0	0
0	1	1	1	0	0	0
1	0	0	0	0	0	0
1	0	1	0	0	0	0
1	1	0	0	0	0	0
1	1	1	1	1	1	1

3. $(p' \wedge q) \vee (p \wedge q) \vee (p \wedge q') = ((p' \vee p) \wedge q) \vee (p \wedge q')$
$$= q \vee (p \wedge q')$$
$$= (q \vee p) \wedge (q \vee q')$$
$$= q \vee p$$
$$= p \vee q$$

5. a. $m(p, q, r) = p'qr'$
   b. $m(p, q, r, s) = p'qr's$

9. $f_{12} \wedge f_7 = f_4$

11. $p'q'r's' \vee p'qrs \vee pq'r's \vee pqrs$

13. a. $f_8' = f_7$

b. $f_{12} \wedge (f_2 \vee f_5) = f_{12} \wedge f_7 = f_4$

15. Let $f(x, y, z) = x \wedge (y \vee z)' \vee (x' \wedge y \vee z)$

The truth table for $f(x, y, z)$ is

x	y	z	$f(x, y, z)$
0	0	0	0
0	0	1	1
0	1	0	1
0	1	1	1
1	0	0	0
1	0	1	1
1	1	0	0
1	1	1	1

$$f(x, y, z) = x'y'z \vee x'yz' \vee x'yz \vee xy'z \vee xyz$$

17.

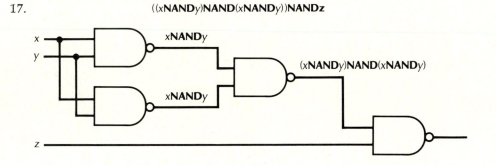

$$((x\textbf{NAND}y)\textbf{NAND}(x\textbf{NAND}y))\textbf{NAND}z$$

21. The expression represented by the Karnaugh map in Figure 7.42 is $rq \vee pqs' \vee p'q'rs$.

23. The Karnaugh map for the expression $x \vee (y' \vee z)'$ is

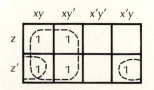

The simplified expression is $x \vee yz'$. The inverse of $x \vee (y' \vee z')$ is

$$
\begin{aligned}
(x \vee (y' \vee z)')' &= (x \vee (y' \wedge z'))' \\
&= x' \wedge (y \wedge z')' \\
&= x' \wedge y' \vee z \\
&= x'y' \vee z
\end{aligned}
$$

25. The simplified expression represented by the Karnaugh map in Figure 7.43 is $xz \vee y'z \vee x'y'$.

27. $C(16, 2) = 120$ and $C(16, 3) = 560$

29. a. $x' \vee xqz'$ b. always true

31.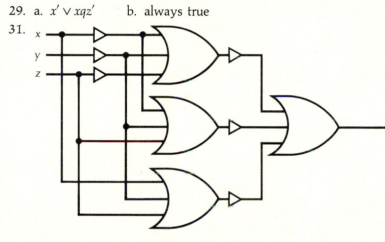

35. Use **AND** gates with five inputs. Negate all possible two-input combinations. Use an **OR** gate to combine the outputs of the **AND** gates.

37. $(r'st \vee rst \vee rst')'$

*39. Use the information in the solution for exercise 8 in Exercises 7.2.

CHAPTER 8

Exercises 8.1

1. To be a semigroup $(\mathbf{N}, \Delta)$ must be associative. Since $(a \triangle b) \triangle c = \min(\min(a, b), c) = \min(a, b, c) = \min(a, \min(b, c)) = a \triangle (b \triangle c)$, $(\mathbf{N}, \Delta)$ is associative.

   To be a monoid $(\mathbf{N}, \Delta)$ must also have an identity element. Assume that e is the identity element. Then $a \triangle e = e \triangle a$ for all a in $\mathbf{N}$. This implies that $\min(a, e) = a$ for all a in $\mathbf{N}$, which in turn implies that e is an upper bound for $\mathbf{N}$. But $\mathbf{N}$ has no upper bound. Thus there is no identity element for $(\mathbf{N}, \Delta)$.

3. $\mathscr{P}(S)$ with $\cup$ (union) is not a group. There is no inverse for an arbitrary element. $\mathscr{P}(S)$ with $\cap$ (intersection) is not a group. There is no inverse for an arbitrary element.

  N with $+$ (addition) is not a group. The inverse of $x \in$ **N** is not in **N**.

  N with $\cdot$ (multiplication) is not a group. The inverse of $x \in$ **N** is not in **N**.

  Z is an abelian group with respect to $+$ (addition).

  Z with $\cdot$ (multiplication) is not a group. The inverse of $x \in$ **Z** is not in **Z**.

  $\mathbf{R}^+$ is not a group with respect to $+$ (addition). There are no inverses.

  $\mathbf{R}^+$ is an abelian group with respect to $\cdot$ (multiplication).

  $R(S)$ is not a group with respect to $\circ$ (composition). There are no inverses.

  $F(S)$ is not a group with respect to $\circ$ (composition). There are no inverses.

  M_n is not a group with respect to matrix multiplication. There are no inverses.

  B is not a group with respect to conjunction. The element 0 has no inverse.

  B is not a group with respect to disjunction. The element 1 has no inverse.

5. 3^9

7. a. $2 +_6 2 = 4$
      $2 +_6 4 = 0$
      $2 +_6 0 = 2$
   b. $f(x) = 2x$
      $f(f(x)) = 2(2x) = 2^2 x$
      $f(f(f(x))) = 2(2(2x)) = 2^3 x$

      $\vdots$

      nth power of $f(x) = 2^n x$

9. Both of the following groups of order 4 are abelian:

$\cdot$	a	b	c	d
a	a	b	c	d
b	b	c	d	a
c	c	d	a	b
d	d	a	b	c

$\cdot$	a	b	c	d
a	a	b	c	d
b	b	a	d	c
c	c	d	a	b
d	d	c	b	a

*11. Assume first that G is a group, and let a and b represent arbitrary elements of G. Now a^{-1} (the inverse of a) is in G, and so are $x = a^{-1}b$ and $y = ba^{-1}$. With these choices for x and y, we have

$$ax = a(a^{-1}b) = (aa^{-1})b = eb = b$$

and

$$ya = (ba^{-1})a = b(a^{-1}a) = be = b$$

Thus G contains solutions to $ax = b$ and $ya = b$.

Suppose now that the equations always have a solution in G. We first show that G has an identity element. Let a represent an arbitrary but fixed element in G. The equation $ax = a$ has a solution in G, say $x = u$. We shall show that u is a right identity for every element in G. To do this, let b be an arbitrary element of G. With z a solution to $ya = b$, we have

$$za = b \quad \text{and} \quad bu = (za)u = z(au) = za = b$$

Thus u is a right identity for every element in G. In a similar fashion, we can show that there exists a left identity, say v. Then $vu = v$ and $vu = u$. Thus $e = u = v$.

For any a in G, let z be a solution to $ax = e$, and let w be a solution to $ya = e$. Combining these equations, we have

$$z = ez$$
$$= waz$$
$$= we$$
$$= w$$

Thus $z = w$ is an inverse for a. This proves G is a group.

13. a.

+	000	001	010	011	100	101	110	111
000	000	001	010	011	100	101	110	111
001	001	010	011	100	101	110	111	000
010	010	011	100	101	110	111	000	001
011	011	100	101	110	111	000	001	010
100	100	101	110	111	000	001	010	011
101	101	110	111	000	001	010	011	100
110	110	111	000	001	010	011	100	101
111	111	000	001	010	011	100	101	110

This represents a group.

b.

+	000	001	010	011	100	101	110	111
000	000	001	010	011	100	101	110	111
001	001	010	011	100	101	110	111	100
010	010	011	100	101	110	111	100	101
011	011	100	101	110	111	100	101	110
100	100	101	110	111	100	101	110	111
101	101	110	111	100	101	110	111	100
110	110	111	100	101	110	111	100	101
111	111	100	101	110	111	100	101	110

This is a monoid, not a group.

Exercises 8.2

1. Use Example 8.13 to show R is a congruence relation. The induced quotient monoid is

$+_n$	$[0]$	$[1]$	$[2]$	$[3]$	$\cdots$	$[n-1]$
$[0]$	$[0]$	$[1]$	$[2]$	$[3]$	$\cdots$	$[n-1]$
$[1]$	$[1]$	$[2]$	$[3]$	$[4]$	$\cdots$	$[0]$
$[2]$	$[2]$	$[3]$	$[4]$	$[5]$	$\cdots$	$[1]$
$\vdots$	$\vdots$	$\vdots$	$\vdots$	$\vdots$		$\vdots$
$[n-1]$	$[n-1]$	$[0]$	$[1]$	$[2]$	$\cdots$	$[n-2]$

Tables for $(\mathbf{Z}_2, +_2)$ and $(\mathbf{Z}_3, +_3)$ are

$+_2$	$[0]$	$[1]$
$[0]$	$[0]$	$[1]$
$[1]$	$[1]$	$[0]$

$+_3$	$[0]$	$[1]$	$[2]$
$[0]$	$[0]$	$[1]$	$[2]$
$[1]$	$[1]$	$[2]$	$[0]$
$[2]$	$[2]$	$[0]$	$[1]$

3. The product monoid $(\mathbf{Z}_2 \times \mathbf{Z}_5, +)$ consists of ordered pairs where the first element is from $\{0, 1\}$ and the second element is from $\{0, 1, 2, 3, 4\}$. The first elements are added modulo 2, and the second components are added modulo 5.

5. $\mathbf{Z}_2 \times \mathbf{Z}_2$ is not cyclic since every element has order 2.

7. A permutation of a set X is a bijection of X onto X. Consider a row (column) labeled g in a group operation table. To show that this row (column) is a permutation of the elements of the group we show that it is one-to-one and onto. If the row (column) is not a one-to-one map of the elements of the group to the elements of the group, then there is some element of g (say k) that appears at least twice

in the row (column). That is, there must be a_1 and a_2 such that $g \cdot a_1 = g \cdot a_2 = k$. But this is impossible unless $a_1 = a_2$, for if $g \cdot a_1 = g \cdot a_2$, then $a_1 = g^{-1} \cdot g \cdot a_1 = g^{-1} \cdot g \cdot a_2 = a_2$. Therefore, each row (column) is an injection of the elements of the group. To show that each row (column) is onto the elements of the group, we show that each element of the group appears at least once in each row (column). Because for every element, a, of the group $g \cdot (g^{-1} \cdot a) = a$, a appears in row g (column g) in the column (row) labeled $g^{-1} \cdot a$. Hence, each row and column is a permutation of the elements of the group.

9.

$\cdot$	x	y	z
x	x	y	z
y	y	z	y
z	z	y	y

11. $\mathbf{Z}_n = \{[0], [1], [2], \ldots, [n-1]\}, +_n)$ is a cyclic group, since $\mathbf{Z}_n = \langle[1]\rangle$.

*13. Let $(G, \cdot)$ be an abelian group with identity e. Let $A = \{x \in G: x^2 = e\}$. Note that $e \in A$ since $e^2 = e$. To show closure, consider $x, y \in A$. Then $(x \cdot y)^2 = x^2 \cdot y^2 = e \cdot e = e$, so that $x \cdot y \in A$. Let $x \in A$. $(x^{-1})^2 = (x^2)^{-1} = e^{-1} = e$, so that $x^{-1} \in A$. Thus $(A, \cdot)$ is a subgroup of G.

15. Let $(A, \cdot)$ be a cyclic group. Let $x, y \in A$. Then $x = a^{k_1}$ and $y = a^{k_2}$ where a is the group generator. Then $x \cdot y = a^{k_1} \cdot a^{k_2} = a^{k_1 + k_2} = a^{k_2 + k_1} = a^{k_2} \cdot a^{k_1} = y \cdot x$. Thus $(u, \cdot)$ is abelian.

17. Let $b \neq e$. $\langle b \rangle \neq \{e\}$, and $\langle b \rangle$ is a subgroup of G. If the only subgroups of G are the trivial ones, $\langle b \rangle = G$. However, if n is the order of G and $n = mk$, m and $k > 1$, then b^m has order k, so G has a non-trivial subgroup.

19. $H = \{e, p, q\}$ is a subgroup of the group in Example 8.15 since it is closed under the operation $\cdot$. Now we compute xH and Hx where $x \in G$.

$$eH = \{e, p, q\} \quad He = \{e, p, q\}$$
$$pH = \{p, q, e\} \quad Hp = \{p, q, e\}$$
$$qH = \{q, e, p\} \quad Hq = \{q, e, p\}$$
$$rH = \{r, t, s\} \quad Hr = \{r, s, t\}$$
$$sH = \{s, r, t\} \quad Hs = \{s, t\}$$
$$tH = \{t, s, r\} \quad Ht = \{t, r\}$$

The distinct left cosets are eH and rH. The distinct right cosets are He, Hr, Hs, and Ht. Thus $\{e, p, q\}$ is not a normal subgroup.

Exercises 8.3

1. a. $R = \{(p, p), (p, q), (q, p), (q, q), (r, r), (r, s), (s, r), (s, s)\}$
   The quotient monoid is

	$[p]$	$[r]$
$[p]$	$[p]$	$[r]$
$[r]$	$[r]$	$[r]$

   The homomorphism is $f(p) = [p]$, $f(q) = [p]$, $f(r) = [r]$, $f(s) = [r]$.
   b. $R = \{(e, e), (a, a), (b, b), (b, c), (b, d), (c, b), (c, c), (c, d), (d, b), (d, c), (d, d)\}$
   The quotient monoid is

	$[e]$	$[a]$	$[b]$
$[e]$	$[e]$	$[a]$	$[b]$
$[a]$	$[a]$	$[b]$	$[b]$
$[b]$	$[b]$	$[b]$	$[b]$

   The homomorphism is $f(e) = [e]$, $f(a) = [a]$, $f(b) = [b]$, $f(c) = [b]$, $f(d) = [b]$.

3. Let $f: \mathbf{Z}_5 \to \mathbf{Z}_5$ be defined by $f([x]) = [2x]$ for $x \in \{0, 1, 2, 3, 4\}$. Consider $f([x + y]) = [2(x + y)]$. $f([x]) + f([y]) = [2x] + [2y]$. Since $[2(x + y)] = [2x + 2y] = [2x] + [2y]$, f is a group homomorphism.

5. Let $(G, \cdot)$ and $(G', \cdot)$ be groups, and let H denote the product group $G \times G'$. Consider $\gamma: H \to G$ defined as $\gamma((g, g')) = g$. γ is a homomorphism since $\gamma((g_1, g') * (g_2, g'')) = \gamma((g_1 \cdot g_2, g' \cdot g'')) = g_1 \cdot g_2$ and $\gamma((g_1, g') * (g_2, g'')) = \gamma(g_1, g') \cdot \gamma(g_2, g'') = g_1 \cdot g_2$. The kernel, K, is $\{(e_1, g'): e_1$ is the identity in G and $g' \in G'\}$. K is isomorphic to G'. Use $f(e_1, g') = g'$ as the isomorphism.

7. Let $(G, \cdot)$ be a finite abelian group. Let r be a positive integer, $r \leq |G|$. Define $f: G \to G$ by $f(x) = x^r$. Consider $f(x \cdot y) = (x \cdot y)^r = x^r \cdot y^r$. Now consider $f(x) \cdot f(y) = x^r \cdot y^r$. Therefore, f is a group homomorphism.

Review Problems

1. a. $(\mathbf{R}^+, +)$, where $+$ represents addition, is a semigroup since addition of real numbers is associative.
   b. $(B, \wedge)$, where $\wedge$ represents conjunction, is a semigroup since conjunction is associative.

3. We can easily verify that concatenation is associative by considering words such as $((aab)(abb))(aab) = (aaa)(aab) = aaa$ and $(aab)((abb)(aab)) = (aaa)(abb) = aaa$. Note that the answer will always be determined by the first three letters of the leftmost word.

5. a. Monoid
   b. Not monoid—no identity and not closed
   c. Not monoid—no identity
   d. Monoid

7. a. The operation table for this system is

θ	1	2	3	4	5
1	1	1	1	1	1
2	1	2	2	2	2
3	1	2	3	3	3
4	1	1	1	4	4
5	1	2	3	4	5

   b. (S, θ) is a commutative monoid. θ is associative since minimum $(a,$ minimum $(b, c))$ is equal to minimum (minimum $(a, b), c)$. The operation is commutative since minimum (a, b) is equal to minimum (b, a). The identity element is 5.

9. The operation is associative. The identity is 0. The inverse of 0 is 0. The inverse of 1 is 4. The inverse of 2 is 3. The inverse of 3 is 2. The inverse of 4 is 1. This is a group.

11. a. Abelian group   d. Monoid
    b. Commutative semigroup   e. Group
    c. Commutative monoid

13. All the permutations of the set $\{a, b\}$ are

$$P_1 = \{(a, a), (b, b)\}$$
$$P_2 = \{(a, b), (b, a)\}$$

The operation table of this group of permutations is

$\cdot$	P_1	P_2
P_1	P_1	P_2
P_2	P_2	P_1

15. Since $P_1{}^2 = P_1$, one subgroup is $(P_1, \cdot)$. Consider the subgroup generated by P_2. $P_2{}^2 = P_3, P_2{}^3 = P_1$. Therefore another subgroup is $(\{P_1, P_2, P_3\}, \cdot)$. This is also the subgroup generated by P_3. Consider the group generated by P_4. Since $P_2{}^2 = P_1$, another subgroup is $(\{P_1, P_4\}, \cdot)$. Similarly, the remaining subgroups are $(\{P_1, P_5\}, \cdot)$ and $(\{P_1, P_6\}, \cdot)$.

17. a.

·	I	H	V	R
I	I	H	V	R
H	H	I	R	V
V	V	R	I	H
R	R	V	H	I

b. We can easily verify the group properties. The identity element is I. Each element is its own inverse. Commutativity can be observed in the symmetry of the operation table. To verify associativity, all we need to do is compute expressions such as $(H \cdot V) \cdot R = R \cdot R = I$ and $H \cdot (V \cdot R) = H \cdot H = I$.

19. a.

·	1	2	3	4	5	6
1	1	2	3	4	5	6
2	2	4	6	1	3	5
3	3	6	2	5	1	4
4	4	1	5	2	6	3
5	5	3	1	6	4	2
6	6	5	4	3	2	1

b. $2^2 = 4$ and $6^2 = 1$

c. $3^1 = 3, 3^2 = 2, 3^3 = 6, 3^4 = 4, 3^5 = 1, 3^6 = 1$. The order of the subgroup is 6.

d. $(G, \cdot_7)$ is cyclic. A generating element is 1.

21. We clearly have a group since every element is its own inverse. To show that the group is abelian, consider $(x*y) * (x*y) = e$, $x*(y*x)*y = e$, $x*x*(y*x)*y = x*e$, $(y*x)*y = x$, $(y*x)*y*y = x*y$, $x*y = x*y$.

23. Consider $(a \cdot b) \cdot (a \cdot b)^{-1} = e$. Multiplying by a^{-1}, we have $a^{-1} \cdot (a \cdot b) \cdot (a \cdot b)^{-1} = a^{-1} \cdot e$. Because $\cdot$ is associative, this is equivalent to $(a^{-1} \cdot a) \cdot b) \cdot (a \cdot b)^{-1} = (a^{-1} \cdot e)$. Using the definitions of inverse and identity elements, we have $b \cdot (a \cdot b)^{-1} = a^{-1}$. Now multiplying by b^{-1} and simplifying, we get $(a \cdot b)^{-1} = b^{-1} \cdot a^{-1}$, which is the desired result.

25. H is a subgroup of index 2 in G. Then $G = H \cup Hx = H \cup xH$. Since $H \cap H_x = \emptyset = H \cap xH$ we must have $xH = Hx$ for all $x \in G$. Therefore, $(H, \cdot)$ is a normal subgroup of $(G, \cdot)$.

27. a. The operation $+$ is associative since the operation on each coordinate is associative. The identity element is 00000. Each element is its own inverse. Therefore $(B^5, +)$ is a group.

b. The subgroups are of the form: $(\{00000, b_1b_2b_3b_4b_5\}, +)$, where $b_1b_2b_3b_4b_5$ is an arbitrary element in B^5.

29. The function $h: G \to G$ defined by $h(x) = e$ for all $x \in G$ is a homomorphism because $h(x_1 : x_2) = e$ and $h(x_1) \cdot h(x_2) = e \cdot e = e$.

31. Consider the function f defined as $f(0) = 1$, $f(1) = 2$, $f(2) = 4$, and $f(3) = 3$. The function is an isomorphism.

33. a. This is a group code with identity 00000 and each element its own inverse.

b. The coset table is

$$0 \to 00000 \ 00001 \ 00010 \ 00011 \ 00100 \ 00101 \ 00110 \ 00111 \ 01000 \to 0$$
$$1 \to 11111 \ 11110 \ 11101 \ 11100 \ 11011 \ 11010 \ 11001 \ 11000 \ 10111 \to 1$$

$$0 \to 01001 \ 01010 \ 01100 \ 10000 \ 10001 \ 10010 \ 10100 \ 11000 \to 0$$
$$1 \to 10110 \ 10101 \ 10011 \ 01111 \ 01110 \ 01101 \ 01011 \ 00111 \to 1$$

This is a double-error correcting code. As an example, consider the transmission of 0 as 00000 and the reception of 00110 with two errors. It would decode as 0.

CHAPTER 9

Exercises 9.1

3. The FSA in Figure 9.9 accepts any sequence of letters and digits that ends in a space and whose initial symbol is a letter. For example, it accepts *a*1, *hello*, and *class*1*unit*, but it does not accept 12, *$amount*, or 1*stclass*.

*5. Let $M = (S, A, T, s_0, F)$. Let $x \in L(M)$ such that $|x| > |S|$. Let $x = x_1x_2 \ldots x_k$ and let $s_0, s_1, \ldots, s_k$ be the path traversed through the states when the input word x is recognized. Since $k = |x| > |S|$, there must be a cycle $s_i, \ldots, s_j$ in the path. Removing the symbols $x_{i+1}, \ldots, x_j$ from x generates a word v whose length is less than k but greater than or equal to 1. Therefore,

1. $x = u \cdot v \cdot w$ where $v = x_{i+1} \ldots x_j$, $u = x_1 \ldots x_i$ and $w = x_{j+1} \ldots x_k$.
2. As we already saw, $|v| \geq 1$.
3. It is easy to see that $u \cdot v^i \cdot w \in L(M)$ for $i = 1, 2, \ldots$ since all such strings end as x does and $x \in L(M)$.

7. Let $L = \{a^i: i = j^2 \text{ for an integer } j\}$. Clearly L is infinite. Suppose that M is an FSA recognizing L, and M has p states. Let $x = x_1x_2 \ldots x_k$ be a word in L of length $k \geq p$. Let $y = x_{i+1} \ldots x_j$, $i < j$ such that the input symbols $x_{i+1}, \ldots, x_j$ cause a cycle in the state machine diagram. Let $Z = \{z = x_1x_2 \ldots y \ldots yx_{j+1}x_k$: y is repeated one or more times$\}$. Every word $z \in Z$ is in L, since they all end as x does. We now must show that Z contains words not of the form a^i, where i is a perfect square ($i = j^2$ for an integer j).

We know that for any $z \in Z$, $|z| = |x| + n|y|$ where $n \geq 1$. We also know that $|x| = k = r^2$ for some integer r. If $z \in L$, then $|z| = m = s^2$ for some integer s. Thus $r^2 + n|y|$ must be of the form s^2 for all $n \geq 1$. This would require that the number of cycles always be $(s^2 - r^2)/|y|$. For any other value of n the word z is not of the required form.

Exercises 9.2

1.

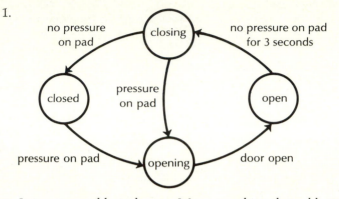

3. It is not possible to have a Moore machine that adds 1 to an integer but accepts the inputs with high order first (i.e., from left to right) because the high-order bits of the sum are not determined by the high-order bits of the summands.

Exercises 9.3

1. a. $(\lambda, 0, aaabb) \rightarrow (A, 1, aabb) \rightarrow (Aa, 1, abb) \rightarrow$
   $(Aaa, 1, bb) \rightarrow (Aa, 2, aBb) \rightarrow (A, 2, aaBb) \rightarrow$
   $(\lambda, 2, AaaBb) \rightarrow (A, 0, aaBb) \rightarrow (AA, 1, aBb) \rightarrow$
   $(AAa, 1, Bb) \rightarrow (AAaB, 1, b) \rightarrow (AAa, 2, BB) \rightarrow$
   $(AA, 2, aBB) \rightarrow (A, 2, AaBB) \rightarrow (AA, 0, aBB) \rightarrow$
   $(AAA, 1, BB) \rightarrow (AAAB, 1, B) \rightarrow (AAABB, 1, \Delta) \rightarrow$
   Halts

   b. $(\lambda, 0, aabbb) \rightarrow (A, 1, abbb) \rightarrow (Aa, 1, bbb) \rightarrow$
   $(A, 2, aBbb) \rightarrow (\lambda, 2, AaBbb) \rightarrow (A, 0, aBbb) \rightarrow$
   $(AA, 1, Bbb) \rightarrow (AAB, 1, bb) \rightarrow (AA, 2, BBb) \rightarrow$
   $(AAB, 2, Bb) \rightarrow (AABB, 2, b) \rightarrow$ Halts

   c. $(\lambda, 0, aaabbb) \rightarrow (A, 1, aabbb) \rightarrow (Aa, 1, abbb) \rightarrow$
   $(Aaa, 1, bbb) \rightarrow (Aa, 2, aBbb) \rightarrow (A, 2, aaBbb) \rightarrow$
   $(\lambda, 2, AaaBbb) \rightarrow (A, 0, aaBbb) \rightarrow (AA, 1, aBbb) \rightarrow$
   $(AAa, 1, Bbb) \rightarrow (AAaB, 1, bb) \rightarrow (AAa, 2, BBb) \rightarrow$
   $(AA, 2, aBBb) \rightarrow (A, 2, AaBBb) \rightarrow (AA, 0, aBBb) \rightarrow$
   $(AAA, 1, BBb) \rightarrow (AAAB, 1, Bb) \rightarrow (AAABB, 1, b) \rightarrow$
   $(AAAB, 2, BB) \rightarrow (AAABB, 2, B) \rightarrow (AAABBB, 1, \Delta) \rightarrow$
   Halts

3. a. The following Turing machine concatenates two nonempty words by shifting the second word back one cell to the left so that only the result is left on the tape.

state	a	[ ]	Δ
0	$(0, a, R)$	$(1, [\ ], R)$	$(0, \Delta, R)$
1	$(1, a, L)$	$(2, a, R)$	—
2	$(1, [\ ], R)$	—	—
3	—	—	—

The input alphabet is $\{a, [\ ]\}$, the tape alphabet is $\{a, [\ ], \Delta\}$, and the start state is $s_0 = 0$. When there is no next move defined for a given state and input, the table entry is just —.

As an example, consider the input string $aa[\ ]aaa$:

$$(\lambda, 0, aa[\ ]aaa) \rightarrow (a, 0, a[\ ]aaa) \rightarrow (aa, 0, [\ ]aaa) \rightarrow$$
$$(aa[\ ], 1, aaa) \rightarrow (aa, 1, [\ ]aaa) \rightarrow (aaa, 2, aaa) \rightarrow$$
$$(aaa[\ ], 1, aa) \rightarrow (aaa, 1, [\ ]aa) \rightarrow (aaaa, 2, aa) \rightarrow$$
$$(aaaa[\ ], 1, a) \rightarrow (aaaa, 1, [\ ]a) \rightarrow (aaaaa, 2, \lambda) \rightarrow$$

Halts

b. The following Turing machine concatenates two nonempty words by creating a new word for the result. The words and the result are left on the tape.

state	a	[ ]	X	$*$	Δ
0	$(1, X, R)$	$(0, [\ ], R)$	—	$(5, *, L)$	—
1	$(1, a, R)$	$(1, [\ ], R)$	—	$(2, *, R)$	$(2, *, R)$
2	$(2, a, R)$	—	—	—	$(3, a, L)$
3	$(3, a, L)$	—	—	$(4, *, L)$	—
4	$(4, a, L)$	$(4, [\ ], L)$	$(0, X, R)$	—	—
5	—	$(5, [\ ], L)$	$(5, a, L)$	—	$(6, \Delta, R)$
6	—	—	—	—	—

The input alphabet is $\{a, [\ ]\}$, the tape alphabet is $\{a, [\ ], X, *, \Delta\}$, and the start state is $s_0 = 0$. When there is no next move defined for a given state and input, the table entry is just —.

As an example, consider *aa*[ ]*a* as an input tape. The following sequence of ID's describes the behavior of the machine:

$(\lambda, 0, aa[\ ]a)$ → $(X, 1, a[\ ]a)$ → $(Xa, 1, [\ ]a)$ →

$(Xa[\ ], 1, a)$ → $(Xa[\ ]a, 1, \Delta)$ → $(Xa[\ ]a*, 2, \Delta)$ →

$(Xa[\ ]a, 3, *a)$ → $(Xa[\ ], 4, a*a)$ → $(Xa, 4, [\ ]a*a)$ →

$(X, 4, a[\ ]a*a)$ → $(\lambda, 4, Xa[\ ]a*a)$ → $(X, 0, a[\ ]a*a)$ →

$(XX[\ ], 1, [\ ]a*a)$ → $(XX[\ ], 1, a*a)$ → $(XX[\ ]a, 1, *a)$ →

$(XX[\ ]a*, 2, a)$ → $(XX[\ ]a*a, 2, \Delta)$ → $(XX[\ ]a*, 3, aa)$ →

$(XX[\ ]a, 3, *aa)$ → $(XX[\ ], 4, a*aa)$ → $(XX, 4, [\ ]a*aa)$ →

$(X, 4, X[\ ]a*aa)$ → $(XX, 0, [\ ]a*aa)$ → $(XX[\ ], 0, a*aa)$ →

$(XX[\ ]X, 1, *aa)$ → $(XX[\ ]X*, 2, aa)$ → $(XX[\ ]X*a, 2, a)$ →

$(XX[\ ]X*aa, 2, \Delta)$ → $(XX[\ ]X*a, 3, aa)$ → $(XX[\ ]X*, 3, aaa)$ →

$(XX[\ ]X, 3, *aaa)$ → $(XX[\ ], 4, X*aaa)$ → $(XX[\ ]X, 0, *aaa)$ →

$(XX[\ ], 5, X\Delta aaa)$ → $(XX, 5, [\ ]a\Delta aaa)$ → $(X, 5, X[\ ]a\Delta aaa)$ →

$(X, 5, Xa[\ ]a\Delta aaa)$ → $(\lambda, 5, aa[\ ]a\Delta aaa)$ → $(\lambda, 5, \Delta aa[\ ]a\Delta aaa)$ →

Halts

c. The machine of part (a) will work if either of the words is empty.

5. The machine determines if there are more *b*'s than *a*'s in the input string.

7. Assume that there exists a one-to-one correspondence between $F = \{f_1, f_2, f_3, \ldots\}$, the set of functions from **N** to $\{0, 1\}$, and **N**. Assume that a specific tabulation of this correspondence has been made. A possible example of this tabulation might be

N	Function
1	$f_1 = \{(1, 0), (2, 0), (3, 1), (4, 0), \ldots\}$
2	$f_2 = \{(1, 1), (2, 0), (3, 1), (4, 0), \ldots\}$
3	$f_3 = \{(1, 0), (2, 1), (3, 1), (4, 0), \ldots\}$
4	$f_4 = \{(1, 0), (2, 0), (3, 0), (4, 0), \ldots\}$
$\vdots$	$\vdots$

Now consider the function $f: \mathbf{N} \to \{0, 1\}$ defined as $f(i) = 0$ if $f_i(i) = 1$ and $f(i) = 1$ if $f_i(i) = 0$. Thus, for the above example,

$$f = \{(1, 1), (2, 1), (3, 0), (4, 1), \ldots\}$$

This function differs from f_1 since $f_1(1) \neq f(1)$. It differs from f_2 since $f_2(2) \neq f(2)$. In general, it differs from f_i since $f_i(i) \neq f(i)$. The function f, therefore, does not appear anywhere on the list. Thus, contrary to our assumption, F cannot be put into one-to-one correspondence with N and, therefore, it is not enumerable.

Exercises 9.4

1.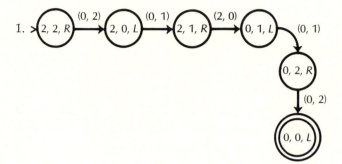

Review Problems

1. It accepts all words containing 01 as a substring.

3. a. The input is -12, and the transitions are

> 0 to 1 on token "$-$"
> 1 to 2 on token "1"
> 2 to 2 on token "2"
> 2 to 4 on token "."

State 4 is an accepting state.

b. The input is 12.34, and the transitions are

> 0 to 2 on token "1"
> 2 to 2 on token "2"
> 2 to 4 on token "."
> 4 to 4 on token "3"
> 4 to 4 on token "4"

State 4 is an accepting state.

c. The input is $+.0001$, and the transitions are

0 to 1 on token "$+$"

1 to 3 on token "."

3 to 4 on token "0"

4 to 4 on token "0"

4 to 4 on token "0"

4 to 4 on token "1"

State 4 is an accepting state.

d. The input is 00005, and the transitions are

0 to 2 on token "0"

2 to 2 on token "0"

2 to 2 on token "0"

2 to 2 on token "0"

2 to 2 on token "5"

State 2 is an accepting state.

5. The words must contain an odd number of b's.

7. A word must start with a letter. The letter is followed by zero or more letters or digits.

9.

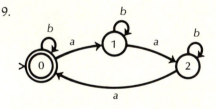

11. Suppose the language $L = \{A^n: n$ a prime$\}$ is recognized by some FSA. Since L is infinite, by the Pumping Lemma (see exercise 5 of Exercises 9.1), we have words x, y, and z such that $xy^n z \in L$ for each $n \geq 0$ with $|y| \geq 1$. Then $x = a^p$, $y = a^q$, and $z = a^r$, where $p, r \geq 0$ and $q > 0$. Then $a^{p+nq+r} \in L$ for each $n > 0$. But this is impossible since $p + nq + r$ is not a prime for all $n > 0$. For, let $n = p + 2q + r + 2$. Then $p + nq + r = (q + 1)(p + 2q + r)$.

13. a. Input 1001: output *aabaa*.   b. Input 010101: output *aaababa*.

c.

Current state	Next state		Output
	Input		
	0	1	
0	0	1	a
1	2	1	a
2	0	1	b

15.

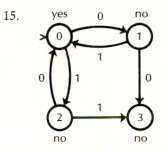

17.

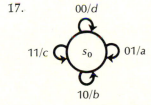

19. Modify the machine given in the solution for exercise 4 of Exercises 9.3 to recognize all words of the form $a^n b c^n$.

CHAPTER 10

Exercises 10.1

1. $p(3 \text{ heads}) = 1/8$; $p(1 \text{ head and } 2 \text{ tails}) = 3/8$

3. Let (S, p) be a probability space. Let A, B, C be subsets of S. Let $D = B \cup C$ and note that $A \cap D = A \cap (B \cup C) = (A \cap B) \cup (A \cap C)$. Thus

$p(A \cup B \cup C)$
$= p(A \cup D) = p(A) + p(D) - p(A \cap D)$
$= p(A) + p(B) + p(C) - p(B \cap C) - [p(A \cap B) + p(A \cap C) - p(A \cap B \cap C)]$
$= p(A) + p(B) + p(C) - p(B \cap C) - p(A \cap B) - p(A \cap C) + p(A \cap B \cap C)$

5. a. The odds of drawing a jack from a standard 52-card deck are 1/12.
   b. The odds of drawing a jack from a deck consisting of face cards only are 1/2.
   c. The odds of drawing a jack from a deck consisting of hearts only are 1/12.

7. Let $n = d_1 d_2$ where d_1 and d_2 are the digits from 0 to 9. $p(n) = p(d_1) \cdot p(d_2)$ for $00 \leq n \leq 99$. Since d_1 and d_2 are independent, $p(d_1) = p(d_2) = 1/10$. Therefore, $p(n) = 1/100$ for any n from 00 to 99.

Exercises 10.2

1. Let R = the ball is red, W = the ball is white, and B = the ball is blue. First we want $p(R \mid W')$: $p(R/W') = p(R \cap W')/p(W') = (3/9)/(7/9) = 3/7$. Next we want $p(B \mid W')$: $p(B \mid W') = p(B \cap W')/p(W') = (4/9)/(7/9) = 4/7$.

3. Let X = is a member of fraternity X, Y = is a member of fraternity Y, Z = is a member of fraternity Z, and Pr = is on probation. We are looking for the conditional probabilities $p(X \mid Pr)$, $p(Y \mid Pr)$, and $p(Z \mid Pr)$:

$$p(X \mid Pr) = p(X \cap Pr)/p(Pr) = (2/100)/(11/100) = 2/11$$
$$p(Y \mid Pr) = p(Y \cap Pr)/p(Pr) = (3/100)/11/100) = 3/11$$
$$p(Z \mid Pr) = p(Z \cap Pr)/p(Pr) = (6/100)/(11/100) = 6/11$$

Note that $p(X \mid Pr) + p(Y \mid Pr) + p(Z \mid Pr) = 1$.

5. Let A = applicant is qualified, and let B = applicant passes the test. We are seeking $p(A \mid B)$:

$$p(A \mid B) = [p(B \mid A)p(A)]/[p(A)p(B \mid A) + p(A')p(B \mid A')]$$
$$= [(0.8) \cdot (0.6)]/[(0.6) \cdot (0.8) + (0.4) \cdot (0.2)] = 0.48/0.56 = 0.86$$

7. Let S be the sample space consisting of the results of tossing a coin four times. Let A be the event that the first two tosses yield heads. Any event involving the last two tosses, such as "the third toss yields a tail," is independent of A.

Exercises 10.3

1. a. Expected value $= 1 \cdot p(1) + (-2) \cdot p(2) + 3 \cdot p(3) +$
$$(-4) \cdot p(4) + 5 \cdot p(5) + (-6) \cdot p(6)$$
$$= -\$0.50$$

   b. If the die is tossed five times you would expect to lose 0.50 on each toss. You would, therefore, lose $2.50.

3. We will show that $\sigma^2 = np(1 - p)$. From the definition of variance, we have

$$\sigma^2 = \sum_{x=0}^{n} x^2 \, B(n, x; p) - \mu^2$$

or

$$\sigma^2 = \sum_{x=0}^{n} x^2 \, C(n, x)p^x(1 - p)^{x-1} - \mu^2$$

This can also be expressed as

$$\sigma^2 = \sum_{x=1}^{n} npx \, C(n - 1, x - 1)p^{x-1} \, (1 - p)^{(n-1)-(x-1)} - \mu^2$$

Then, since $x = (x - 1) + 1$, we can write

$$\sigma^2 = np \sum_{x=1}^{n} (x - 1) \, C(n - 1, x - 1)p^{x-1}(1 - p)^{(n-1)-(x-1)} - \mu^2$$

$$+ np \sum_{x=1}^{n} C(n - 1, x - 1)p^{x-1} \, (1 - p)^{(n-1)-(x-1)} - \mu^2$$

$$= np \cdot (n - 1)p \cdot ((p - 1) + p)^{n-2} + np \cdot ((p - 1) + p)^{n-1} - \mu^2$$

$$= np \cdot (n - 1)p + np = (n^2 - n)p^2 + np - \mu^2$$

Therefore

$$\sigma^2 = np \cdot (n - 1)p + np = (n^2 - n)p^2 + np - (np)^2$$
$$= n(p - p^2) = np(1 - p)$$

5. $\mu = (-5)(0.3) + (-3)(0.2) + (0)(0.1) + (2)(0.2) + (4)(0.1) + (8)(0.1) = -0.5$
   $\sigma^2 = (-5)^2(0.3) + (-3)^2(0.2) + (0)^2(0.1) + (2)^2(0.2) + (4)^2(0.1) + (8)^2(0.1) - \mu^2$
   $= 17.85$
   $\sigma = \sqrt{\sigma^2} = 4.22$

Review Problems

1. There are $C(52, 5)$, or 2,598,960, ways to draw five cards from 52 cards. There are $C(4, 2)$, or 6, ways to draw two jacks from four jacks, and there are $C(48, 3)$, or 17,296, ways to draw three cards that are not jacks. Thus

$$p(2 \text{ jacks}) = (17296 \cdot 6)/2598960 = 0.04$$

3. a. $p(\text{sum} = 4) = 3/36 = 1/12$ b. $p(\text{sum} > 5) = 26/36 = 13/18$
5. Let $A = \text{sum} > 7$, and let $B = \text{one die shows a 2}$. We need $p(B|A)$:

$$p(B|A) = p(A \cap B)/p(A) = (1/18)/(30/32) = 2/15$$

7. a. $p(\text{red}) = (1/4)(3/7 + 1/3 + 1/3 + 5/8) = 0.43$
   b. $p(\text{not blue}) = (1/4)(5/7 + 6/6 + 2/3 + 6/8) = 0.78$
   c. $p(\text{Urn} = 4 \,|\, \text{ball} = \text{blue}) = 0.22$

9. $(1/4) + (1/4) + (1/4) + (1/12) + (1/12) + p(f) = 1$.
   Therefore, $p(f) = 1/12$.

11. $(10/25) \cdot (9/24) \cdot (8/23) = 720/13800$

13. $p(A) = 1/2$ and $p(B) = 1/4$; $p(A \cup B) = 5/16$. Therefore, $p(A \cap B) = 7/16$.

15. An event independent of the event "the second and third tosses are heads" is any event involving the first and/or the fourth tosses.

17. The probability distribution is

X_i	1	2	3	4	5	6
p_i	$\dfrac{1}{36}$	$\dfrac{3}{36}$	$\dfrac{5}{36}$	$\dfrac{7}{36}$	$\dfrac{9}{36}$	$\dfrac{11}{36}$

$$\mu = 1(1/36) + 2(3/36) + 3(5/36) +$$
$$4(7/36) + 5(9/36) + 6(11/36)$$
$$= 161/36 = 4.47$$
$$\sigma^2 = 1(1/36) + 4(3/36) + 9(5/36) + 16(7/36) + 25(9/36) + 36(11/36) -$$
$$\mu^2 = 1.971$$

19. There are 512 outcomes with head on the first toss. Of these $C(9, 5)$, or 126, have five additional heads. Thus $p(6 \text{ heads}\,|\,\text{head on first toss}) = 126/512$. Given a head on the first toss, there are $C(9, 6)$, or 84, outcomes with six tails. Thus $p(6 \text{ tails}\,|\,\text{head on first toss}) = 84/512$.

*21. Observe that for 100 tosses, the distribution function is $B(x, 100; 0.5)$ with mean $\mu = 50$ and variance $\sigma^2 = 25$. We are seeking $P = p(|x - \mu| \le 2\sigma)$.
   Now

$$\sigma^2 = \sum_{\{x:\, |x - \mu| \le 2\sigma\}} (x - \mu)^2 B(x, 100; 0.5) + \sum_{\{x:\, |x - \mu| > 2\sigma\}} (x - \mu)^2 B(x, 100; 0.5)$$

Hence

$$\sigma^2 > \sum_{\{x:\, |x - \mu| > 2\sigma\}} (x - \mu^2) B(x, 100; 0.5)$$
$$> 4\sigma^2(1 - p), \text{ so}$$
$$1 - p < 1/4 \quad \text{and} \quad p > 3/4$$

Index